COASTAL FLOOD RISK REDUCTION

COASTAL FLOOD RISK REDUCTION

The Netherlands and the U.S. Upper Texas Coast

Edited by

SAMUEL BRODY

Institute for a Disaster Resilient Texas, Texas A&M University, College Station, TX, United States
Department of Marine and Coastal Environmental Science, Texas A&M University, Galveston Campus, Galveston, TX, United States

YOONJEONG LEE

Institute for a Disaster Resilient Texas, Texas A&M University, College Station, TX, United States
Department of Marine and Coastal Environmental Science, Texas A&M University, Galveston Campus, Galveston, TX, United States

BAUKJE BEE KOTHUIS

Department of Hydraulic Engineering, Faculty of Civil Engineering and Geosciences, Delft University of Technology, Delft, The Netherlands
Netherlands Business Support Office, Houston, TX, United States

Elsevier
Radarweg 29, PO Box 211, 1000 AE Amsterdam, Netherlands
The Boulevard, Langford Lane, Kidlington, Oxford OX5 1GB, United Kingdom
50 Hampshire Street, 5th Floor, Cambridge, MA 02139, United States

Copyright © 2022 Elsevier Inc. All rights reserved.

No part of this publication may be reproduced or transmitted in any form or by any means, electronic or mechanical, including photocopying, recording, or any information storage and retrieval system, without permission in writing from the publisher. Details on how to seek permission, further information about the Publisher's permissions policies and our arrangements with organizations such as the Copyright Clearance Center and the Copyright Licensing Agency, can be found at our website: www.elsevier.com/permissions.

This book and the individual contributions contained in it are protected under copyright by the Publisher (other than as may be noted herein).

Notices

Knowledge and best practice in this field are constantly changing. As new research and experience broaden our understanding, changes in research methods, professional practices, or medical treatment may become necessary.

Practitioners and researchers must always rely on their own experience and knowledge in evaluating and using any information, methods, compounds, or experiments described herein. In using such information or methods they should be mindful of their own safety and the safety of others, including parties for whom they have a professional responsibility.

To the fullest extent of the law, neither the Publisher nor the authors, contributors, or editors, assume any liability for any injury and/or damage to persons or property as a matter of products liability, negligence or otherwise, or from any use or operation of any methods, products, instructions, or ideas contained in the material herein.

ISBN: 978-0-323-85251-7

For information on all Elsevier publications
visit our website at https://www.elsevier.com/books-and-journals

Publisher: Candice Janco
Acquisitions Editor: Louisa Munro
Editorial Project Manager: Naomi Robertson
Production Project Manager: Kumar Anbazhagan
Cover Designer: Mark Rogers

Typeset by STRAIVE, India

Contents

Contributors   *xiii*

1. Introduction to the Coastal Flood Risk Reduction Program   1
Samuel Brody, Baukje Bee Kothuis, and Yoonjeong Lee

2. Mixing tulips with tacos: Flood prevention practices and policies— A comparison of north Texas coastal communities and the Netherlands   5
William Merrell

Introduction	5
Dutch and American approaches to flood risk reduction	5
History and evolution of Dutch and US flood policy	7
Brief review of present flood defenses on the Dutch central coast and the upper Texas coast	10
Using Dutch principles to protect the Galveston Bay region	11
Proposed Texas plan	13
Future collaboration between Dutch and Texas researchers	14
References	15
Further reading	16

SECTION I: Predicting the floods: Environmental/physical underpinnings

3. Storm surge modeling in the Gulf of Mexico and Houston-Galveston regions   19
Bruce A. Ebersole, Thomas W. Richardson, and Robert W. Whalin

Likelihood of extreme water levels	19
Storm surge generation in the northwestern Gulf	21
State of engineering practice—Modeling hurricane storm surge	23
Different applications of storm surge models	24
Relevance of JSU (2018) research to flood risk	25
Comparison with the Netherlands situation	28
Conclusions	29
Acknowledgments	30
References	30

4. Modeling the movement of water and sediment in coastal environments — 33
Jens Figlus

Introduction — 33
Combining coastal hydrodynamics, sediment transport, and morphodynamics — 34
Utilizing sediment (transport) to mitigate flooding — 38
Conclusions — 42
References — 43

5. Urban flood modeling: Perspectives, challenges, and opportunities — 47
Antonia Sebastian, Andrew Juan, and Philip B. Bedient

Introduction — 47
Pluvial flooding as a critical area of research — 48
Modeling flood hazards in urban areas — 49
Modeling the risks associated with urban flooding — 55
Key research priorities — 55
Summary — 56
References — 57

6. Using machine learning to predict flood hazards based on historic damage — 61
William Mobley and Russell Blessing

The need for alternative flood hazard models and visualizations — 61
Establishing a machine learning workflow for flood hazard estimation — 63
A potential application for the Netherlands — 72
Summary and conclusions — 73
References — 74

7. Compound flooding — 77
Antonia Sebastian

Introduction — 77
Modeling coastal flood hazards using numerical and statistical approaches — 78
Managing coastal flood hazards — 82
Conclusions — 86
References — 86

SECTION II: Paying the price: Socioeconomic and political underpinnings of flood risk

8. Cost-benefit analysis of a proposed coastal infrastructure for reducing storm surge-induced impact in the Upper Texas Coast — 91
Meri Davlasheridze and Kayode O. Atoba

Introduction	91
Modeling approach	93
Dutch approach to coastal infrastructure BCA and lessons for the United States	100
References	104

9. The role of insurance in facilitating economic recovery from floods — 109
Samuel Brody, Wesley E. Highfield, and Russell Blessing

Introduction	109
The US NFIP as the basis for household economic recovery	109
Record of historical loss through the NFIP	113
Beyond the NFIP: Emergence of private insurance markets	115
Flood risk reduction in the Netherlands: A counter approach	115
Conclusions	117
References	117

10. Behavioral insights into the causes of underinsurance against flood risks: Experimental evidence from the Netherlands — 119
Peter J. Robinson and W.J. Wouter Botzen

Introduction	119
Insufficient demand for flood insurance	121
Results	126
Policy recommendations	131
References	133

11. Assessing economic risk, safety standards, and decision-making — 137
Matthijs Kok

Introduction	137
Risk approaches	137
Economic optimization	140
Safety standards	140
Examples	145
Comparison and discussion	147
Concluding remarks	148
References	148

SECTION III: Place-based design and the built environment

12. Infrastructure impacts and vulnerability to coastal flood events — 151
Jamie E. Padgett, Pranavesh Panakkal, and Catalina González-Dueñas

- Introduction — 151
- International case studies of coastal flood impacts on infrastructure — 152
- Envisioning the future of coastal infrastructure design and management — 158
- Conclusions — 161
- References — 162

13. Understanding the impacts of the built environment on flood loss — 167
Samuel Brody, Wesley E. Highfield, and Russell Blessing

- Putting more people in harm's way — 167
- Spread of impervious surfaces — 168
- Built environment as obstacles — 171
- Inadequate and aging infrastructure — 172
- Looking to the Dutch for solutions — 172
- Summary and conclusions — 174
- References — 175

14. Plan evaluation for flood-resilient communities: The plan integration for resilience scorecard — 177
Matthew Malecha, Siyu Yu, Malini Roy, Nikki Brand, and Philip Berke

- Transatlantic application — 179
- Feijenoord, Rotterdam — 180
- Nijmegen — 183
- De Staart neighborhood, Dordrecht — 185
- Conclusions — 187
- References — 189
- Further reading — 190

15. Dreaming about Houston and Rotterdam beyond oil and ship channels — 193
Han Meyer

- Introduction — 193
- Building the dream of the modern industrial urban landscape — 194
- Cracks and fractures in the dream — 198
- Creating new perspectives: The ship channel as a leverage — 202
- Conclusions — 207
- References — 207

16. A new nature-based approach for floodproofing the Metropolitan Region Amsterdam — 209
Anne Loes Nillesen

Introduction — 209
Conclusions — 223
Reflection of applicability in the Houston context — 224
References — 224

17. Green infrastructure-based design in Texas coastal communities — 227
Galen Newman and Dongying Li

Introduction — 227
The shift toward green infrastructure for flood mitigation in coastal Texas — 228
LID and GI as flood mitigation tools — 229
Planning and design promoting GI — 231
Application project of GI in a community design along the Texas coast — 232
Moving forward — 238
Acknowledgments — 239
References — 239

18. Integrated urban flood design in the United States and the Netherlands — 241
Fransje Hooimeijer, Yuka Yoshida, Andrea Bortolotti, and Luca Iuorio

Introduction — 241
Spatial design approach — 242
The spatial design potentials of the risk approach — 243
Case study: Vlissingen (Flushing) — 246
Case study: Galveston — 249
Discussion — 250
Conclusions — 252
References — 253

SECTION IV: Resilient solutions for flood risk reduction—Convergence of knowledge

19. Flood risk reduction for Galveston Bay: Preliminary design of a coastal barrier system — 257
Sebastiaan N. Jonkman and Erik. C. van Berchum

Introduction — 257
Setting the scene: Risk-based evaluation of strategies — 258
Preliminary design of a coastal spine system — 259
Closing discussion — 267
References — 268

20. Design, maintain and operate movable storm surge barriers for flood risk reduction   271
Marc Walraven, Koos Vrolijk, and Baukje Bee Kothuis

Introduction	271
Movable storm surge barriers	272
Specific characteristics and their implications	276
How reasoned design could enable more efficient MMO: Three cases	280
Conclusions	284
Acknowledgments	285
References	286

21. Designing and implementing coastal dunes for flood risk reduction   287
Jens Figlus

Introduction	287
Natural dunes	287
Dunes and storm impacts	289
Engineered dunes	292
Implementation of engineered dunes in Texas and the Netherlands	294
Conclusions	298
References	299

22. A proactive approach for the acquisition of flood-prone properties in advance of flood events   303
Kayode O. Atoba

Property acquisition for flood resiliency in the United States	303
Contextual differences between buyouts in the United States and the Netherlands	305
A model for prioritizing ecological gains for property acquisition in the United States	307
Discussions and conclusions	312
References	313

23. Wetlands as an ecological function for flood reduction   317
Wesley E. Highfield

Introduction	317
Early comparative research	317
Simulation-based research	318
Observational research	319
Recent advances in identifying the type and shape of wetlands in reducing flood loss	323
Comparing the Dutch experience	325
Conclusions	325
References	326
Further reading	327

24. Designing and building flood proof houses — 329
Anne Loes Nillesen

Introduction	329
Flood proof housing types	330
Conclusions	337
Reflection	338
References	339

25. Risk communication tools: Bridging the gap between knowledge and action for flood risk reduction — 341
Samuel Brody and William Mobley

The role of risk perception	341
Tools that enhance communication of risk	342
Examples of data-driven web communication tools	344
Conclusion: Challenges and opportunities	346
References	347

SECTION V: Immersive place-based learning through convergence approach

26. How to design a successful international integrative research and education program — 351
Yoonjeong Lee and Baukje Bee Kothuis

Introduction	351
Background theories and concepts of the program design	352
Program design incorporating convergence	354
Lessons learned	362
Conclusions	363
Acknowledgments	363
References	363

27. Measuring the educational effects of problem- and place-based research education programs: The student survey — 365
Yoonjeong Lee and Baukje Bee Kothuis

Introduction	365
Transformative and authentic learning and education	366
Methods	368
Results	370
Discussion	372
Conclusions	372
Appendix: Scoring Rubric for NSF PIRE CFRRP Student Survey	373
References	376

28. A specific transdisciplinary co-design workshop model to teach a multiple perspective problem approach for integrated nature-based design377
Jill H. Slinger and Baukje Bee Kothuis

Introduction	377
Theoretical background	379
Method	381
Results	382
Learning outcomes	389
Concluding remarks	393
Acknowledgment	393
References	393

29. Flood risk assessment of storage tanks in the Port of Rotterdam397
Sabarethinam Kameshwar

Introduction	397
AST inventory analysis	399
Storm surge hazard data	400
Vulnerability analysis	401
Dutch flood risk management philosophy	404
Impact of PIRE program	407
Summary	407
References	408

30. Experiences on place-based learning and research outcomes from the perspective of a student411
Alaina Parker-Belmonte

Background	411
Introduction	411
Program methodology	413
Case study research and results	413
NSF PIRE and beyond	420
References	421

31. Conclusions423
Samuel Brody, Baukje Bee Kothuis, and Yoonjeong Lee

Index *425*

Contributors

Kayode O. Atoba
Institute for a Disaster Resilient Texas, Texas A&M University, College Station, TX, United States

Philip B. Bedient
Department of Civil and Environmental Engineering, Rice University, Houston, TX, United States

Philip Berke
Department of City and Regional Planning, University of North Carolina at Chapel Hill, Chapel Hill, NC, United States

Russell Blessing
Institute for a Disaster Resilient Texas, Texas A&M University, College Station, TX, United States

Andrea Bortolotti
Delta Urbanism Research Group, Department of Urbanism, Faculty of Architecture and the Built Environment, Delft University of Technology, Delft, The Netherlands

W.J. Wouter Botzen
Department of Environmental Economics, Institute for Environmental Studies (IVM), VU University Amsterdam, Amsterdam; Utrecht University School of Economics (U.S.E.), Utrecht University, Utrecht, The Netherlands; Risk Management and Decision Processes Center, The Wharton School, University of Pennsylvania, Philadelphia, PA, United States

Nikki Brand
Department of Strategic Development, Delft University of Technology, Delft, The Netherlands

Samuel Brody
Institute for a Disaster Resilient Texas, Texas A&M University, College Station; Department of Marine and Coastal Environmental Science, Texas A&M University, Galveston Campus, Galveston, TX, United States

Meri Davlasheridze
Department of Marine and Coastal Environmental Science, Texas A&M University, Galveston Campus, Galveston, TX, Unites States

Bruce A. Ebersole
Department of Civil and Environmental Engineering, Jackson State University, Jackson, MS, United States

Jens Figlus
Department of Ocean Engineering, Texas A&M University, College Station, TX, United States

Catalina González-Dueñas
Department of Civil and Environmental Engineering, Rice University, Houston, TX, United States

Wesley E. Highfield
Institute for a Disaster Resilient Texas, Texas A&M University, College Station; Department of Marine and Coastal Environmental Science, Texas A&M University, Galveston Campus, Galveston, TX, United States

Fransje Hooimeijer
Delta Urbanism Research Group, Department of Urbanism, Faculty of Architecture and the Built Environment, Delft University of Technology, Delft, The Netherlands

Luca Iuorio
Delta Urbanism Research Group, Department of Urbanism, Faculty of Architecture and the Built Environment, Delft University of Technology, Delft, The Netherlands

Sebastiaan N. Jonkman
Department of Hydraulic Engineering, Faculty of Civil Engineering and Geosciences, Delft University of Technology, Delft, The Netherlands

Andrew Juan
Department of Civil and Environmental Engineering, Rice University, Houston, TX, United States

Sabarethinam Kameshwar
Department of Civil and Environmental Engineering, Louisiana State University, Baton Rouge, LA, United States

Matthijs Kok
Department of Hydraulic Engineering, Faculty of Civil Engineering and Geosciences, Delft University of Technology, Delft, The Netherlands

Baukje Bee Kothuis
Department of Hydraulic Engineering and Flood Risk, Faculty of Civil Engineering and Geosciences, Delft University of Technology, Delft, The Netherlands; Netherlands Business Support Office, Houston, TX, United States

Yoonjeong Lee
Institute for a Disaster Resilient Texas, Texas A&M University, College Station; Department of Marine and Coastal Environmental Science, Texas A&M University, Galveston Campus, Galveston, TX, United States

Dongying Li
Department of Landscape Architecture and Urban Planning, Texas A&M University, College Station, TX, United States

Matthew Malecha
Department of Landscape Architecture and Urban Planning, Texas A&M University, College Station, TX, United States

William Merrell
Department of Marine and Coastal Environmental Science, Texas A&M University at Galveston, Galveston, TX, United States

Han Meyer
Department of Urbanism, Delft University of Technology, Delft, The Netherlands

William Mobley
Institute for a Disaster Resilient Texas, Texas A&M University, College Station, TX, United States

Galen Newman
Department of Landscape Architecture and Urban Planning, Texas A&M University, College Station, TX, United States

Anne Loes Nillesen
Urban and Rural Climate Adaptation, Defacto Urbanism, Rotterdam, The Netherlands

Jamie E. Padgett
Department of Civil and Environmental Engineering, Rice University, Houston, TX, United States

Pranavesh Panakkal
Department of Civil and Environmental Engineering, Rice University, Houston, TX, United States

Alaina Parker-Belmonte
Professional Landscape Architect, TX, United States

Thomas W. Richardson
Department of Civil and Environmental Engineering, Jackson State University, Jackson, MS, United States

Peter J. Robinson
Department of Environmental Economics, Institute for Environmental Studies (IVM), VU University Amsterdam, Amsterdam, The Netherlands

Malini Roy
Department of Landscape Architecture and Urban Planning, Texas A&M University, College Station, TX, United States

Antonia Sebastian
Department of Earth, Marine and Environmental Sciences, The University of North Carolina at Chapel Hill, Chapel Hill, NC, United States

Jill H. Slinger
Faculty of Technology, Policy and Management, Delft University of Technology, Delft, Netherlands; Institute for Water Research, Rhodes University, Makhanda, South Africa

Erik. C. van Berchum
Department of Hydraulic Engineering, Faculty of Civil Engineering and Geosciences, Delft University of Technology, Delft, The Netherlands

Koos Vrolijk
Ministry of Infrastructure and Water Management, Rijkswaterstaat, Rotterdam, Netherlands

Marc Walraven
Ministry of Infrastructure and Water Management, Rijkswaterstaat, Rotterdam, Netherlands

Robert W. Whalin
Department of Civil and Environmental Engineering, Jackson State University, Jackson, MS, United States

Yuka Yoshida
Delta Urbanism Research Group, Department of Urbanism, Faculty of Architecture and the Built Environment, Delft University of Technology, Delft, The Netherlands

Siyu Yu
Department of Landscape Architecture and Urban Planning, Texas A&M University, College Station, TX, United States

CHAPTER 1

Introduction to the Coastal Flood Risk Reduction Program

Samuel Brody[a,b], Baukje Bee Kothuis[c,d], and Yoonjeong Lee[a,b]
[a]Institute for a Disaster Resilient Texas, Texas A&M University, College Station, TX, United States
[b]Department of Marine and Coastal Environmental Science, Texas A&M University, Galveston Campus, Galveston, TX, United States
[c]Department of Hydraulic Engineering and Flood Risk, Faculty of Civil Engineering and Geosciences, Delft University of Technology, Delft, The Netherlands
[d]Netherlands Business Support Office, Houston, TX, United States

Floods are the most deadly, disruptive, and costly natural hazard worldwide. Not a day goes by without news of some type of flood-induced disaster being reported across multiple media outlets. From chronic deluges of rainfall to 1000-year storm surge events, the toll of persistent inundation, especially in low-lying coastal areas, continues to mount. Increasing physical risk combined with rapid land use change and development in flood-prone areas has amplified the adverse economic and human impacts in recent years. Never before have the repercussions from storm events driven by both surges and rainfall been so damaging to local communities to the point that curbing their impacts has become, in many countries, a national priority.

Nowhere is the growing threat of floods to the economic well-being of society more apparent than in the upper Texas coast of the United States. In 2017, Hurricane Harvey brought record rainfall to this region causing catastrophic losses covering 49 counties, an area roughly the size of New England. Over 150,000 residential structures were inundated by floodwaters and damage estimates range in the hundreds of billions of dollars. Harvey simply followed a continuous string of storms delivering at least 6 in of rain, peppered with gigantic events with names such as Memorial Day, Ike, Rita, and Allison, and so on. For many households in the Houston area, floodwaters in one's home is an annual occurrence. Each successive flood episode inflicting the Texas coast with millions of dollars in losses exposes the underlying risk of placing millions of people in harm's way and sends a signal to other large metropolitan areas in similar situations that the problem is only getting worse.

In 2015, years before Hurricane Harvey broke every rainfall record in the book, a team of collaborators from universities in the Netherlands and the United States came together and hatched a 5-year research plan to address growing flood problems in the Houston-Galveston area as an example to the rest of the world. After all, the Dutch have long been renowned for their gritty ability to not just survive, but also thrive in the face of a persistent threat of floods. After their national flood disaster in 1953, the country made it the national priority to protect the welfare and safety of its population in perpetuity. The Netherlands and the upper Texas coast of the United States also make for a very useful comparison. Both regions are extremely flood-prone and have experienced continual adverse impacts throughout their histories. And, both are approximately the same size with similar populations, economies, and economic growth patterns. However, each country has responded to the risk of floods in fundamentally different ways, providing important points of comparison for those interested in becoming more resilient over the long term.

This international team of researchers, holed up for the day in a windowless conference room situated on a small barrier island clinging to the Texas coast, were working under several assumptions: (1) unlike some other natural disasters, floods are often exacerbated or even entirely created by human development decisions such as roadways, rooftops, parking lots, etc.; (2) floods problems persist in roughly the same place under the same conditions year after year. Their predictability makes them more fixable; (3) the rising cost of floods is not solely a function of changing weather patterns, sea-level rise, or a problem that can be solved through engineering solutions alone. Rather, flood risk and associated losses can only be understood and eventually reduced through integrated investigation across multiple disciplines, cultures, and international boundaries.

All of the researchers in the room that day, from engineers to architects, agreed that there is a critical need for program-level, trans-disciplinary inquiry in science and engineering that will lay a foundation for decision-making aimed at increasing the flood resiliency of communities in the United States and around the world. And, that any successful study must combine a diversity of physical and social science data, methods, and analytical techniques to form a more comprehensive understanding of flood risk.

This brainstorming session became the National Science Foundation Partnerships for International Research and Education (PIRE)—Coastal Flood Risk Reduction Program. Now in its sixth year, this collaborative research initiative activates an integrated, place-based education, and student exchange between multiple institutions in the Netherlands and the US Researchers from Texas A&M University, Rice University, Jackson State University, TU Delft, and VU Amsterdam, and others converge every year in the Netherlands to learn about cutting edge flood risk reduction techniques and then transport this knowledge to the Upper Texas Coast. Working with and learning from the brightest minds from a country that has set the benchmark for flood mitigation and protection is one of the best ways to address our national problem in the United States. So far, over 50 students from multiple universities have had transformative educational experiences in the Netherlands. Moreover, the program has spawned innovative research methods, improved understanding of flood risk reduction, and generated mitigation techniques that can be applied to Houston and other coastal urban areas.

The pages of this book are filled with insights, findings, and recommendations from dozens of experts formally participating in the PIRE program. Multiple case studies integrating the fields of engineering, hydrology, landscape architecture, economics, and planning address the following broad research questions: (1) what are the underlying characteristics of physical flood risks and how can they be predicted; (2) why are human communities and the associated built environment so vulnerable to flood impacts; (3) how are physical, social, and built-environment variables interrelate to exacerbate flood risk; and (4) which mitigation techniques, both structural and nonstructural, are most effective in reducing the adverse impacts of floods? Chapters cover a range of issues, from engineering flood gates and landscape design, to risk communication and authentic learning practices. They provide a rare opportunity to look at flood problems from an international, comparative perspective over a multiyear time frame. Together, they represent a transdisciplinary and holistic approach to addressing flood problems in the Netherlands and the United States.

This edited book is outlined by five sections representing major thematic areas associated with flood risk reduction. Section I focuses on predicting floods based on environmental and physical underpinnings. The authors examine the factors often setting the boundary conditions for flooding and associated impacts. Special attention is paid to predicting risk via inundation from precipitation and tidal-based events, as well as other nonphysics-oriented methods for predicting flooding. Section II shifts to investigating the socioeconomic and political drivers of flood risk. Authors highlight economic impacts, cost–benefit analysis, insurance-related issues, and risk perceptions. Section III examines the role of planning, design and the built environment. Here, authors from both countries emphasize how development patterns, buildings, and critical infrastructure contribute to the degree of flood impacts. Nature-based planning, green infrastructure, and urban design are also explored as a way to address increasing flood risk over the long term. Section IV brings together themes of previous chapters to focus on finding solutions that enhance flood resiliency. Multiple mitigation strategies are presented and evaluated in detail, including coastal barriers, dune systems, property acquisition, flood-proofing, and risk communication. Finally, Section V is dedicated to assessing the impact of placed-based and authentic learning—a major goal of the PIRE program. The authors describe how this style of learning was implemented for students participating in the international multidisciplinary flood risk reduction research and education program. Specific examples and observational data are presented on the effectiveness of the educational approach over the 5-year time frame, including descriptions of (1) multidisciplinary and transformative education approaches and student-centered place-based learning; (2) the student exchange and specific outcomes/examples of work; and (3) observational impacts of place-based and problem-based on learning and resilient thinking. The book concludes with lessons learned from the large body of work produced and experiences undertaken throughout the PIRE program and sets an agenda for future collaborative work that will benefit flood risk reduction in both countries.

We hope this book provides critical knowledge and pathways to more flood resilient futures by comparing and contrasting the pursuits of two leading countries dealing with increasing flood risk. It is a valuable resource for decision-makers, researchers, experts, students, and others interested in reducing the adverse impacts of floods over the long term.

CHAPTER 2

Mixing tulips with tacos: Flood prevention practices and policies— A comparison of north Texas coastal communities and the Netherlands

William Merrell
Department of Marine and Coastal Environmental Science, Texas A&M University at Galveston, Galveston, TX, United States

Introduction

The powerful Dutch/Texas research and education collaborations evidenced in this book grew in a remarkably short time. After Hurricane Ike, Texas researchers sought answers to storm surge flooding and reached out to Dutch experts. This was no accident because the Dutch are the best and most established engineers and scientists studying water-related issues in the world. The resulting collaboration provided access to Dutch researchers' long-term stable base of knowledge and experience. This cooperation was particularly welcome because the citizens of Texas continue to lead all the US states in flood-related deaths as well as experience millions of dollars per year in flood damages.

The newly formed Texas-based research effort, led by Texas A&M University's Institute for a Disaster Resilient Texas (IDRT), focuses on the disasters and floods affecting people and property in Texas. Because Dutch and Texas researchers working on flood risk reduction found the relationship mutually beneficial, numerous and significant collaborations developed quickly, much to the benefit of Texas's nascent program. This development has allowed the newly formed Institute to jump-start work on flood risk reduction strategies and practices of national and international importance as well as those focused on Texas.

In this chapter, we will explore the similarities and occasional profound differences between the Netherlands and the upper Texas coast that have influenced and guided our joint research and education initiatives and delve into both regions' respective approaches to flood risk reduction.

Dutch and American approaches to flood risk reduction

There are two quite different strategies that significantly reduce flood impacts on the built environment:

1. Prevention—prevent the flood damages by providing protection for human infrastructure and settlements and/or by purposefully choosing to settle and build out of harm's way.
2. Recovery—accept failure to prevent flood damages, but have policies in place that encourage rapid rebuilding and recovery.

To date, the Netherlands has embraced prevention as its national strategy while the United States has relied mostly on recovery.

In general, prevention schemes are aimed at reducing the vulnerability of the human-built infrastructure of a region to storm surge damage. This can be accomplished by preventing the presence of a human infrastructure and reducing it where it exists; or by building comprehensive structures designed specifically to protect existing human settlements or, less often, valuable environmental resources. Both approaches, although quite different, decrease flood damages by reducing the vulnerability of human settlements or environmental resources to flooding.

However, attempts at restricting, building, or encouraging retreat, although popular in concept, have generally failed in the face of strong incentives for people to live and work near water and coasts. Increases in population in coastal areas have been attributed to numerous economic and environmental factors. In general, centers for the transportation of goods have emerged in coastal areas because of the economic advantages of transporting goods on water. As people move to the area to fill available jobs in industrial complexes, residential infrastructure and businesses emerge to support these populations and in turn create more jobs and residents. Also, people are attracted to the recreational opportunities and the welcome view fields that coasts provide. As human populations and the infrastructure continue to grow in coastal areas, the human footprint negatively affects the amount of natural space and often the quality of that space. Thus, when a natural disaster strikes, both the human population and natural environments experience greater risk.

To protect against coastal surges in the Netherlands, the Dutch have relied on a combination of reducing or avoiding human settlements very near their coast and on an extensive and comprehensive engineered protection system. On their northern coasts, Dutch towns and cities are usually located behind enormous natural coastal dune systems. But the central coast is protected by a massive engineered project, the Delta works, a combination of barriers, fortified sand dunes, and gates. For riverine flooding, the Dutch traditionally used protection from levees to confine their rivers but recently have seen the need to reduce human structures near rivers with their innovative Room for the River initiative.

Much of the Netherlands behind its coastal barrier is at or below sea level and has great economic importance. Therefore, Dutch coastal barriers must be designed and built to extreme standards to minimize the risk of failure. Thus, the levels of protection enjoyed in the Netherlands are huge. In the low-lying, economically productive region near Rotterdam, protective measures on Dutch coasts are designed with a probability of exceedance of only once in 10,000 years. This level of protection is reduced to a one in

4000 years on the more agriculturally oriented southern coast. Dutch rivers were protected to a one in 3000 years occurrence but that has more recently been changed to an exceedance of once in 1250 years.

On the other hand, in the United States, where recovery is the dominant flood response policy, standards of protection from flooding are low. The United States Army Corps of Engineers (USACE) relies on a low universal standard of 1% or one exceedance in 100 years protection. From a practical standpoint, this is a moving target because of increases in Sea Level Rise (SLR), and rainfall rates attributed to global warming are rapidly changing what a 100-year protection means.

A recent report by the United States National Academies National Research Council titled "Reducing Coastal Risks on the East and Gulf Coasts" states there is no basis for the USACE's 1% standard being applied universally. Further, the report states that the benefit/cost analysis augmented by environmental and social dimensions provides a reasonable national framework for evaluating coastal risk management Investments. It also notes that hard structures are likely to become increasingly important to protect cities along our coasts.

The United States, with its lower protection standards, expects and experiences significant flood damages, essentially a planned failure to protect. The US policy encourages approaches aimed at fostering recovery; for examples, see the widely available and encouraged federal flood insurance for individual residences and businesses and nearly universal replacement of damaged municipal infrastructure through federal grants. However, recovery often takes many years and the most vulnerable populations, the poor and the elderly, often lack the means and energy to recover.

The US policy of allowing flooding but encouraging building back also demands frequent evacuation of coastal residents, especially those who live in low-lying areas or in wind or flood vulnerable structures. Evacuation has its own issues exemplified by the Rita evacuation in Texas that caused many more deaths than the storm itself.

The US flood policy is now moving more toward protection because the strategy of allowing failure (flooding) and then recovering has become increasingly expensive and socially unfair. A real change toward adopting a regional protection-based surge prevention strategy is evident with the construction of the greater New Orleans storm surge barrier. In response to the devastation caused by Hurricane Katrina, over $14 billion has been spent on a network of structures in the New Orleans area designed to protect the City from a hurricane-induced storm surge. The project is designed to reduce storm flooding in parts of New Orleans to the 100-year flood plain level, although when the ability to retain floodwaters after overflow is included, the overall protection is closer to one occurrence in 500 years.

History and evolution of Dutch and US flood policy

The prevalent policies and practices in both the Netherlands and the United States are products of past experiences.

The Dutch have a long and ultimately successful history of using structures to deal with flooding and surge suppression. As early as the 13th century, the inhabitants of the Netherlands were using dikes to protect the region from flooding. A ruling by Count Floris V in 1280 determined that "the monastery, the knight, the priest, the common man, everybody alike" was responsible for funding dike maintenance (Bijker, 2007).

The Rijkswaterstaat (RWS), founded in 1728 as the Bureau voor den Waterstaat, is part of the Ministry of Infrastructure and Water Management of the Netherlands and provides overall responsibility for Dutch water management. RWS is responsible for the management of major waters, the rivers, estuaries, and seas. RWS ensures that the government authorities responsible are alerted to floods or potential storm surges. In addition, RWS maintains dikes, dams, weirs, and storm surge barriers. Furthermore, RWS protects the coast and gives more room to rivers by deepening floodplains and constructing secondary channels. Thus, in the Netherlands, strong water management including flood risk reduction has been centralized at the national level for centuries.

RWS interfaces with district water boards that are responsible for regional waters, such as canals and smaller waterways. The district water boards also help protect the country from flooding and ensure that farmers have sufficient water for their crops. Furthermore, they are responsible for wastewater purification and ensuring that the water is clean enough to keep fish stocks healthy.

Until the Disaster of 1953, protection from flooding in the Netherlands was dominated by polders, "donut dikes," designed to protect individual cities from coastal flooding, and containment dikes protecting against river overflow.

On the night of January 31, 1953, a severe storm with hurricane strength winds caused a massive tidal surge in the North Sea which coincided with high tide. The water level rose by more than five meters above sea level, inundating Dutch sea defenses and causing sudden and widespread flooding in the regions of Zuid-Holland, Zeeland, and Noord-Brabant.

Over 1800 people died in the disaster and serious high-level talks began on how to protect the country from future flood disasters. These talks led to the development of a comprehensive Delta Plan and the implementation of the ambitious Delta Works project. This famous and protracted project, designed to protect the Dutch coast and estuaries of the Rhine, Meuse, and Scheldt, began in 1958 and was at least temporarily completed with a storm surge barrier near Rotterdam in 1998. Flood reduction features of the present Delta Works are described later in this chapter.

In addition to building more walls and dikes, the Dutch have begun applying a more long-term, holistic perspective to riverine flooding using the idea that instead of keeping water at bay, space can be allocated to safely accommodate flooding. This holistic approach led to Room for the River, a $3 billion project initiated in 2006 that involves some 40 different infrastructure projects along Dutch rivers and waterways.

Almost seemingly embarrassed by their engineering successes, the Dutch have recently embraced the concept of working with nature as a guiding concept. An experimental project in the Netherlands that could be applied to the upper Texas coast is the Zandmotor or Sand Engine—a 1-km-long sand structure that was built in 2011 along a stretch of coastline of South Holland. Dredgers collected 21.5 million cubic meters of sand, which was then deposited in a hook-shaped sand island offshore. By design, waves and coastal currents slowly disperse the sand and naturally replenish nearby beaches. This experiment has performed as designed in slowing and preventing beach erosion. It has also been more efficient and economical, as well as less disruptive to the beach and to wildlife, than traditional approaches such as regular sand replenishment.

Unlike RWS in the Netherlands, there is no central US authority that controls national water policy and practice. Instead, there are highly fragmented authorities to governing aspects of water, with the USACE designated the agency responsible for building flood protection structures but having very limited authority over water quality or marine ecosystem health. Fragmented responsibilities are also common in the states', including Texas's, management of water resources. This is in stark contrast to the clear inclusive authority given to Dutch water boards at the regional level.

Remarkably, the Ike Dike protection barrier for the upper Texas coast will be designed and built by the USACE but they will have no role in its operation and maintenance. The State of Texas has passed legislation forming a five-county Gulf Coast Protection District,[a] which will provide matching funds for the USACE project and then, upon completion, take over responsibility for the Ike Dike's operation and management. This is quite a different approach than the concentration of authority and responsibility at the federal level in the Netherlands, which covers all aspects of a major flood defense barrier's existence.

Natural disasters, particularly hurricanes, storm surges, and floods, have played a key role in the practices of American coastal engineering (Bijker, 2007). Because of the adverse effects on lives and property, natural disasters typically result in elevated public awareness of regional vulnerability (Wiegel & Saville Jr, 1996). Historically, the damage caused by specific hurricanes has motivated several limited coastal barrier projects: for example, the New England Hurricane and Coastal Storm Barriers (1960s) and the Galveston Seawall (early 1900s).

On the upper Texas coast, the 1900 storm, also known as the Great Storm, which killed as many as 8000 people, led the City of Galveston to pay for construction of a protective seawall. (Ironically, Dutch dredges were used in its construction, so even in the early 1900s, Texas coastal construction was dependent on Dutch technologies.) The seawall was designed to prevent storm surge from entering the City from the Gulf of

---

[a] The district encompasses five counties: Harris, Galveston, Jefferson, Chambers, and Orange counties.

Mexico. It did its job in the 1915 Hurricane preventing surge from the Gulf but did allow back surge into Galveston from the Bay side.

More recently, Hurricane Ike in 2008 caused major surges around Galveston Bay. Large losses were experienced because of the enormous growth in port and industrial facilities as well as business and residential infrastructure since 1900. Ike-generated experiences and damages caused the upper Texas Coast to consider more holistic solutions to stop hurricane-induced surges both at the coast and in Galveston Bay.

The United States has also recommended a more holistic approach to riverine flooding. Notably, the Galloway Report was commissioned after the 1993 floods and published in 1995 and authored by retired Corps Gen. Gerald E. Galloway, Jr. The report recommended pulling some levees back from the Mississippi river to create more flood storage. The Report also proposed a set of research topics. One of them was whether federal policies create "moral hazards" of greater flood damages by protecting floodplains for development and agriculture.

Brief review of present flood defenses on the Dutch central coast and the upper Texas coast

The Dutch principles of shortening and strengthening the coast have led to the Delta Works Project, a massive network of surge suppression devices designed to stop coastal surge at the coast while controlling river flooding (see Fig. 1).

To shorten and strengthen the coast, early aspects of the Delta Plan included closing the tidal outlets of the Maas and Rijn rivers. Fig. 1 shows the locations of these early-constructed solid coastal barriers, the Brouwersdam and the Haringvlietdam. These solid barriers are effective against coastal surge but have transformed formerly saltwater estuaries into freshwater lakes.

Later in the implementation of the Delta Plan, coastal defenses evolved to use gates to allow for saltwater circulation and ecological function in the bays. The Dutch use of circulation and navigation gates has transformed thinking about approaches to surge protection because coastal barriers can now be constructed to not only effectively suppress surge when the gates are closed but also to allow the passage of large ocean-going vessels as well as appropriate estuarine circulation when the gates are open. The Deltaworks contains examples of both navigation and circulation gates. Fig. 1 shows the Maeslantkering, a very large navigation gate near Rotterdam, and the Oosterscheldekering, a series of gates which allow tidal flow in the Eastern Scheldt estuary. The Oosterscheldekering has been declared one of the modern Seven Wonders of the World by the American Society of Civil Engineers.

The upper Texas coast has no comprehensive protection. The Galveston seawall provides limited protection for direct Gulf surges over its length but no protection for back surge from the Bay. Most of the Gulf side of the Bolivar Peninsula and Galveston Island

Fig. 1 The Delta works, a complex system in the Netherlands to protect the human population from flooding. The straight lines and dotted lines are gates and dams within the network. *(Image courtesy of OpenStreetMap.org.)*

does not have a seawall and is only protected by low sand dunes. The shores of Galveston Bay, which hold the largest concentration of petrochemical facilities in the United States, are largely unprotected. What might a comprehensive protection scheme for the upper Texas coast look like?

Using Dutch principles to protect the Galveston Bay region

The Dutch lead the world in surge protection with their Deltaworks designed to protect inland areas from North Sea storms. The coastal spine, also called the "Ike Dike" approach to protecting the entire Galveston Bay region, relies on the Dutch principles of shorting and strengthening the coast as its primary flood defense strategy and employs Dutch technology in its design.

Much of the Netherlands behind its coastal barrier is at or below sea level and has great economic importance. Therefore, the Dutch coastal barrier must be designed and built to extreme standards to minimize the risk of failure in the low-lying, economically productive region near Rotterdam.

However, the stakes are high in Texas too; the Port of Houston is the United States' busiest export port and the shores of Galveston Bay and the Houston Ship channel contain the largest petrochemical complex in the country. As the sea level rises, more of the Galveston Bay area will be more susceptible to flooding and will eventually be below or near sea level.

While each region's primary common threat is coastal surge, there are important differences between the surge and flooding threats in each region. In the Netherlands, major rivers flow into the coastal barrier and, if the flow is restricted too long by closing the coastal surge barrier, the rivers will flood. So controlling the river runoff is very important during North Sea storms which can cause intense rainfall over a wide region as well as significant surge. This has led to a sophisticated Dutch barrier system where gates are closed in careful sequence and to a policy of waiting as long as possible before closing the barrier to surge. The latter is achieved by simply not closing the barrier until the local surge reaches a predetermined height. In Texas, for typical surge-causing hurricanes, there is no significant river runoff or local rainfall threat in Galveston Bay because of its large surface area. So the barrier's closing sequence and timing is dominated by other considerations.

An important contribution of the threat in the Houston/Galveston region is the forerunner often generated by a hurricane traversing the Gulf. Because of the Coriolis force, the hurricane's intense cyclonic winds create an Ekman wave that is forced ahead of the hurricane by its motion. Over a wide continental shelf, the Ekman wave can push considerable water ahead of the hurricane's arrival. This is commonly referred to as a forerunner or fore-surge. Forerunners in Galveston Bay can arrive as many as 18 h ahead of the hurricane landfall and be several feet high, especially in the western portion of the Bay. This additional water depth in the forerunner-engorged Bay allows the Bay to support significantly higher internal surges when the hurricane's winds do arrive. Therefore, instead of waiting for a predetermined local surge height to close the Galveston barrier, the likely strategy would be to keep the Bay's depth as shallow as possible, which would in turn argue to close the barrier at the low tide that occurs just before the forerunner arrives.

A Texas coastal barrier including surge gates built to 17 ft. above sea level, the height of the Galveston seawall constructed after the 1900 storm, would offer a high level of protection from any storm that has made landfall near Galveston over the past 150 years.

However, there is a possibility of surges over 20 ft. from very intense but also very rare Gulf of Mexico storms, such as Camille (1969) and the 1934 Labor Day Hurricane. If protection from these intense storms is desired, much more robust defenses would be needed, which leads to the obvious possibility of raising the height of the coastal barrier or to the more complex strategy of creating secondary lines of defense in Galveston Bay. These secondary protections in Galveston Bay might allow a lower coastal barrier, so total protection becomes a complex trade-off between the coastal barrier and in-bay measures.

Although Galveston Bay's presence complicates Texas design considerations, it is intriguing that the Bay's large surface area and shallow depth might not only cause problems but might also offer significant benefits. The Bay's greatest problem is that it can have internal surges that increase water levels around its boundaries when significant winds blow over it. Fortunately, frictional effects limit the height of these surges to about the depth of the Bay (12 ft). Also, unlike the Netherlands or New Orleans, the land behind the Galveston Bay barrier is not below sea level. Thus, the Galveston barrier can accept considerable leakage, either by design such as allowing overtopping or by barrier failure. In this situation, some surge flows past the barrier into the Bay but, because of Galveston Bay's large surface area, the inflow only causes a small increase in the depth before the hurricane winds arrive. Moreover, because intense storms usually move quickly northward, they soon depart the region. These strategies take advantage of the Bay's capacity as a huge retention pond.

Proposed Texas plan

While both regions need surge gates to shorten the coast, one is immediately struck that the principal navigation gate (Maeslantkering) and the principal environmental gate (Oosterscheldekering) in the Netherlands not only have much different functions but also serve separate geographical areas. On the other hand, Bolivar Roads, the major pass into Galveston Bay that serves the Ports of Houston, Texas City, and Galveston as well as allows at least 80% of the water exchange between Galveston Bay and the Gulf of Mexico, would require the presence of both navigation and environmental gates.

Fig. 2 in Jonkman and Van Berchum in this volume (Chapter 19), captioned "Visualization of the coastal spine system (Jonkman et al., 2015)," presents the results of proof of concept work completed at the Technical University of Delft in the Netherlands and Texas A&M's Galveston campus on the coastal spine, "Ike Dike," that would protect the entire Houston/Galveston region from storm surge.

Fig. 2 (Chapter 19) presents an overview of the system and its main elements. The barrier includes storm surge barriers in Bolivar Roads and San Luis Pass and land barriers to protect Galveston Island and Bolivar Peninsula designed to stop or limit overland flow into the bay.

The total length of the proposed coastal spine in Fig. 2 is 58.5 miles (94 km) and consists of 56 miles (90 km) of land barriers and two storm surge barriers with a combined length of 2.5 miles (4 km).

The Dutch/Texas design studies leading to the Ike Dike concept presented in Jonkman and Van Berchum's chapter were undertaken between 2012 and 2017. More recently, a conceptual design of a similar coastal barrier has been published by the USACE and the Texas General Land Office (GLO) (USACE, 2020). Although both concepts use the coastal spine strategy to stop surge at the coast, there are notable differences between the two approaches. These differences include the following. The USACE/GLO plan includes dunes, but much lower and smaller dunes than those proposed in Jonkman and Van Berchum's chapter, thus offering smaller levels of risk reduction. Also, the USACE adopted a different storm surge barrier in the navigational section in the form of paired floating sector gates which may be vulnerable to back surge and may limit large vessel traffic. In the USACE plan, there are no gates proposed for San Luis pass. Moreover, the USACE plan estimates higher cost estimates for shared features such as the Bolivar Roads storm surge barrier. Finally, the USACE plan includes a ring levee around Galveston and several in-bay features for local risk reduction.

One crucial aspect that is still being developed is the management, maintenance, and funding of the new Texas system. In the Netherlands, the large storm surge barriers are managed by the federal government (Rijkswaterstaat, the Dutch equivalent of the USACE) and levees are mostly managed by local water boards.

As mentioned earlier in this chapter, the Ike Dike protection barrier for the upper Texas coast will be designed and built by the USACE, but they will have no role in its operation and maintenance. Instead, those responsibilities will be fulfilled by a five-county Gulf Coast Protection District that will provide matching funds for the USACE project and then, upon completion, take over responsibility for the Ike Dike's operation and management. Appointments are now being made to the District's board, which will have taxing authority, but only in the five county regions: Chambers, Galveston, Harris, Jefferson, and Orange County.

Future collaboration between Dutch and Texas researchers

Arrivals at the Amsterdam airport are greeted by signage informing them that they are 11 ft. below sea level. This brings home that the potential damages from massive coastal flooding are catastrophic in the Netherlands while usually "only" disastrous in the United States As evident from the technical section of this book, hydraulic infrastructure design is strongly influenced by the nature of the flooding threat and the geography of the region one wishes to protect. However, structure design is also influenced by national, regional, and local public policy and people's attitudes toward flooding risks. The collaboration of Texas and Dutch scientists and engineers has taught us not only the necessity of a comprehensive approach but also to recognize that the solution has to be tailored to the specific region as well as the nature of the threat. Ultimately, technical and scientific understanding have to be folded into complex political and social environments.

There are strong social values at work in both regions. The Dutch are a more orderly, risk-averse people who largely trust their government. This is reflected in the strong levels of protection against coastal storm events being codified in national law. Contrast that with the Americans, especially Texans, who not only tolerate a high level of risk but also largely mistrust the government at all levels.

The United States' mindless national concept of forcing a one in 100-year protection on all regions that require protection regardless of their importance or vulnerability is slowly giving way to more flexible approaches. Obviously, protection levels influence the size, strength, and even presence of protective hydraulic Infrastructure in the two countries. But that difference is collapsing as the US national policy of mainly relying on recovery from flooding has started its gradual demise with a move toward more regional protection schemes as evidenced in the construction of the greater New Orleans Barrier and the aggressive federal response to Hurricane Sandy (2012).

Political and social influences cause details of flood risk reduction planning to be different in Texas and the Netherlands. But from a technical viewpoint, both the Dutch homeland and the Houston/Galveston region benefit from a strategy of stopping the surge at the coast. The coastal spine, "Ike Dike," approach employs the Dutch strategy adopted after their 1953 coastal flooding catastrophe to both shorten the coast and strengthen surge defenses. The major differences between the two regions that affect hydraulic infrastructure design are a direct result of the geography behind the barrier in each country and subtle but important differences in the surge and flooding threats.

Sea level rise will eventually make the Texas coastal sea levels higher than the elevations of large inland areas, similar to the situation that the Dutch deal with now. Obviously, we in Texas will continue to learn from Dutch experiences and practices as they are our precursors in dealing with climate change. The Dutch will have the opportunity to benefit from Texas's aggressive program of new construction and new approaches as both SLR and more intense rainfall events put an unprecedented strain on Texas coastal flood defenses. We look forward to continuing our fruitful collaborations, knowing that there are many lessons to be learned on both sides of the Atlantic Ocean.

References

Bijker, W. E. (2007). American and Dutch coastal engineering: Differences in risk conception and differences in technological culture. *Social Studies of Science, 37*(1), 143–151.

Jonkman, S. N., Lendering, K. T., van Berchum, E. C., Nillesen, A., Mooyaart, L., de Vries, P., et al. (2015). *Coastal spine system-interim design report.* TUDelft, IV-infra, Royal HaskoningDHV, Texas A&M University, Defacto.

USACE. (2020). *Coastal Texas protection and restoration feasibility study, coastal Texas study draft report.* U.S. Army Corps of Engineers Galveston District. https://www.swg.usace.army.mil/Business-With-Us/Planning-Environmental-Branch/Documents-for-Public-Review/.

Wiegel, R. L., & Saville, T., Jr. (1996). History of coastal engineering in the USA. In *History and heritage of coastal engineering* (pp. 513–600). ASCE. https://doi.org/10.1061/9780784401965.015.

Further reading

OpenStreetMap Contributors. (2008). *Deltawerken*. https://commons.wikimedia.org/wiki/File:Deltawerken_na.png.

SECTION I

Predicting the floods: Environmental/physical underpinnings

CHAPTER 3

Storm surge modeling in the Gulf of Mexico and Houston-Galveston regions

Bruce A. Ebersole, Thomas W. Richardson, and Robert W. Whalin
Department of Civil and Environmental Engineering, Jackson State University, Jackson, MS, United States

Likelihood of extreme water levels

Hurricanes are the principal weather events that lead to severe coastal flooding around the periphery of the Gulf of Mexico. Gulf hurricanes have their origins over the warm tropical waters in the Atlantic Ocean, the Caribbean Sea, or within the Gulf itself. Storm surge is the primary contributor to coastal flooding. Heavy rainfall can accompany hurricanes and it can be a significant contributor to flooding as well. Storm surge can exacerbate inland flooding by retarding drainage of rainfall off the watershed and toward the coast. Locally, heavy rainfall can exacerbate flood levels during high surge events.

Both the storm surge and the astronomical tide influence total water levels. Compared to the peak surge levels that occur during major land-falling hurricanes, water level changes associated with the astronomical tide are quite small throughout the Gulf. At Galveston, the hurricane surge can exceed 6 m, whereas, the mean tide range on the open coast is only 0.36 m. The tide is mostly diurnal, and the rise and fall of a hurricane's peak surge is usually a short-duration event, lasting between 12 and 24 h. Therefore, it is less likely that a hurricane surge peak coincides with high tide.

The vulnerability of an area to flooding is often characterized by the annual exceedance probability of extreme water levels. Table 1 shows the estimated water levels that are associated with different return periods (i.e., the inverse of annual exceedance probability) for several locations in the Houston-Galveston region. Fig. 1 shows the listed locations. Water level estimates are from JSU (2018); they were made using the models described in the section "State of engineering practice—Modeling hurricane storm surge" for computing hurricane storm surge. Cited water levels correspond to mean statistical values. Water levels associated with long return periods are quite high. Water levels within Galveston Bay can be much higher than open coast levels. This is due to the following process: (1) surge propagation from the Gulf into the Bay via tidal passes, (2) inundation of low-lying western Galveston Island and Bolivar Peninsula, and

Table 1 Water surface elevation (in meters, NAVD88) for selected return periods, Houston-Galveston region, USA.

Return period (years)	City of Galveston (Gulf side)	City of Galveston (bay side)	Clear Lake entrance	Bayport entrance	Upper Houston Ship Channel
50	2.7	2.7	3.1	3.1	3.9
100	3.3	3.2	3.6	3.7	4.6
500	4.3	4.3	4.8	4.8	6.0
1000	4.9	4.6	5.2	5.1	6.5

subsequent overflow of water into the bay, and (3) the strong and rapid response of water to hurricane-force winds within this very shallow, semienclosed bay. More discussion of storm surge generation in this region of the Gulf follows in the section "Storm surge generation in the northwestern Gulf".

Fig. 1 Location map for the Houston-Galveston region, and footprint of the Ike Dike coastal spine concept.

Storm surge generation in the northwestern Gulf

The wind is the primary driver of storm surge in the Gulf of Mexico. In the northern hemisphere, hurricanes have a counterclockwise rotating wind circulation about the eye. Both the weaker far-field winds and the much stronger core winds that wrap around the eye are important in generating the storm surge. Wind-driven surge is primarily generated on the gently sloping continental shelf, in the nearshore coastal region and in bays and estuaries, because the shallower the water depth, the more effective the wind is in dragging the water around. Regions, where the continental shelf is widest, are most prone to the formation of storm surge. The "quarter-circle" configuration of the shoreline and continental shelf in the northwest corner of the Gulf is one such region (see the inset bathymetry map in Fig. 2). In the northwestern part of the Gulf, there are two

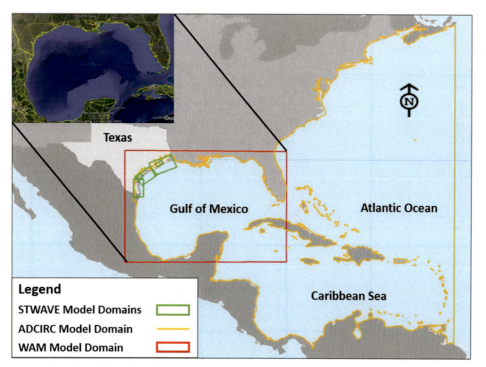

Fig. 2 Gulf of Mexico bathymetry and domains of the coupled ADCIRC/WAM/STWAVE models used to simulate storm surges and waves. *Adapted from Federal Emergency Management Agency (FEMA), 2011. Flood Insurance Study: Coastal Counties, Texas. Intermediate Submission 2: Scoping and Data Review. Federal Emergency Management Agency Region 6. Retrieved from https://docplayer.net/41290857-Flood-insurance-study-coastal-counties-texas-intermediate-submission-2-scoping-and-data-review-federal-emergency-management-agency-region-6.html. Accessed 18 September 2014. (Note: digital report).*

important aspects of wind-driven surge generation: one is the formation of a hurricane surge forerunner, and the other is the formation of the peak surge by the hurricane's core winds.

As a hurricane approaches the north Texas coast, its far-field winds blow toward the west and southwest over the Louisiana and Texas continental shelves, respectively, causing the development of a surge forerunner. The forerunner is an Ekman wave, water moving along these shelves, east to west to southwest, in response to the wind. Water movement also is directed onshore, forced to the right in the northern hemisphere by the Coriolis force associated with the earth's rotation, where it is "stacked" against the coast. Forerunner formation begins when the eye is in the central Gulf, far from land. Forerunner development can commence several days before landfall. The forerunner is revealed as a slow steady rise in water level, with the highest increase occurring at the coastline. Kennedy et al. (2011) described and documented the forerunner during Hurricane Ike in 2008, through measurements and storm surge modeling. During Ike, the water surface elevation reached 2 m above normal at Galveston, 12 h prior to landfall. As the forerunner builds, it forces water into the bays. It also leads to early onset of inundation of the beach berm, subsequent exposure of sand dunes to the direct action of waves, and to dune degradation and possible breaching.

As the hurricane eye moves across the continental shelf toward landfall, the core winds push this accumulated volume of water on the shelf toward the coast, leading to the surge peak. In forming the peak surge, the water surface slope on the inner shelf is nearly balanced by the effective onshore-directed surface wind stress.

Within semienclosed, very shallow water bodies such as Galveston Bay (average water depth of 3 m) and West Bay, the water surface responds to both filling and a tilting action caused by the hurricane-force winds as landfall occurs. Filling arises from forerunner and peak surge propagation through Bolivar Roads and San Luis passes that connects the Gulf to the Bays, and from flow over the low Galveston Island and Bolivar Peninsula once they become inundated. The large volume of water that accumulates inside the bays is dragged from one side of the bay to the other in response to the hurricane's core winds, subjecting areas along the Bay's periphery to high storm surge levels and flooding. The dynamics of surge forerunner and peak surge generation along the north Texas coast and inside the bays are described in much more detail in JSU (2018).

Atmospheric pressure gradients associated with the hurricane also contribute to storm surge, albeit as a lesser contributor. Pressure gradients force water from regions of higher atmospheric pressure toward regions of lower atmospheric pressure; i.e., from the periphery of the hurricane toward its eye. This process creates a build-up in the water surface under the eye, a dome of water. This pressure-induced increase in water surface elevation for severe hurricanes is usually 0.5–1 m, approximately.

Hurricane winds also force the growth of short-period gravity waves in both deep and shallow waters of the Gulf, and within the bays. Short waves contribute to the storm surge near the coastline, with a contribution of approximately 1 m in severe hurricanes. This contribution to surge is less than that associated with the direct effect of wind on the water column.

In the northwestern Gulf, there are large areas of sparsely populated and vegetated terrain near the coast, as well as more densely populated urban areas and industrial areas. The terrain, in both inhabited and uninhabited areas, is relatively flat and low in elevation. An extreme hurricane storm surge can penetrate inland for miles over the flat terrain, flooding large areas of the Texas coast.

State of engineering practice—Modeling hurricane storm surge

The state of engineering practice for modeling hurricane storm surge is the application of individual computer models for simulating hurricane winds and pressures, waves, and tides and storm surge. Individual models are applied as a coupled system in order to address several important aspects of feedback between modeled processes. The following set of models was set up and applied by FEMA (2011) to compute storm surge for the entire Texas coast.

(a) Tropical cyclone planetary boundary layer (PBL) model [Cardone, Greenwood, & Greenwood, 1992; Cardone, Cox, Greenwood, & Thompson, 1994; Thompson & Cardone, 1996; Cardone & Cox, 2009] was applied to provide wind and atmospheric pressure input to both the wave and surge models,

(b) Deep water wave model (WAM) [WAMDI Group, 1988; Komen et al., 1994; Günther, 2005; Smith, Jensen, Kennedy, Dietrich, & Westerink, 2011],

(c) Shallow water wave model (STWAVE) [Smith, Sherlock, & Resio, 2001; Smith, 2007; Smith et al., 2011; Massey, Anderson, Smith, Gomez, & Jones, 2011], and.

(d) Storm surge model (ADCIRC) [Luettich Jr., Westerink, & Scheffner, 1992; Westerink, Luettich, Baptista, Scheffner, & Farrar, 1992; Dietrich et al., 2010].

JSU (2018) adapted the same models for its research into hurricane storm surge in the Houston-Galveston region.

Fig. 2 shows the domains for the ADCIRC, WAM, and STWAVE models utilized by both FEMA (2011) and JSU (2018). The large domain for ADCIRC, which includes the entire Gulf of Mexico, parts of the Atlantic Ocean and the Caribbean Sea, was adopted in order to most accurately simulate a number of important processes relevant to stormwater level generation. These include (1) astronomical tide generation and tide- and hurricane-induced water movement between the Gulf of Mexico and the larger Atlantic/Caribbean basins, (2) storm surge generation on the continental shelf regions around the periphery of the Gulf, and (3) propagation of surge from the Gulf into shallow bays and estuaries through tidal inlets, such as Galveston and West Bays. The domain for the storm surge model also must consider the potential inundation of low-lying terrain by the most severe hurricanes that are to be simulated. These requirements necessitated the use of highly detailed model resolution along the north Texas coast. Hurricane winds and pressures were modeled using the PBL model for a Gulf-wide domain.

A large Gulf-wide WAM model domain is required to accurately simulate short wave growth and propagation outward from the hurricane as it enters and transits the Gulf. Nested local STWAVE model domains along the Texas coast, having higher resolution, were utilized to simulate the growth and transformation of short waves in shallow coastal areas and bays, where wave interactions with the seabed are important. The WAM model provides boundary conditions for the STWAVE model domains. Hurricane wind input to the wave models also came from the PBL model.

Different applications of storm surge models

The complexity of the surge modeling approach, choice of the model domain(s), and the degree of model resolution can vary depending on the purpose of the modeling. FEMA (2011) applied the models mentioned above as part of the FEMA Risk MAP project to update coastal flood risk maps for the Texas coast. Generally, for both flood risk quantification and engineering design purposes, the storm surge model tends to require a high degree of resolution ranging from hundreds to tens of meters. High resolution is required in those areas near the coastline that act as pathways for, or impediments to, the propagation of storm surge as well as in critical inland areas that are at risk of flooding. Generally, the higher the resolution the greater the model accuracy and the greater the computational effort required to apply the model.

Storm-surge forecasting is performed using some of these same models. However, forecasting has its own set of constraints, primarily the need to repeat a hurricane surge forecast every few hours. This generally necessitates a simpler modeling approach and less model resolution in order to reduce computer run time sufficiently to fit within the available forecast time window. For example, the Coastal Emergency Risk Assessments (CERA, 2020) web-based mapping tool utilizes the ADCIRC Surge Guidance System (ASGS) to provide a continuous storm surge forecast for the US Atlantic and Gulf coasts. The ASGS utilizes a large domain ADCIRC model for simulating tides and storm surge, a domain that is quite similar to that shown in Fig. 2, but with less detailed resolution near the coast, in bays and estuaries, and in areas where the terrain can be inundated. The CERA/ASGS system does not consider short wave processes, to reduce the run time. Wind and pressure inputs to the surge model are derived from US National Oceanic and Atmospheric Agency (NOAA) models and from NOAA National Hurricane Center measurements as they become available during actual hurricanes.

The purpose of the storm surge modeling also dictates what types of storms are simulated. FEMA (2011) simulated a large number of hypothetical hurricanes that were defined to represent the full probability space of what is possible along the Texas coast in terms of hurricane intensity, size, forward, speed, and track. Prior to running the full hypothetical storm set, model calibration and validation was performed using actual historical hurricanes. For each historical hurricane, storm meteorologists developed the wind and pressure fields using a synthesis of model results and measurements, to achieve the best possible inputs to the surge and wave models. As is typically done, the predictive skill of the models was assessed via comparisons between model results and measured wave and water level data for each of the historical storms.

Surge forecasting considers only the actual storm that is occurring, using the best available wind and pressure field input to make the simulation.

Relevance of JSU (2018) research to flood risk

Storm-surge modeling is a powerful tool for understanding surge dynamics within complex coastal systems; and it can provide valuable insights regarding the effectiveness of measures taken to reduce flood risk. JSU (2018) adapted the models applied by FEMA (2011) to examine the storm surge dynamics (forerunner and peak surge development and evolution) in the northwestern Gulf of Mexico; and they used the models to assess the effectiveness of the Ike Dike coastal spine concept (Houston Chronicle, 2009) to reduce storm surge in the Houston-Galveston region. Lessons learned can be applied to this region, others situated around the periphery of the Gulf of Mexico, or in similar geographical settings in other parts of the world.

The importance and generation of the forerunner were discussed previously. Both the forerunner and peak surge are relatively unattenuated by the deep tidal pass, Bolivar Roads that connects the Gulf to Galveston Bay, with surge propagation occurring far up the Houston Ship Channel, all the way to the City of Houston (see location in Fig. 2) with relatively little attenuation. More attenuation of the forerunner occurs through the smaller and shallower San Luis Pass.

The relationships between forerunner amplitude and peak surge, both outside and inside Galveston Bay, and the dependence between peak surge and hurricane forward speed, are illustrated in Fig. 3. It also shows results for three hurricanes that approach the Houston-Galveston region from the southeast and make landfall at Freeport, which is 15 miles to the southwest of San Luis Pass. Each hurricane has a different forward speed but the same characteristics otherwise, all of which were held constant during the simulation: central pressure of 930 mb, maximum wind speed of 50 m/s, radius-to-maximum-winds of 33 km. Fig. 3 shows results for an open-coast location at Galveston (Gulf side), and three locations inside Galveston Bay: the City of Galveston (bayside), entrance to Clear Lake, and the upper Houston Ship Channel at the City of Houston.

It is well known that peak surge on the open coast increases with increasing storm intensity and with increasing storm size. The commonly held notion is that peak surge on the open Gulf coast increases slightly with increasing forward speed. Peak surge results for the City of Galveston (Gulf side) change relatively little with increasing forward speed

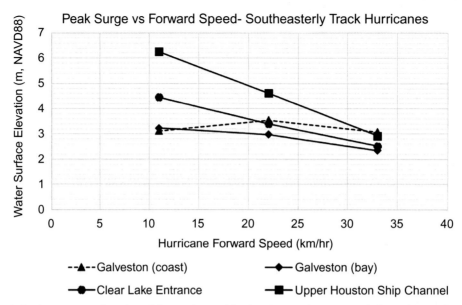

Fig. 3 Peak surge as a function of forward speed for hurricanes that approach from the southeast.

(see the dashed curve), and no clear trend is evident. However, inside Galveston Bay, peak surge increases sharply with decreasing forward speed; and a clear trend is evident. Peak surge in the upper Houston Ship Channel can be twice as great as peak surge on the open coast for slow-moving hurricanes. This arises due to the role of the hurricane surge forerunner. The slower a hurricane moves, the higher the forerunner and the greater the volume of water that enters the bay, which can then be driven up into the ship channel by the strong hurricane winds. Fast-moving hurricanes produce less of a forerunner and there is less time for the forerunner to fill the bays. Research also showed that hurricanes that approach from the south generate slightly higher forerunners than the same hurricane from the southeast, which leads to higher peak surge inside the bay.

JSU (2018) also performed storm-surge modeling to examine the effectiveness of the Ike Dike coastal spine concept for reducing hurricane flood risk in the Houston-Galveston area, and the benefits of the three barrier sections that comprise the spine (see the spine footprint in Fig. 1). The middle section includes a surge gate system at Bolivar Roads pass; and the western section includes a gate system at San Luis Pass. The land barriers have a crest elevation of 5.2 m.

A series of "proxy storms" were identified, individual hypothetical hurricanes whose spatial distribution of peak surge within Galveston Bay matched well the spatial distribution of the statistical 10-, 100-, and 500-yr peak surge values computed from the very large set of hurricanes simulated by FEMA (2011). Surge model simulations for this small set of proxy storms, and Hurricane Ike, enabled considerable insights to be gained concerning the performance of different configurations of the Ike Dike, without the need to run each barrier configuration through a much larger set of hypothetical hurricanes.

With the Ike Dike concept, which "defends" at the coast, hurricanes can still generate a sizable storm surge within Galveston Bay because of its large size. Model results indicated that the Ike Dike concept is extremely effective in reducing peak storm surge throughout Galveston and West Bays for all the storms that were considered, including Hurricane Ike and the 100-yr and 500-yr proxy storms. The Ike Dike at an elevation of 5.2 m served as an effective means to reduce surge generated within the Bay to manageable levels.

Results suggest that the eastern section (see Fig. 1) might not be cost-effective in terms of flood-risk-reduction benefits to the Houston-Galveston region. The eastern section provides relatively little benefit in reducing peak surge at the City of Galveston, along the densely developed western shoreline of Galveston Bay, and in the industrialized upper reaches of the Houston Ship. These three areas have a higher potential for damage/loss than do others in the region. This conclusion regarding the eastern section was reached by closely examining time series of water surface elevation throughout Galveston Bay for hurricanes that have the greatest potential to cause flood damage along the western shoreline. For these storms, in the absence of an eastern section, when the peak surge is occurring along the western Bay shoreline, surge waters that flank the middle section enter into the eastern part of the Bay that is experiencing strong wind set down. Therefore, the added water does not contribute much at all to increasing peak surge levels on the western side. The added water does contribute to some increase in peak surge on the eastern side of the bay in the latter stages of the hurricane's transit through the bay.

Surge modeling showed the degree to which the absence of a western section, which includes a gate system at San Luis Pass, allows considerable storm surge associated with a hurricane's forerunner and core winds to propagate into West Bay. The absence of the western section also allows flow over the barrier island south of the Pass once it is inundated. Modeling revealed that even with Bolivar Roads Pass closed, the forerunner propagates through San Luis Pass into West Bay and via the Port of Galveston navigation channel into Galveston Bay, albeit with a reduced but still significant amplitude. For Hurricane Ike, with Bolivar pass closed but leaving San Luis Pass open, surge propagation via this pathway can increase peak surges inside Galveston Bay by 1–2 ft. for future sea level. Penetration of the forerunner increased for a higher sea level.

In the light of the importance of forerunner penetration into Galveston Bay, and its important role in dictating peak surge levels within the Bay, a reliable hurricane surge-forecasting model that accurately simulates long wave propagation into the bays through gated and ungated passes would be a very useful predictive tool. The tool could provide valuable guidance for decision-making concerning the optimal time for closure of surge gates included in the Ike Dike concept.

Comparison with the Netherlands situation

Along the coast of the Netherlands, extratropical storm events that transit the North Sea are the principal weather events that dictate coastal flood risk. Unlike severe hurricanes that only occasionally affect the northwestern Gulf coast, several extratropical storms effect the Netherlands each year. The most severe hurricanes in the Gulf have higher wind speeds (thus higher storm surge generating potential) than wind speeds encountered in these extratropical storms; although, extreme extratropical storm winds can exceed the threshold speed for defining a hurricane. The extratropical storms tend to be larger and longer-duration events than hurricanes, often spanning a few days. Astronomical tide fluctuations are significantly greater along the Netherlands coast compared to those in the Gulf of Mexico. For example, the mean tide range at Hoek van Holland is 1.7 m. The large tide range contributes significantly to elevated total water levels during storms, because high surge levels can occur at one or more high tides.

Table 2 summarizes the likelihood of extreme water levels at Hoek van Holland from Voorendt (2016). The author cites two sources of probabilistic water levels: (1) one is based on measured water level data acquired between 1859 and 1958, and (2) the second is based on measured data acquired between 1863 and 2013. Results are quite similar despite the 50-yr difference in data record length.

Along the Netherlands coast, 50- and 100-yr water levels are greater than values along the north Texas coast. This is attributed to the greater frequency of extratropical storm events, the large tide range, and the much wider continental shelf off the Netherlands coast. The water depth in the shallow North Sea increases to 60 m over a distance of 340–400 km from the coastline, a much wider shelf than found in the northwestern Gulf of Mexico. The extensive shallow shelf, the high wind speeds that can accompany extratropical storms, and the longer duration of extratropical storms combine to make this region highly conducive to the formation of storm surge. For the 500-yr and 1000-yr return periods, water levels at Hoek van Holland are generally less than water levels in the Houston-Galveston region. This is attributed to the greater surge-producing potential of the most severe hurricanes that are possible in the Gulf (much higher wind speeds).

Table 2 Water surface elevation (in meters, NAP) for selected return periods, open coast at Hoek van Holland, Netherlands.

Return period (years)	Data—1859–1958	Data—1863–2013
50	3.3	3.4
100	3.5	3.6
500	4.0	4.1
1000	4.2	4.2

Significant forerunner generation along the Netherlands coast is unlikely. This is due to the configuration of the coastline and shelf and to the prevalence of strong winds from the north during severe extratropical storms. Winds from the north tend to force the peak surges. Such winds do not lend themselves to forerunner formation because the Coriolis force would force the north-to-south-to-west water movement along the Netherlands coast toward the right, away from the coast.

In contrast to the Houston-Galveston region, much of the Netherlands coast is protected by a large system of massive dunes and levees and by storm surge barriers across rivers and estuaries to keep the storm surge out of inhabited areas. These projects were constructed during the years 1954–97 as part of the Delta Works (2020). This extensive system of projects was built to greatly reduce the risk that extreme coastal storm surges can propagate into estuaries and rivers, breach coastal dikes and levees, and inundate large expanses of inhabited land in the Netherlands, which has a considerable area below sea level. The USACE (2020) examines the feasibility of a system for the Houston-Galveston region to reduce flood risk.

The Netherlands has a 150-yr long record of measured water level data at Hoek van Holland. Because of the long record length and a large number of extratropical storm events included in that record, the measured data are mostly used to compute exceedance probabilities for severe stormwater levels along the Netherlands coast, assess flood risk, and they were used in the design of most of the Delta Works along the coast as discussed in Voorendt (2016).

Storm surge modeling is a crucial component of storm preparedness in the Netherlands. Storm surge forecasts for the Netherlands coast involve the use of a different, but similar, set of meteorological, tide and storm surge, and wave models, along with assimilation of measured water level data. The Rijkswaterstaat, jointly with the Netherlands' Royal Meteorological Institute, issues surge forecasts. Wind and pressure inputs are provided from different meteorological models, depending upon the nature of the forecast, i.e., a short-range versus a medium-range forecast. de Kleermaeker, Verlaan, Kroos, and Zijl (2012) discuss recent developments in operational forecasting of storm surge in the Netherlands.

Conclusions

In the Houston-Galveston region of the north Texas coast, both the hurricane surge forerunner and peak surge are extremely important in understanding flood risk. Along the Netherlands coast, the storm surge threat originates from different types of storms, having some different characteristics, notably longer durations, and the storm surge forerunner is not an important process. The threat of severe storm surge is higher in the Netherlands for shorter return periods, but higher along the north Texas coast for longer return periods due to the greater potential for higher wind speeds during infrequent hurricanes.

In both the Netherlands and the United States, a system of similar coupled models that simulate storm atmospheric pressures and winds, and storm surge and wave generation, is the current state of engineering practice for understanding coastal storm surge, quantifying vulnerability to coastal flooding, designing flood risk reduction projects, and storm surge forecasting. Storm surge modeling enabled understanding of the complex dynamics of forerunner and peak surge generation in the northwestern Gulf of Mexico and Houston-Galveston region; and it enabled evaluation and optimization of the performance of the Ike Dike concept to reduce the risk of flooding in this area. The Ike Dike coastal spine concept takes the same flood risk reduction approach that is adopted in the Netherlands, i.e., a strong coastal barrier that shortens the coast and prevents a surge from entering the shallow bays and estuaries.

Acknowledgments

The authors are grateful for the support and technical guidance provided by Dr. William J. Merrell, Dr. Samuel Brody, and other faculty and staff members from the Texas Center for Beaches and Shores, Texas A&M University—Galveston Campus, and for the support provided by the Bay Area Coastal Protection Alliance. The authors also thank Dr. Chris Massey and Margaret Owensby, Coastal & Hydraulics Laboratory, US Army Engineer Research and Development Center, Vicksburg, Mississippi, USA, for their extensive technical support and assistance with storm surge simulations and data from those simulations.

References

Cardone, V.J., & Cox, A. T. (2009). Tropical cyclone wind field forcing for surge models: Critical issues and sensitivities. *Natural Hazards, 51*, 29–47.

Cardone, V.J., A.T. Cox, J.A. Greenwood, and E.F. Thompson. (1994). Upgrade of tropical cyclone surface wind field model. Misc. Paper CERC-94-14, Vicksburg, MS: U.S. Army Corps of Engineers Waterways Experiment Station. https://apps.dtic.mil/dtic/tr/fulltext/u2/a283530.pdf. (Note: digital report).

Cardone, V.J., C.V. Greenwood, and J.A. Greenwood (1992). Unified program for the specification of tropical cyclone boundary layer winds over surfaces of specified roughness. Contract Rep. CERC 92-1. U.S. Army Corps of Engineers Waterways Experiment Station, Vicksburg, Mississippi, USA. https://erdc-library.erdc.dren.mil/jspui/bitstream/11681/2797/1/CR-CERC-92-1.pdf. *(Note: digital report)*.

Coastal Emergency Risk Assessments (CERA). (2020). https://cera.coastalrisk.live/. (Accessed 30 November 2020).

de Kleermaeker, S., Verlaan, M., Kroos, J., & Zijl, F. (2012). A new coastal flood forecasting system for the Netherlands. In *Hydro12—Taking care of the sea, Rotterdam, Netherlands.* , November.

Delta Works. (2020). http://www.deltawerken.com/Deltaworks/23.html. (Accessed 30 November 2020).

Dietrich, J. C., Bunya, S., Westerink, J. J., Ebersole, B. A., Smith, J. M., Atkinson, J. H., et al. (2010). A high-resolution coupled riverine flow, tide, wind, wind wave and storm surge model for southern Louisiana and Mississippi: Part II—Synoptic description and analysis of hurricanes Katrina and Rita. *Monthly Weather Review, 138*, 378–404.

Federal Emergency Management Agency (FEMA), 2011. Flood Insurance Study: Coastal Counties, Texas. Intermediate Submission 2: Scoping and Data Review. Federal Emergency Management Agency Region 6. Retrieved from https://docplayer.net/41290857-Flood-insurance-study-coastal-counties-texas-intermediate-submission-2-scoping-and-data-review-federal-emergency-management-agency-region-6.html. Accessed 18 September 2014. (Note: digital report).

Günther, H. (2005). *WAM cycle 4.5 version 2.0* (p. 38). Geesthacht, Germany: Institute for Coastal Research, GKSS Research Centre.

Houston Chronicle. (2009). Oceanographer: 'Ike Dike' could repel most storm surges, https://www.chron.com/news/houston-texas/article/Oceanographer-Ike-Dike-could-repel-most-storm-1627411.php. Accessed 30 November 2020.

Jackson State University (JSU), 2018. Final Report—Ike Dike Concept for Reducing Hurricane Storm Surge in the Houston-Galveston Region, Jackson, Mississippi, United States, 549pp. URL: https://www.tamug.edu/ikedike/pdf/JSU-Final-Report-Ike-Dike-Concept.pdf (Note: digital report). Accessed 30 November 2020.

Kennedy, A. B., Gravois, U., Zachry, B. C., Westerink, J. J., Hope, M. E., Dietrich, J. C., et al. (2011). Origin of the Hurricane Ike forerunner surge. *Geophysical Research Letters*, L08805. https://doi.org/10.1029/2011GL047090.

Komen, G. J., Cavaleri, L., Donelan, M., Hasselmann, K., Hasselmann, S., & Janssen, P. A. E. M. (1994). *Dynamics and modelling of ocean waves*. Cambridge, UK: Cambridge University Press. 560pp.

Luettich, R. A., Jr., Westerink, J. J., & Scheffner, N. W. (1992). ADCIRC: An advanced three-dimensional circulation model for shelves coasts and estuaries, report 1: Theory and methodology of ADCIRC-2DDI and ADCIRC-3DL, dredging research program technical report DRP-92-6. In *U.S. Army Engineers Waterways Experiment Station, Vicksburg, Mississippi, USA*. https://apps.dtic.mil/dtic/tr/fulltext/u2/a261608.pdf (Note: digital report).

Massey, T.C., Anderson, M.E., Smith, J.M., Gomez, J., and Jones, R. (2011). STWAVE: Steady-state spectral wave model user's manual for STWAVE. Version 6.0. ERDC/CHL SR-11-1, U.S. Army Engineer Research and Development Center, Vicksburg, Mississippi, USA. https://ewn.el.erdc.dren.mil/tools/stwave/STWAVE_manual.pdf. (Note: digital report).

Smith, J. M. (2007). *Full-plane STWAVE with bottom friction: II. Model overview*. CHETN-I-75. Vicksburg, Mississippi, USA: U.S. Army Engineer Research and Development Center. https://apps.dtic.mil/dtic/tr/fulltext/u2/a471582.pdf (*Note: digital report*).

Smith, J. M., Jensen, R. E., Kennedy, A. B., Dietrich, J. C., & Westerink, J. J. (2011, July). Waves in wetlands: Hurricane Gustav. In *Coastal Engineering Proceedings, 1(32), Shanghai, China*. https://doi.org/10.9753/icce.v32.waves.29.

Smith, J.M., A.R. Sherlock, and D.T. Resio. (2001). STWAVE: Steady-state spectral wave model user's manual for STWAVE, Version 3.0. ERDC/CHL SR-01-1. U.S. Army Engineer Research and Development Center. Vicksburg, Mississippi, USA. https://ewn.el.erdc.dren.mil/tools/stwave/STWAVE_manual.pdf. (Note: digital report).

Thompson, E. F., & Cardone, V. J. (1996). Practical modeling of hurricane surface wind fields. *ASCE Journal of Waterway, Port, Coastal and Ocean Engineering*, 122(4), 195–205.

U. S. Army Corps of Engineers (USACE). (2020). Coastal Texas Study 2020 Draft Feasibility Report. URL: https://www.swg.usace.army.mil/Business-With-Us/Planning-Environmental-Branch/Documents-for-Public-Review/. Accessed 30 November 2020. (*Note: digital report*).

Voorendt, M. Z. (2016). *The development of the Dutch flood safety strategy*. Technical Report. Delft University of Technology. ISBN/EAN 978-90-74767-18-7, 70pp.

WAMDI Group. (1988). The WAM model—A third Generation Ocean wave prediction model. *Journal of Physical Oceanography*, 18, 1775–1810.

Westerink, J. J., Luettich, R. A., Baptista, A. M., Scheffner, N. W., & Farrar, P. (1992). Tide and storm surge predictions using finite element model. *ASCE Journal of Hydraulic Engineering*, 118(10), 1373–1390.

CHAPTER 4

Modeling the movement of water and sediment in coastal environments

Jens Figlus
Department of Ocean Engineering, Texas A&M University, College Station, TX, United States

Introduction

Water can enter coastal systems and produce flooding in a variety of ways. Elevated water levels due to storm surge and wave setup, wave overtopping, runoff from heavy rain events, and increased river discharge resulting from upstream dam releases or appreciable rainfall induce flow to low-lying coastal areas with the potential to result in flooding. Apart from the damaging water masses, flood flows can move a considerable amount of sediment. This aspect usually receives less attention than the dynamics and the presence of floodwaters, but the movement of sediment nonetheless affects infrastructure, communities, and even the flood flows themselves. For example, as water overtops or inundates coastal dunes during a storm, the dune sediment is eroded and transported both offshore and onshore by waves, seaward-directed return flows, and landward-directed overwash flows (Sallenger Jr., 2000). If large amounts of the dune are eroded, a breach forms and even more water can penetrate the dune line and enter the landward area, further enlarging the breach. Thus, the erosion of dune sediment by flood flow has effectively led to a positive feedback mechanism through which flooding is further enhanced (e.g., Cañizares & Irish, 2008). In addition, if the dunes are protecting developed landward areas, the mobilized sediment can cover roads and other infrastructure, effectively reducing the capacity for emergency access, causing hazardous conditions, and resulting in removal costs. Not including sediment movement in modeling such a scenario will lead to underprediction of flooding and neglection of other hazards.

Another example of the adverse effects of sediment flow is related to the role of pollutants during flood events. Pollutants are often attached to sediments or buried underneath sediments. The power of flood flows to mobilize and transport sediments also allows for pollutants to be mobilized and transported (e.g., Dellapenna, Hoelscher, Hill, Al Mukaimi, & Knap, 2020). This can manifest itself in elevated levels of pollutant concentrations (e.g., heavy metals) in flood flows and is particularly critical where developed and industrialized areas are close to sensitive ecosystems, as is the case for most coastal systems near population centers.

The ability of hydrodynamic forces to move sediment can also be used to benefit coastal flood protection schemes as presented later in this chapter using examples from the Netherlands (sand engine, mud motor) and Texas (nearshore submerged berm nourishment and bedload collector). Such "Building with Nature" or "Engineering with Nature" concepts in the Netherlands and the United States, respectively, include the beneficial redistribution of placed dredged sediment by natural forcing to offer sustainable and cost-effective solutions to ecosystem restoration and coastal flood protection (e.g., de Vriend, van Koningsveld, Aarninkhof, de Vries, & Baptist, 2015; Temmerman et al., 2013).

The above examples highlight the importance of properly modeling erosion, transport, and deposition of sediment by flood flows as a critical element of the flooding process. The physics governing the movement of sediment by water are complex and have fascinated researchers for decades (e.g., Bagnold, 1946, 1956; Grant & Madsen, 1979, 1982; James, Jones, Grace, & Roberts, 2010; Shields, 1936). Detailed explanations of sediment transport physics are beyond the scope of this chapter. Here, a brief overview of the major physical processes and options to model sediment dynamics of coastal water flows are given with the understanding that chemical, as well as biological processes, can also play a role. The chapter closes with examples of intentional use of coastal sediment transport in flood protection schemes in the Netherlands and Texas. The intent is to further stress the importance of sound understanding of coastal sediment transport in driving flood resilience innovation.

Combining coastal hydrodynamics, sediment transport, and morphodynamics

Modeling approaches to coastal flood flows that treat only the water phase and assume a fixed bed without sediment transport and morphological changes are very common. They tend to produce results with sufficient accuracy for many problems and applications. For example, typical physical model studies of breakwater overtopping rates for various hydrodynamic forcing conditions in wave flumes or wave basins consider only the overtopped water since such structures usually are not constructed with fine granular sediment such as sand that is easily moved by water (e.g., Losada, Lara, Guanche, & Gonzalez-Ondina, 2008). Similarly, numerical model simulations of coastal flood dynamics and inundation during storm impact often do not include sediment movement and morphology changes since the focus is on water movement and inundation over large spatial scales (i.e., grids of hundreds of kilometers horizontal dimensions) over multiple tidal cycles for which detailed simulation of sediment dynamics would introduce high additional computation costs (e.g., Dietrich et al., 2011; Sebastian et al., 2014).

Nonetheless, there are many situations in the coastal flooding context where properly modeling both water movement and sediment dynamics is critical in understanding the system under investigation and the related flooding. In such situations, it becomes important how the model combines water and sediment movement. In the following, some context for physical as well as numerical modeling approaches is provided.

Physical modeling

Physical models created to better understand combined coastal water and sediment movement typically involve medium to large-scale 2D wave flumes as shown in Fig. 1 (e.g., Tomasicchio et al., 2011; van Gent, van Thiel de Vries, Coeveld, de Vroeg, & van de Graaff, 2008) and 3D wave basins (e.g., Kamphuis, 1991; Taqi & Figlus, 2019) as shown in Fig. 2 with the ability to include loose sediment beds and sometimes even currents. Physical models are a great way to incorporate all the physics of a

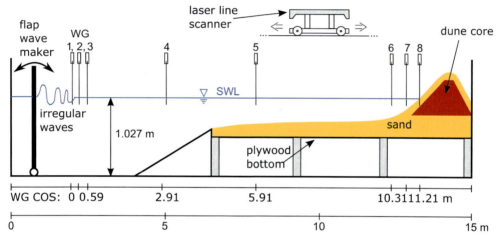

Fig. 1 Schematic representation of 2D wave flume experiment for irregular wave impact on moveable sand bed and core-enhanced dune.

Fig. 2 Photo of 3D wave basin experiment on wave overtopping of sand-covered rubble-mounds at Texas A&M University's Haynes Laboratory.

problem since real fluids and real sediment can be used. However, most experimental facilities are not sized to run experiments at 1:1 scale and the type of fluid and sediment feasible for use are usually limited. This means scaling laws have to be applied to properly interpret model results and translate findings from model to prototype. Van Rijn, Tonnon, Sánchez-Arcilla, Cáceres, and Grüne (2011), for example, studied scaling laws for beach and dune erosion processes using experimental data from three differently sized wave flumes. The problem is that the use of sediment particles subject to flow can introduce conflicting scaling requirements for geometry, kinematics, and dynamics of the system. Modeling coastal mobile bed sediment transport and morphology is perhaps the most difficult out of all physical hydraulic models (Hughes, 1993). Thus, researchers have to carefully choose which laws to prioritize and how to interpret the results. Depending on the objective of the experiment, appropriate scaling laws have to be used based on the similitude of characteristic nondimensional numbers between model and prototype (Heller, 2011). For experiments involving a free water surface and waves, some form of Froude similarity is commonly used to link model and prototype (Hughes, 1993). The Froude number quantifies the relative importance of inertia and gravity on a fluid and can be written as

$$Fr = \frac{u}{\sqrt{gL}} \tag{1}$$

where u is a characteristic flow velocity, g is the acceleration due to gravity, and L is a characteristic length scale. Including a moveable sand bed to the problem makes proper physical model scaling more difficult. Depending on the specific aspect of the problem under investigation, other forms of Froude scaling using a densimetric Froude number or a grain Froude number can be used (e.g., Aguirre-Pe, Olivero, & Moncada, 2003). These are often employed when the initiation of sediment motion, bedforms (i.e., ripples, sand waves) or the mode of sediment transport (i.e., suspended load, bed load) need to be reproduced accurately and other scaling relations can be relaxed.

For example, a 1:20 geometric scale model of a beach and dune system under storm wave impact in a wave flume using freshwater and fine sand can be used to assess the process of wave-induced dune erosion during a storm (see Chapter 24 for more details on dune erosion). Acceleration due to gravity is generally the same in model and prototype. The length scales of the dune and beach can be set based on the 1:20 geometry scale (scale factor $\lambda = 1/20$) as long as the length and height of the flume are adequate. To uphold Froude similarity, model times and velocities need to be scaled with a factor of $\sqrt{\lambda}$. This means that a prototype wave period of 15 s translates to a model wave period of 3.4 s. The problem is that wave breaking processes and turbulence do not scale appropriately since the fluid in both model and prototype is water with a given density and viscosity with only minor differences based on salinity and temperature. Similarly, fine sand used in the model would correspond to much coarser material of the same density in the prototype. Model sediment size representing granular material cannot be reduced much beyond fine sand since finer material may be subject to electrostatic forces (i.e., cohesion) between individual grains, which will have a significant effect on model results. Thus, it is desirable to run such physical model tests at the largest geometric scale possible or use engineered sediment particles with tuned properties such as density and shape that can help fulfill the most critical scaling laws.

Despite the mentioned difficulties, scaled physical model studies have been critical in identifying the most important processes of problems related to the coastal movement of water and sediment and have helped in formulating numerical models that compute coastal sediment transport and morphodynamics based on hydrodynamic forcing from water level changes, currents, and waves.

Numerical modeling

Combined sediment transport and hydrodynamics in coastal systems are difficult to model numerically due to the inherent complexity of the interaction of sediment and flow in addition to the challenges posed by nearshore hydrodynamics such as nonlinear wave and current interactions, wave breaking, surface roller energy dissipation, turbulence, intermittent wetting and drying cycles due to water level fluctuations, and shallow overland flows during flooding events. Furthermore, the characteristics of the sediment vary in space and time and, even at a single location and time step, are not easily described by just one parameter. Usually, a whole distribution of grain sizes is present, made up of grains from different sources, often with somewhat different properties related to density, shape, and roundness. Numerical models have to be chosen or developed based on the specific research or project objective. This usually involves a compromise between resolving small-scale sediment transport processes and detailed hydrodynamics, and simulating large spatial and/or temporal scales. For this reason, a large variety of different numerical modeling approaches exists (Amoudry & Souza, 2011).

Waves in coastal areas display random characteristics that are typically best handled by random wave transformation models. These can be classified into phase-resolving and phase-averaging models depending on whether they resolve individual wave motions or not. Phase-resolving models can describe physical wave transformation processes in coastal areas including shoaling, diffraction, refraction, and dissipation at the intra-wave scale (Nam, Larson, Hanson, & Hoan, 2009) but due to their required fine resolution in space and time are suitable only for small coastal areas (order of tens of kilometers in horizontal grid extents) and short-term (i.e., storm duration) simulations. Phase-averaging models do not resolve individual wave motions but rather describe slowly varying wave quantities such as wave amplitude and wave energy as they propagate through bodies of water. While this comes with certain limitations (e.g., Holthuijsen, Herman, & Booij, 2003), it allows simulations of much larger areas (i.e., hundreds of kilometers in horizontal grid extents) and temporal scales compared to phase-resolving models.

Nearshore currents driven by random waves are critical in moving sediment in coastal areas and have been modeled by many researchers with a variety of modeling approaches (Choi, Kirby, & Yoon, 2015; De Vriend & Stive, 1987; Smit, Zijlema, & Stelling, 2013; Thornton & Guza, 1986). The most sophisticated ones include the effect of tidal motions, wind, rivers, random waves, wave breaking, and surface rollers and are capable of simulating many nearshore phenomena such as edge waves, surf beats, infragravity waves, and longshore currents, all of which are important in coastal sediment transport.

In the intermittently wet and dry area of the coastal system along the border of ocean and land, water depths are very small. This creates difficulties for many numerical hydrodynamic models due to vertical grid resolution and also complicates the application of linked sediment transport formulas. It is in this area, however, where the combination of hydrodynamics and sediment transport are critical for flood generation. During storms, water levels are elevated, wave height increases, and erosion of the beach and dune system including processes such as overwash and breaching can lead to severe flooding of the backdune area. These processes are typically simulated using process-based models that focus on specific dominant processes such as dune scarping by waves and often use empirical equations derived from calibration with limited laboratory and field data to determine wave-driven overtopping flow and associated dune erosion and breaching.

To simulate the flooding initiated by a storm at a coastline with dunes, all of the modeling aspects given above are required to determine the correct amount of water and sediment reaching the areas behind the dune line. This process usually involves the coupling of multiple models in such a way that the changes in sediment transport and morphology induced by the changing hydrodynamics during the storm provide direct feedback to the hydrodynamics computed in the next time step.

As computing power and numerical model sophistication continue to increase, such computations will be feasible for more projects, larger areas, and longer durations. It has to be understood, however, that the inherent difficulties in accurately simulating sediment transport, remain and require continued research. In addition, the general scarcity of field data available for numerical model validation continues to be a concern that should be tackled through intensified field measuring efforts.

Utilizing sediment (transport) to mitigate flooding

Hydrodynamic forcing from tidal currents and waves can be used to move placed sediment material in such a way that it benefits flood resilience of coastal systems. Often, sediment dredged from navigation channels is available for such beneficial use of dredged material (BUDM) projects. Beach nourishment (the placement of sediment directly on the beach) and saltmarsh restoration (the placement of sediment and vegetation along wetland fringes) have been go-to approaches for decades to delay the effects of coastal erosion, restore ecosystems, and provide flood mitigation benefits to nearby communities. There is, however, room for continued improvement and innovation to save costs and harness the power of natural processes when engineering beach and marsh restoration efforts. While approaches to coastal flood mitigation differ in many ways between the Netherlands and Texas as discussed throughout this book, there are concepts related to coastal sediment transport and beneficial use of placed sediment that can be implemented successfully in both locations and should be part of the transatlantic discussion on coastal flood resilience. Some related innovative projects from the Netherlands and Texas are highlighted in the following.

Sand engine

A pilot mega-nourishment (MN) project known as the Delfland sand engine, or Zandmotor in Dutch, has been in existence on the Dutch coast since 2011. The general idea is based on the economy of scale: Rather than implementing many separate small nourishment projects along the coast, one large nourishment project is realized at a single location with the hope that coastal sediment transport processes will redistribute the material over time to feed beaches further up the coast and thus contribute to flood resilience along a substantial stretch of the Dutch coastline.

The sand engine is located on the coast of Holland between Scheveningen and Rotterdam where the prevailing alongshore drift of sediment is to the northeast. The sand engine peninsula was constructed using 17 million m^3 of sand (21.5 million m^3 if the two accompanying shoreface nourishments are included) initially extending approximately 1 km offshore and a bit over 2 km alongshore, soon forming a near-Gaussian shoreline shape (de Schipper et al., 2014). Monthly topographic surveys showed rapid initial redistribution of sediment during the first 1.5 years after construction where material from near the tip spread out to the flanks of the peninsula and adjacent beaches (de Schipper et al., 2016). Fig. 3 shows a photo of the sand engine beach.

Fig. 3 Photo of the sand engine beach looking south toward the port of Rotterdam, the Netherlands.

An MN limits the temporary disturbance during construction to a single location which may be a plus for the overall environment, but to which degree it can reduce other nourishment activities along the coast is still under investigation. The sand engine has not evolved entirely as expected raising issues related to the local ecosystem, recreational and navigational safety, groundwater, and potential changes to nearshore processes (Stive et al., 2013). It is a promising concept but such a large modification of the coastal system requires continued monitoring and detailed modeling to understand the involved complexities and improve predictions of future evolution. The Dutch sand engine example can provide valuable lessons to other locations around the world, including the Texas coast. However, it has to be understood that coastal systems are unique and that the almost infinite nearby offshore sand resources of the Netherlands play a major role in the success of such an undertaking.

Mud motor

The Wadden Sea includes muddy tidal flats and channels along the Dutch mainland coast that are protected from North Sea waves by barrier islands. Harlingen is one of the ports located along that mainland coastline and requires frequent dredging of its approach channels. The mud motor objective is to feed and expand the mainland salt marsh area near Harlingen using dredged sediment and its strategic placement (Baptist et al., 2019; van Eekelen & Baptist, 2019). The idea includes the use of naturally occurring current forcing at different tidal stages to suspend placed fine sediments into the water column and transport them toward an existing marsh area. One of the intended benefits is the reduced flood risk of landward regions due to the extension of the existing salt marsh area over time (Baptist et al., 2017). The method involves disposal of channel dredge material in subtidal placement areas during different tidal elevations and associated current regimes. This is accomplished via small hopper dredges using the bottom release of sediment in predefined water depths and distances to the target marsh area. In general, disposal happens during flood tide when water movement in the tidal channel is directed toward the designated marsh area. This operation occurs during daylight hours in fall and winter to minimize disturbance. The overall concept is that the increased concentration of suspended fine sediments in the water column near the target marsh area will lead to sedimentation that can help vegetation colonize further areas of the mudflats and turn them into salt marsh. Observations have shown that the expected sedimentation has been somewhat hindered by freshwater influence on water column stratification and no seaward expansion of the marsh edge has occurred in a short initial time frame. The extent of marsh area at this location does not seem to be limited by the supply of suspended sediment but by morphological evolution of the bed level (Willemsen et al., 2018), meteorological (mainly wind direction and waves) and ecological (high density of worms influence the establishment of salt marsh vegetation) factors. It has been found, for example, that a higher-than-normal density of a special type of worm species influenced the establishment of pioneer vegetation at the site in a negative way (van Regteren, ten Boer, Meesters, & de Groot, 2017). This highlights the complexities of such projects and the importance of multidisciplinary planning approaches.

Submerged nearshore feeder bar

At South Padre Island (SPI), Texas, a submerged feeder bar located outside the surf zone (9-m water depth) in nearshore waters of the Gulf of Mexico (Work & Otay, 1997) is being tested as an alternative means to supply dredged sediment to the beach by natural transport processes and reduce ongoing coastal erosion and flood risk. While beach nourishment projects can be important components of coastal risk-reduction strategies, they can be intrusive to coastal ecosystems and beach use. Similar to prior examples, the submerged feeder bar system is intended to deliver placed material from nearby ship channel dredging operations to the SPI nearshore and beach by means of natural nearshore circulation. Such sustainable sediment practices that do not use heavy machinery and intrusive beach placement methods are increasingly desirable (e.g., Berkowitz, Piercy, Welp, & VanZomeren, 2019; Brutsché, McFall, Bryant, & Wang, 2019; McFall, Smith, Pollock, Rosati, & Brutsché, 2016 and Welch, Mogren, & Beeney, 2016). Since 1988 SPI has undertaken intermittent submerged bar nourishment projects as part of a BUDM scheme (Aidala, Burke, & McLellan, 1992). Sediment removed by maintenance dredging is regularly placed back into the littoral system, and thus available for cross-shore and alongshore sediment transport to the beaches. Although nearshore placement is typically less expensive than beach placement, quantitative evidence is needed to understand how the material spreads and to determine whether it is eventually delivered to the active surf zone and deposited on the upper beach template. A recent sediment tracer and hydrodynamic field study showed that sediment from the feeder bar can be deposited on nearby beaches but also revealed the complexity of the system and the difficulty in predicting long-term sediment movement in the nearshore (Figlus et al., 2021). Experience and data from such projects are important in the development of numerical models to gain a better understanding of sediment transport dynamics that can aid in the reduction of coastal flood risk.

Bedload collector

The harvesting of sediment for later use in protection schemes against flooding is an important area with room for innovation. Most dredging operations are fairly intrusive owing to their method of picking up sediment from the seabed. Cutter-suction heads or water jets loosen bed material during the dredging process and create high levels of local turbidity in addition to the potential to harm wildlife. Bedload collectors operate on the principle that bedload sediment can be harvested by gravity at the rate of naturally occurring bedload transport. While bedload collectors have been successfully collecting sediment material in riverine settings where the transport is unidirectional, testing of such systems for coastal applications is ongoing (Tucker et al., 2018). The principle of bedload collectors is simple but requires adaptation to each location of interest. In contrast to suspended load, bed load is the mode of sediment transport describing the movement of heavier sediment grains close to the bed of a river or coastal system. Grains can roll, slide, or bounce near the surface driven by currents and wave motions but always maintain some intermittent connection to the bed. Heavier and larger grains are often desired for shore protection projects since they can resist erosion better than fine material. Hence, a device that can collect these types of grains from a system without disturbing the system too much is desirable. The surf zone region of the coast often features strong wave-generated currents moving sediment material along the shoreline (the same process that is redistributing sediment at the sand engine). A bedload collector essentially provides a trap for denser grains to fall in as they travel along the seabed. The width, depth, length, and geometric shape of the gap are parameters that need to be adjusted based on local sediment and transport conditions. Once settled in the trap, the mixture of water and bedload material is pumped onto land for further use. Bedload collectors can be a critical component of sediment back-passing systems where they can replace more intrusive methods of sediment dredging and where bedload material is the desired fraction of sediment to use. Physical model tests of bedload collector configurations in both field and laboratory settings are essential to perfect this sediment harvesting method and make it suitable for use in coastal systems.

Conclusions

Sediment movement in coastal systems is primarily driven by hydrodynamic forcing generated from wind, water level changes, waves, and currents. This chapter highlights the importance of including sediment transport and morphodynamic processes for accurate coastal flood hazard predictions while recognizing the inherent difficulties in modeling combined water and sediment flows together with associated morphological evolution. Both physical and numerical model approaches are discussed in light of their respective contributions and limitations to a better understanding of coastal flooding processes. While spatial and temporal restrictions often limit the applicability of such models, many coastal flood model scenarios require the inclusion of sediment movement to make accurate predictions. Examples include the consideration of dune dynamics during storm impact for accurate flood modeling or the movement of polluted sediments during flood flows. In-depth knowledge and accurate modeling of coastal sediment transport processes, however, does not only help with better flooding predictions. Sediment movement and its accurate modeling can also be harnessed to help reduce flood risk in coastal systems. Project examples from the Netherlands and the United States show that the natural interaction between hydrodynamic forcing and sediment transport can be utilized in a beneficial way to deliver sediment to coastal areas where it is needed for flood protection. It is clear from the presented examples that continued innovation and research are needed to optimize the prediction as well as the beneficial use of combined water and sediment dynamics in coastal systems.

References

Aguirre-Pe, J., Olivero, M. L., & Moncada, A. T. (2003). Particle densimetric froude number for estimating sediment transport. *Journal of Hydraulic Engineering*, *129*(6), 428–437. https://doi.org/10.1061/(ASCE)0733-9429(2003)129:6(428).

Aidala, J. A., Burke, C. E., & McLellan, T. N. (1992). Hydrodynamic forces and evolution of a nearshore berm at South Padre Island, Texas. In *Proceedings of the hydraulic engineering sessions at Water Forum '92* (pp. 1234–1239).

Amoudry, L. O., & Souza, A. J. (2011). Deterministic coastal morphological and sediment transport modeling: A review and discussion. *Reviews of Geophysics*, *49*(2). https://doi.org/10.1029/2010RG000341.

Bagnold, R. A. (1946). Motion of waves in shallow water—Interaction between waves and sand bottoms. *Proceedings of the Royal Society A*, *187*(1008), 1–18. https://doi.org/10.1098/rspa.1946.0062.

Bagnold, R. A. (1956). The flow of cohesionless grains in fluids. *Philosophical Transactions of the Royal Society London A*, *249*(964), 235–297. https://doi.org/10.1098/rsta.1956.0020.

Baptist, M. J., van Eekelen, E., Dankers, P. J. T., Grasmeijer, B., van Kessel, T., & van Maren, D. S. (2017). *Working with nature in Wadden Sea ports*. https://library.wur.nl/WebQuery/wurpubs/524847.

Baptist, M., Vroom, J., Willemsem, P., Puijenbroek, M., van Maren, B., van Steijn, P., et al. (2019). *Beneficial use of dredged sediment to enhance salt marsh development by applying a 'mud motor': Evaluation based on monitoring* (Research Report C088/19 & 1209751). Wageningen Marine Research and Deltares. https://doi.org/10.18174/500109.

Berkowitz, J., Piercy, C., Welp, T., & VanZomeren, C. (2019). *Thin layer placement: Technical definition for U. S. Army Corps of Engineers applications* (p. 9). Engineer Research and Development Center (U.S.) (Technical Note) 10.21079/11681/32283.

Brutsché, K., McFall, B., Bryant, D., & Wang, P. (2019). *Literature review of nearshore berms* (Final Report ERDC/CHL SR-19-2) (p. 59). Engineer Research and Development Center (U.S.). https://doi.org/10.21079/11681/32509.

Cañizares, R., & Irish, J. L. (2008). Simulation of storm-induced barrier island morphodynamics and flooding. *Coastal Engineering*, *55*(12), 1089–1101. https://doi.org/10.1016/j.coastaleng.2008.04.006.

Choi, J., Kirby, J. T., & Yoon, S. B. (2015). Boussinesq modeling of longshore currents in the SandyDuck experiment under directional random wave conditions. *Coastal Engineering*, *101*, 17–34. https://doi.org/10.1016/j.coastaleng.2015.04.005.

de Schipper, M. A., de Vries, S., Ruessink, G., de Zeeuw, R. C., Rutten, J., van Gelder-Maas, C., et al. (2016). Initial spreading of a mega feeder nourishment: Observations of the sand engine pilot project. *Coastal Engineering*, *111*, 23–38. https://doi.org/10.1016/j.coastaleng.2015.10.011.

de Schipper, M. A., De Vries, S., Stive, M., de Zeeuw, R., Rutten, J., Ruessink, G., et al. (2014). Morphological development of a mega-nourishment; first observations at the sand engine. *Coastal Engineering Proceedings*, *1*(34), 73.

De Vriend, H. J., & Stive, M. J. F. (1987). Quasi-3D modelling of nearshore currents. *Coastal Engineering*, *11*(5), 565–601. https://doi.org/10.1016/0378-3839(87)90027-5.

de Vriend, H. J., van Koningsveld, M., Aarninkhof, S. G. J., de Vries, M. B., & Baptist, M. J. (2015). Sustainable hydraulic engineering through building with nature. *Journal of Hydro-Environment Research*, *9*(2), 159–171. https://doi.org/10.1016/j.jher.2014.06.004.

Dellapenna, T. M., Hoelscher, C., Hill, L., Al Mukaimi, M. E., & Knap, A. (2020). How tropical cyclone flooding caused erosion and dispersal of mercury-contaminated sediment in an urban estuary: The impact of Hurricane Harvey on Buffalo Bayou and the San Jacinto Estuary, Galveston Bay, USA. *Science of the Total Environment, 748*. https://doi.org/10.1016/j.scitotenv.2020.141226, 141226.

Dietrich, J. C., Zijlema, M., Westerink, J. J., Holthuijsen, L. H., Dawson, C., Luettich, R. A., et al. (2011). Modeling hurricane waves and storm surge using integrally-coupled, scalable computations. *Coastal Engineering, 58*(1), 45–65. https://doi.org/10.1016/j.coastaleng.2010.08.001.

Figlus, J., Song, Y.-K., Maglio, C. K., Friend, P. L., Poleykett, J., Engel, F. L., et al. (2021). Particle tracer analysis for submerged berm placement of dredged material near South Padre Island, Texas. *Journal of Dredging, 19*(1), 14–30.

Grant, W. D., & Madsen, O. S. (1979). Combined wave and current interaction with a rough bottom. *Journal of Geophysical Research, Oceans, 84*(C4), 1797–1808. https://doi.org/10.1029/JC084iC04p01797.

Grant, W. D., & Madsen, O. S. (1982). Movable bed roughness in unsteady oscillatory flow. *Journal of Geophysical Research, Oceans, 87*(C1), 469–481. https://doi.org/10.1029/JC087iC01p00469.

Heller, V. (2011). Scale effects in physical hydraulic engineering models. *Journal of Hydraulic Research, 49*(3), 293–306. https://doi.org/10.1080/00221686.2011.578914.

Holthuijsen, L. H., Herman, A., & Booij, N. (2003). Phase-decoupled refraction–diffraction for spectral wave models. *Coastal Engineering, 49*(4), 291–305. https://doi.org/10.1016/S0378-3839(03)00065-6.

Hughes, S. A. (1993). *Physical models and laboratory techniques in coastal engineering. Vol. 7.* World Scientific. https://doi.org/10.1142/2154.

James, S. C., Jones, C. A., Grace, M. D., & Roberts, J. D. (2010). Advances in sediment transport modelling. *Journal of Hydraulic Research, 48*(6), 754–763. https://doi.org/10.1080/00221686.2010.515653.

Kamphuis, J. W. (1991). Alongshore sediment transport rate. *Journal of Waterway, Port, Coastal, and Ocean Engineering, 117*(6), 624–640. https://doi.org/10.1061/(ASCE)0733-950X(1991)117:6(624).

Losada, I. J., Lara, J. L., Guanche, R., & Gonzalez-Ondina, J. M. (2008). Numerical analysis of wave overtopping of rubble mound breakwaters. *Coastal Engineering, 55*(1), 47–62. https://doi.org/10.1016/j.coastaleng.2007.06.003.

McFall, B. C., Smith, J., Pollock, C. E., Rosati, J. I., & Brutsché, K. E. (2016). *Evaluating sediment mobility for siting nearshore berms* (Technical Note ERDC/CHL CHETN-IV-108) (p. 11). Engineer Research and Development Center.

Nam, P. T., Larson, M., Hanson, H., & Hoan, L. X. (2009). A numerical model of nearshore waves, currents, and sediment transport. *Coastal Engineering, 56*(11), 1084–1096. https://doi.org/10.1016/j.coastaleng.2009.06.007.

Sallenger, A. H., Jr. (2000). Storm impact scale for barrier islands. *Journal of Coastal Research, 16*(3), 890–895.

Sebastian, A., Proft, J., Dietrich, J. C., Du, W., Bedient, P. B., & Dawson, C. N. (2014). Characterizing hurricane storm surge behavior in Galveston Bay using the SWAN+ADCIRC model. *Coastal Engineering, 88*, 171–181. https://doi.org/10.1016/j.coastaleng.2014.03.002.

Shields, A. (1936). Anwendung der Ähnlichkeitsmechanik und der Turbulenzforschung auf die Geschiebebewegung. In *Vol. 26. Mitteilungen der preußischen Versuchsanstalt für Wasserbau und Schiffbau, Berlin* (p. 26).

Smit, P., Zijlema, M., & Stelling, G. (2013). Depth-induced wave breaking in a non-hydrostatic, near-shore wave model. *Coastal Engineering, 76*, 1–16. https://doi.org/10.1016/j.coastaleng.2013.01.008.

Stive, M. J. F., de Schipper, M. A., Luijendijk, A. P., Aarninkhof, S. G. J., van Gelder-Maas, C., van Thiel de Vries, J. S. M., et al. (2013). A new alternative to saving our beaches from sea-level rise: The sand engine. *Journal of Coastal Research, 290*, 1001–1008. https://doi.org/10.2112/JCOASTRES-D-13-00070.1.

Taqi, A., & Figlus, J. (2019). A 3D physical model study of reinforced dune evolution during storm conditions. In *Proceedings of coastal structures 2019. Coastal Structures 2019, Hannover, Germany*.

Temmerman, S., Meire, P., Bouma, T. J., Herman, P. M. J., Ysebaert, T., & De Vriend, H. J. (2013). Ecosystem-based coastal defence in the face of global change. *Nature, 504*(7478), 79–83. https://doi.org/10.1038/nature12859.

Thornton, E. B., & Guza, R. T. (1986). Surf zone longshore currents and random waves: Field data and models. *Journal of Physical Oceanography, 16*(7), 1165–1178. https://doi.org/10.1175/1520-0485(1986)016<1165:SZLCAR>2.0.CO;2.

Tomasicchio, G. R., Sánchez-Arcilla, A., D'Alessandro, F., Ilic, S., James, M. R., Sancho, F., et al. (2011). Large-scale experiments on dune erosion processes. *Journal of Hydraulic Research, 49*(sup1), 20–30. https://doi.org/10.1080/00221686.2011.604574.

Tucker, R., Trevion, R., Welp, T., Maglio, C., Moya, J., Tyler, Z., et al. (2018). Evolution of an innovative dredging technology for harvesting coarse-grained sediment; from riverine to marine applications. In *Western Dredging Association Summit. WEDA Summit & Expo, Norfolk, VA.*

van Eekelen, E., & Baptist, M. J. (2019). The mud motor: A beneficial use of dredged sediment to enhance salt marsh development. *Terra et Aqua, 155*(4), 28–42.

van Gent, M. R. A., van Thiel de Vries, J. S. M., Coeveld, E. M., de Vroeg, J. H., & van de Graaff, J. (2008). Large-scale dune erosion tests to study the influence of wave periods. *Coastal Engineering, 55*(12), 1041–1051.

van Regteren, M., ten Boer, R., Meesters, E. H., & de Groot, A. V. (2017). Biogeomorphic impact of oligochaetes (Annelida) on sediment properties and Salicornia spp. Seedling establishment. *Ecosphere, 8*(7). https://doi.org/10.1002/ecs2.1872, e01872.

Van Rijn, L. C., Tonnon, P. K., Sánchez-Arcilla, A., Cáceres, I., & Grüne, J. (2011). Scaling laws for beach and dune erosion processes. *Coastal Engineering, 58*(7), 623–636. https://doi.org/10.1016/j.coastaleng.2011.01.008.

Welch, M., Mogren, E. T., & Beeney, L. (2016). *A literature review of the beneficial use of dredged material and sediment management plans and strategies. Vol. 34* (p. 45). Portland State University, Hatfield School of Government, Center for Public Service.

Willemsen, P. W. J. M., Borsje, B. W., Hulscher, S. J. M. H., der Wal, D. V., Zhu, Z., Oteman, B., et al. (2018). Quantifying bed level change at the transition of tidal flat and salt marsh: Can we understand the lateral location of the marsh edge? *Journal of Geophysical Research. Earth Surface, 123*(10), 2509–2524. https://doi.org/10.1029/2018JF004742.

Work, P. A., & Otay, E. N. (1997). Influence of nearshore berm on beach nourishment. *Coastal Engineering, 1996*, 3722–3735. https://doi.org/10.1061/9780784402429.287.

CHAPTER 5

Urban flood modeling: Perspectives, challenges, and opportunities

Antonia Sebastian[a], Andrew Juan[b], and Philip B. Bedient[b]

[a]Department of Earth, Marine and Environmental Sciences, The University of North Carolina at Chapel Hill, Chapel Hill, NC, United States
[b]Department of Civil and Environmental Engineering, Rice University, Houston, TX, United States

Introduction

As demonstrated by recent extreme events, including the TX-LA Floods (2016), Hurricane Harvey (2017), and Tropical Storm Imelda (2019), urban flooding driven by extreme precipitation can have devastating consequences for urban communities. This type of flooding—also known as pluvial flooding or flash flooding—occurs when the capacity of the natural or engineered drainage infrastructure is exceeded by intense rainfall (Carter et al., 2015; Falconer et al., 2009). In this case, inundation may manifest as ponding, street flooding, or structural inundation as floodwaters seek an overland flow pathway to the nearest body of water or major drainage channel. Typically, pluvial flooding occurs rapidly over a small geographic area and recedes within a matter of minutes to hours; however, in extreme cases, miles of infrastructure can remain inundated over long periods of time disrupting businesses and impacting recovery.

Despite their prevalence, pluvial flood hazards have been historically overlooked by the flood modeling community (Rosenzweig et al., 2018). Riverine or coastal floods tend to have larger footprints, higher flow velocities, and deeper inundation, and have historically caused more damage on a per-event basis than pluvial events (Rözer et al., 2016). Moreover, they are assumed to be managed through existing stormwater infrastructure. As a result, advances in flood-hazard modeling and risk assessment in urban contexts have focused primarily on riverine and coastal flooding, relegating pluvial hazards to nuisance flooding and assuming that they cause minimal socioeconomic impact (Van De Ven et al., 2011). However, several recent studies suggest that flooding driven by intense rainfall is more widespread than previously acknowledged and a significant contributor to socioeconomic risk (ASFPM, 2019; Galloway et al., 2018; NASEM, 2019).

In the following sections, we discuss pluvial flooding as a critical area of research. We then provide a brief overview of the types of hydrodynamic models used in urban flood studies and outline recent advances in modeling the hazard associated with two of the main drivers of urban flooding: pluvial and fluvial flooding, using physics-based methods, and the incorporation of data-driven approaches together with physical models to map flood risk.

We will specifically comment on the limitations of existing hydrodynamic models in the context of capturing flood processes in the built environment and highlight the need for more comprehensive risk mapping across multiple flood drivers and their impacts. The model overview and associated discussion are aimed at a general audience. We end the chapter with a section highlighting several key research priorities.

Pluvial flooding as a critical area of research

Flooding is the costliest natural disaster worldwide (UNISDR, 2015). While the vast majority of flood damage occurs as a result of river and coastal flooding brought on by severe tropical cyclones and mesoscale convective systems, the occurrence of relatively small, localized extreme precipitation events can lead to substantial cumulative losses over time. In the United States, evidence points to mounting losses in areas outside of the FEMA Special Flood Hazard Areas (SFHAs) which is used as the primary indicator of risk, but only accounts for riverine or coastal hazards (Blessing, Sebastian, & Brody, 2017). Similarly, flooding caused by "cloudbursts"—or intense rainfall—is an increasing focus of Dutch water management (Dai, Wörner, & van Rijswick, 2018; Van De Ven et al., 2011). While there is no database isolating the contribution of pluvial flooding to global losses, reports suggest that in some countries they may account for between 30% and 50% of all flood impacts (Schanze, 2006).

Pluvial flooding occurs naturally in geographies characterized by poorly infiltrated soils (e.g., clays) and little topographic relief. However, pluvial flooding may also occur as a result of land surface changes, like the proliferation of impervious surfaces associated with urban sprawl (e.g., Odense, Denmark; Skougaard Kaspersen et al., 2015), soil compaction and subsidence (e.g., the Netherlands; Van De Ven et al., 2011), and loss of headwater streams (e.g., Baltimore, USA; Elmore & Kaushal, 2008), as well as atmospheric changes, like increasing extreme precipitation as a result of global warming (e.g., Houston, USA; Van Oldenborgh et al., 2017) and urban heat island (UHI) effects (e.g., Atlanta, USA; Debbage & Marshall Shepherd, 2019). This latter phenomenon emerges as the result of complex land surface-atmospheric dynamics which drive precipitation (Marelle et al., 2020). Across several recent case studies, urban development has been shown to exacerbate precipitation observed during convective rainfall (Niyogi, Lei, Kishtawal, Schmid, & Shepherd, 2017) and tropical cyclone events (Zhang, Villarini, Vecchi, & Smith, 2018).

While pluvial flooding is generally classified as an "urban" problem, the challenges associated with extreme rainfall are not limited to large metropolitan areas. Instead, pluvial flooding is prevalent across all types of human settlements (e.g., urban, suburban, exurban, and rural), particularly in those where existing civil infrastructure is insufficient to handle the magnitude and intensity of storms experienced. For example, in Houston, the storm sewer network has been historically designed to contain a 2-year storm

(COH Infrastructure Design Manual, 2018), however, more severe storms as a result of anthropogenic climate change, coupled with aging infrastructure and changing land uses has demonstrated the limitations of existing built environment systems to accommodate extreme precipitation (Sebastian, Gori, Blessing, van der Wiel, & Bass, 2019).

As the costs associated with managing flooding through centralized infrastructure continue to grow and resources (and space) are increasingly strained, managing flood risks will become increasingly important for community resilience. Flood management will, in part, require (1) accurately modeling the frequency and magnitude of flooding associated with extreme precipitation events, (2) delineating human and infrastructure exposure and vulnerability to flooding, and (3) quantifying present and future risks to optimize investments in infrastructure and identify critical adaptation pathways. In this context, the following section discusses the current state of hydrodynamic modeling and their combination with data-driven approaches to delineate flood hazards and their impacts, particularly within urban and other developed areas.

Modeling flood hazards in urban areas

Hydrodynamic models, in the context of flood inundation modeling, apply physics-based equations to describe and simulate the movement of stormwater within an area. These models vary widely in terms of their complexity and computational expense, and the choice of model depends on the scope of the problem (Bulti & Abebe, 2020) and the extent to which hydrologic processes and/or urban infrastructure (e.g., channels, bridges, storm sewers) influence the hydrodynamic behavior of the system.

In general, hydrodynamic models all solve some form of the Navier-Stokes equations. These equations represent fluid flow in three dimensions (i.e., the planar directions: x and y, and the vertical direction, z) and are based on the principles of conservation of mass and momentum. The equations can be simplified by reducing the number of dimensions of flow or neglecting terms. Based on the numerical representation of the fluid flow and dictated by the dominant flow direction(s), hydrodynamic models are typically categorized as 1D, 2D, and 3D. For example, in applications where fluvial processes dominate, 1D models are used; where pluvial (e.g., extreme precipitation, local drainage) or coastal (e.g., storm surge, tsunamis) processes dominate, 2D or 3D models are better suited.

Table 1 lists several widely used hydrodynamic models and provides selected studies where they have been applied.

1D models

One-dimensional (1D) models (e.g., HEC-RAS 1D, EPA-SWMM, TUFLOW 1D) represent flow in one direction: either upstream or downstream along the length of a river or channel reach. These models solve the Saint-Venant equations, which conserve mass and momentum in one direction. In most 1D models, rivers and channels, including

Table 1 Select models/software used in urban flood studies with examples.

Model/Software	Developer	1D, 2D, 3D	Status	Link	Example studies
HEC–RAS	US Army Corps of Engineers	1D, 2D, 1D/2D	Freeware, not open source	https://www.hec.usace.army.mil/software/hec-ras/	Bass, Juan, Gori, Fang, and Bedient (2016); Gori, Blessing, Juan, Brody, and Bedient (2019); Juan, Gori, and Sebastian (2020)
EPA-SWMM	Environmental Protection Agency	1D	Freeware, not open source	https://www.epa.gov/water-research/storm-water-management-model-swmm	Rabori and Ghazavi (2018); Bai, Zhao, Zhang, and Zeng (2018)
Delft 3D	Deltares	3D	Open source	https://oss.deltares.nl/web/delft3d	Elias, Walstra, Roelvink, Stive, and Klein (2020); Sao (2008)
Infoworks ICM	Innovyze	1D, 2D	Commercial	https://www.innovyze.com/en-us/products/infoworks-icm	Rubinato, Shucksmith, Saul, and Shepherd (2013)
LISFLOOD-FP	University of Bristol	1D, 2D	Research	http://www.bristol.ac.uk/geography/research/hydrology/models/lisflood/	Neal et al. (2012); Wing et al. (2018)
MIKE	DHI	1D, 2D, 3D	Commercial	https://www.mikepoweredbydhi.com/products/mike-flood	Seyoum, Zoran, Price, and Sutat (2012); Xie et al. (2017)
Telemac	Telemac-Mascaret	1D, 2D, 3D	Open source	http://www.opentelemac.org/	Brière, Abadie, Bretel, and Lang (2007); Villaret, Hervouet, Kopmann, Merkel, and Davies (2013)
TUFLOW	BMT	1D, 2D, 3D	Commercial	https://www.tuflow.com/	Phillips, Yu, Thompson, and de Silva (2005)
SOBEK	Deltares	1D, 2D	Commercial	https://www.deltares.nl/en/software/sobek/	Tarekegn, Haile, Rientjes, Reggiani, and Alkema (2010)

Fig. 1 HEC-RAS 1D geometric cross-sectional model of Brays Bayou in Houston, TX.

complex in-stream structures such as bridges, dams, and levees, are explicitly represented as a series of cross sections or river nodes (see Fig. 1). Flow, average velocity, and water levels are computed at each cross section and floodplains are generated by interpolating the values between adjacent cross sections. Flood inundation maps can then be created by subtracting the underlying terrain from the computed and interpolated water surface elevations (WSE). Where fluvial flooding is the primary driver of inundation, the 1D assumption is often sufficient to characterize flood flow and its impacts near the channel (see, e.g., Juan et al., 2020). In fact, the majority of the US Federal Emergency Management Agency (FEMA) Special Flood Hazard Areas (i.e., floodplains) for inland areas are products of steady-state 1D models. A major advantage of 1D models compared to their 2D and 3D counterparts is that computational runtimes tend to be small. This means that applications that require running thousands of simulations, such as in a probabilistic flood hazard analysis, are feasible with a 1D model.

While 1D models can be used to capture the impacts of heavy rainfall-runoff processes when combined with hydrologic models or river gauge data, the resulting floodplains are limited to the extents of the cross sections in the model domain. In other words, while 1D models are able to simulate the effects of riverine flooding (i.e., flooding from channel overtopping its banks), they are not well-suited for capturing more complex processes that may occur in overland areas within the floodplain due to localized rainfall (e.g., pluvial flooding), or in areas hydrologically disconnected from the river system. This is especially true in mild or flat-sloped regions where flow is not limited to any particular direction or in urban environments where local drainage infrastructure may become overwhelmed.

2D models

Two-dimensional models (e.g., HEC-RAS 2D, Infoworks ICM, MIKE FLOOD, TUFLOW 2D) represent flow propagation in the planar directions such that flood waves are allowed to propagate laterally as well as longitudinally within the model domain. In a 2D model, the hydrodynamic response is computed at every storm-impacted cell as stormwater propagates throughout the model domain. Water surface elevation and flux is calculated based on a numerical approximation of the shallow water equations in which

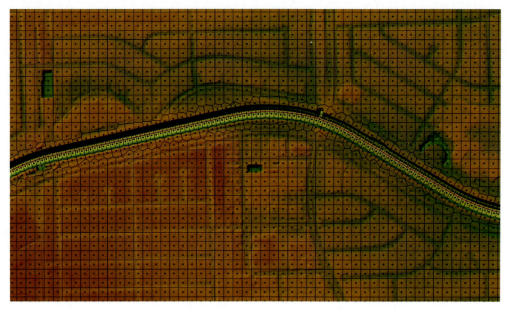

Fig. 2 HEC-RAS 2D model setup of Brays Bayou watershed in Houston, TX.

both mass and momentum are conserved within a plane and flow in the vertical direction is assumed to be shallow or negligible relative to the other two dimensions. Flood inundation can then be estimated based on the difference between total water surface elevation and the underlying topography. The study area may be represented as a single mesh or multiple meshes of interconnected grid cells (see Fig. 2). Based on their spatial representation, these meshes can be structured (i.e., square grids), unstructured (e.g., irregular triangles or quadrilaterals), or flexible (i.e., a combination of structured and unstructured meshes). Unstructured meshes tend to better represent the physical characteristics of the terrain, whereas structured meshes tend to be easier for both pre- and postprocessing.

One major advantage 2D models have over 1D approaches is that they allow for better representation of pluvial flooding and interbasin transfers since flow is not constrained within particular channels or water bodies. This advantage is especially valuable for modeling pluvial flooding within the built environment. In a recent study, Garcia, Juan, and Bedient (2020) used a 2D model, HEC-RAS 2D, to simulate the flood impacts of extreme precipitation, interbasin transfers, and reservoir releases of several watersheds in west Houston during Hurricane Harvey in 2017. Another advantage of 2D models is their ability to leverage high-resolution geospatial data such as LiDAR-derived Digital Elevation Models (DEMs) (i.e., topography), land use/land cover (LULC), and raster-based soil surveys.

While 2D models have existed for over two decades, the adoption of 2D models in practice (i.e., by the general engineering community and governmental agencies) has been slow, in part because the models are more computationally intensive than their 1D counterparts. Historically, this made building a regional 2D model on a personal computer infeasible. As a result, most watershed and regional floodplain studies have used a combination of hydrologic models and 1D hydraulic models to estimate flood extent. In recent years, however, the computational burden of building and running 2D models has decreased substantially due to advances in parallelization, vectorization, and multicore computing, making it increasingly feasible to use 2D models in large-scale regional studies (Courty, Pedrozo-Acuña, & Bates, 2017). For example, Neal et al. (2012) optimized the 2D hydraulic model, LISFLOOD-FP, improving its single-core and multicore performances that resulted in a dramatic reduction in model run times compared to previous iterations of the model.

In May 2021, USACE released a HEC-RAS version 6.0 in which it addressed many of the shortcomings of the previous versions of the 2D model. HEC-RAS 6.0 allows for spatially distributed rainfall and infiltration, complex bridge hydraulics and other flood control structures, as well as wind forces, to better represent coastal hydrodynamics. These new features and capabilities would undoubtedly improve the usability of 2D models in numerous urban applications (Fig. 3).

Coupled models

In some cases, 2D models have been coupled to their 1D counterparts. These hybrid 1D/2D models usually represent the channels and the corresponding riverine floodplain in

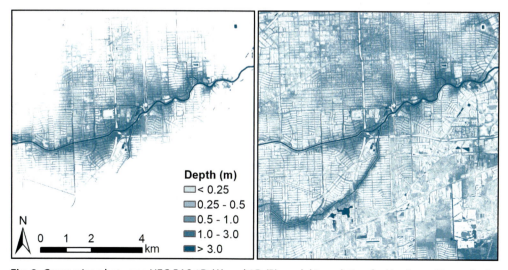

Fig. 3 Comparison between HEC-RAS 1D (A) and 2D (B) model inundation for Hurricane Harvey in the Brays Bayou watershed in Houston, TX.

1D and overland areas in 2D (Gori et al., 2019). The advantage of these models is the ability to delineate the channel and associated infrastructure while still capturing the hydrodynamic process in areas that may be disconnected from the main channel. The second type of 1D/2D models are those which couple 2D surface models with 1D subsurface models. The ability to model surface-subsurface dynamics is particularly important in urban areas where one may want to represent the contributions of underground stormwater infrastructure to inundation at the surface. In many cases, stormwater infrastructure is only designed to capture and route floodwaters during small to medium-sized events. During large events, storm sewers may become overwhelmed or even overflow, further exacerbating flooding. As a result, these processes are overlooked in flood hazard mapping because the positive contribution of these infrastructure systems to reducing inundation is negligible during the most extreme events.

3D models

Three-dimensional (3D) hydrodynamic models (e.g., Delft3D, Telemac 3D, TUFLOW 3D) are seldom used in floodplain applications except for a few specific cases where modeling vertical flow, velocity, and turbulence is important. Some examples include studies of dam breaks and extreme erosion or scour due to major flash floods and/or storm surge. Some 3D models are derived from the Navier-Stokes equations that describe fluid motion in three dimensions (e.g., Ye & McCorquodale, 1998), whereas other models employ the shallow water equations to solve for 2D horizontal flow and use a pseudo-3D solver to compute velocities in the vertical profile to reduce computational cost (e.g., Casulli & Stelling, 1998). Because most applications to mapping flood hazards in urban areas involve shallow water flow, 3D models are seldom used for floodplain mapping.

Alternative approaches

The application of full physics-based, hydrodynamic models is often limited by scale due to computational requirements. This is true for most 2D and 3D models, but can also apply to 1D models as well. To address this issue, simple raster-based methods have been explored as an alternative to physics-based models to obtain reasonable estimates of inundation while leveraging high-resolution terrain data and modeled or observed streamflow data and stage-discharge relationships. One example is the Height Above Nearest Drainage (HAND)-based method in which flood inundation is estimated based on the terrain's relative gravitational potential and stage-discharge relationships (Nobre et al., 2011). The HAND method has been integrated with the US National Water Model to map flooding in real time; however, its accuracy is poor in areas with small streams and low topographic relief (Johnson, Munasinghe, Eyelade, & Cohen, 2019).

In addition, the use of data-driven models that apply statistical tools and machine learning (ML) techniques in flood risk and flood hazard studies has increased dramatically. The popularity of these approaches lies in the fact that conventional sources of flood-related data (e.g., stream gages and surveyed high watermarks) are often limited in their availability, distribution, and reliability. Even when available, these data sources are usually located within the vicinity of a channel, which makes them ill-suited for capturing pluvial flooding and its associated hazard and risk. To address this shortcoming, various ML studies have explored the utilization of alternative sources of information, including rescue requests (e.g., Mobley, Sebastian, Highfield, & Brody, 2019), flood insurance claims (e.g., Knighton et al., 2020), and social media and other crowdsourced data platforms (See, 2019) together with underlying information about the physical environment and/or socioeconomic characteristics of the community to estimate the extent of flooding.

Modeling the risks associated with urban flooding

Conventionally, losses from extreme flood events are considered in terms of their direct, tangible economic impacts. However, in the context of urban flooding, indirect or intangible losses (e.g., infrastructure and business disruptions) may be equally, if not more important to the overall impact of the event (ten Veldhuis, 2011). In general, research focused on improving the assessment of risks from urban flooding has focused on three primary areas: (1) direct household damage (e.g., Merz et al., 2014; Rözer et al., 2019), (2) infrastructure disruption (e.g., Coles, Yu, Wilby, Green, & Herring, 2017; Dong, Yu, Farahmand, & Mostafavi, 2020; Pregnolato, Galasso, & Parisi, 2015), and (3) recovery (e.g., Murdock, de Bruijn, & Gersonius, 2018). In the simplest case, models relating depth of water to expected damage are applied to estimate losses; however, the complexity of the urban environment and limited data with which to validate these approaches leads to considerable uncertainty in these estimates. To overcome these limitations, there has been an increase in research using advanced statistical tools to assess damages and a rapidly expanding area of research that relies on citizen science and crowdsourcing to fill these knowledge gaps (Ten Veldhuis & Clemens, 2010).

Key research priorities

Emerging from the discussion of recent advances in modeling urban flood hazards and associated risk are many opportunities for additional research in support of urban flood resilience. We highlight three key priority areas below:

(1) *Improve tools used to model flood hazards in the built environment and expand their application to regional scales.* Hydrodynamic models that can accurately represent the complexities of the physical processes that drive urban flooding will lead to better stormwater management, development planning, and flood policy. Numerical approximations based on a diffusive wave scheme have enabled researchers to address issues of scale, numerical instability, and high computational time in 2D overland models (Hammond, Chen, Djordjević, Butler, & Mark, 2015). However, more work is needed to integrate stormwater infrastructure both on the surface (e.g., channels, culverts) and below ground (e.g., pipes; Leandro, Schumann, & Pfister, 2016) into these models in order to accurately simulate the urban environment across the city and regional scales.

(2) *Advance scientific understanding of the interactions between climate, land use/land cover, development, and the hydrologic cycle at local and regional scales.* Sound adaptation policy and spatial planning rely on the ability to understand and accurately model flood hazards and how they coevolve with development and climate change across spatial and temporal scales (Pelletier et al., 2015). Recent studies have demonstrated that climate and land use change are both significant drivers of flooding, where their combined impact is greater than that of the isolation (Sebastian et al., 2019; Zhang et al., 2018). However, these dynamic interactions across human-natural systems are still poorly understood and application of these models in practice remains limited. Neglecting to consider these dynamics could lead to decisions that may be maladaptive over longer time scales (Di Baldassarre et al., 2013).

(3) *Increase capacity to assess and manage flood risks in urban areas under current and future conditions.* Data-driven approaches for estimating urban flood risks are becoming increasingly sophisticated, ranging from modeling flood hazards in near real time (e.g., Mobley et al., 2019) to estimating structural damage (e.g., Rözer et al., 2019), to predicting infrastructure disruption, downtime, and recovery (e.g., Gori et al., 2020). Continuing scientific advancement in these areas necessitates detailed data collection and accessibility (e.g., high-resolution infrastructure databases, crowdsourced data (Gaitan & Ten Veldhuis, 2015) and damage information (Spekkers, Clemens, & Ten Veldhuis, 2015; Wing, Pinter, Bates, & Kousky, 2020) to build and validate new models as well as to assess the cost-benefits of proposed flood risk reduction (FRR) strategies.

Summary

Urban flooding is the leading cause of global flood losses and can arise as a result of pluvial, fluvial, or coastal flooding. Of the three flood hazard types, pluvial flooding—flooding that occurs as a result of heavy rainfall and inadequate stormwater infrastructure or limited infiltration capacity—has been historically overlooked because it is assumed to be managed and is perceived to cause less damage on average. However, increasing evidence suggests that pluvial flooding is a significant driver of cumulative damage over time, and that the risks of these events are growing due to changes in hazard exposure, aging infrastructure, urbanization, and global warming. As a result, urban flooding driven by extreme rainfall has emerged as a critical area of research, particularly within the modeling community.

This chapter provides an overview of the various types of models used in urban flood hazard studies as well as their advantages and limitations. The models can be categorized as 1D, 2D, or 3D based on their numerical solution to the Navier-Stokes equations. The 1D approach provides an estimate of flooding when riverine processes dominate and have been widely applied in regulatory flood hazard studies. However, they cannot simulate flooding that may occur due to complex surface processes within the floodplain or in areas disconnected from the primary channel. A downside of 1D models is that they require detailed information about the channel bathymetry and must be linked to a hydrologic model or river gage data to provide hydrologic (i.e., rainfall-runoff) input. In contrast, 2D models can leverage high resolution regional and global geospatial datasets and can incorporate rainfall directly onto the grid. The 2D models capture flood processes that occur in the planar directions where elevation is represented as either a structured or unstructured grid. The primary limitation of 2D models is their computational expense. 3D models are seldom used in urban flood studies as flow in the vertical direction is negligible.

Advances in Desktop and high-performance computing have enabled the uptake of more complex flood hazard models in practice; however, their widespread application is still limited by the availability of resources, validation data, and scale. As a result, the popularity of analytical approaches which take advantage of high-resolution geospatial data as well as point-level data to project the spatial extent of flood hazards has grown dramatically. Together with historical data on flood losses, these data-centric approaches hold promise in areas where detailed information about infrastructure or resources is limited. Nevertheless, much work is still needed to extend both hydrodynamic models and analytical tools to simulate risks under current and projected conditions to address future urban flood resilience.

References

ASFPM. (2019). *Urban flooding: Moving towards resilience*. Washington, DC: ASFPM.

Bai, Y., Zhao, N., Zhang, R., & Zeng, X. (2018). Storm water management of low impact development in urban areas based on SWMM. *Water (Switzerland), 11*. https://doi.org/10.3390/w11010033.

Bass, B., Juan, A., Gori, A., Fang, Z., & Bedient, P. B. (2016). 2015 Memorial day storm flood impacts for changing watershed conditions in Houston, TX. *Natural Hazards Review*. https://doi.org/10.1061/(ASCE)NH.1527-6996.0000241.

Blessing, R., Sebastian, A., & Brody, S. D. (2017). Flood risk delineation in the United States: How much loss are we capturing? *Natural Hazards Review, 18*, 1–10. https://doi.org/10.1061/(ASCE)NH.1527-6996.0000242.

Brière, C., Abadie, S., Bretel, P., & Lang, P. (2007). Assessment of TELEMAC system performances, a hydrodynamic case study of Anglet, France. *Coastal Engineering, 54*, 345–356. https://doi.org/10.1016/j.coastaleng.2006.10.006.

Bulti, D. T., & Abebe, B. G. (2020). A review of flood modeling methods for urban pluvial flood application. *Modeling Earth Systems and Environment*, *6*, 1293–1302. https://doi.org/10.1007/s40808-020-00803-z.

Carter, J. G., Cavan, G., Connelly, A., Guy, S., Handley, J., & Kazmierczak, A. (2015). Climate change and the city: Building capacity for urban adaptation. *Progress in Planning*, *95*, 1–66. https://doi.org/10.1016/j.progress.2013.08.001.

Casulli, V., & Stelling, G. S. (1998). Numerical simulation of 3D quasi-hydrostatic, free-surface flows. *Journal of Hydraulic Engineering*, *124*. https://doi.org/10.1061/(ASCE)0733-9429(1998)124:7(678).

COH Infrastructure Design Manual. (2018). https://houstonrecovers.org/wp-content/uploads/2018/06/IDM-CH-9-Show-changes.pdf.

Coles, D., Yu, D., Wilby, R. L., Green, D., & Herring, Z. (2017). Beyond 'flood hotspots': Modelling emergency service accessibility during flooding in York, UK. *Journal of Hydrology*, *546*, 419–436. https://doi.org/10.1016/j.jhydrol.2016.12.013.

Courty, L. G., Pedrozo-Acuña, A., & Bates, P. D. (2017). Itzï (version 17.1): An open-source, distributed GIS model for dynamic flood simulation. *Geoscientific Model Development*, *10*, 1835–1847.

Dai, L., Wörner, R., & van Rijswick, H. F. M. W. (2018). Rainproof cities in the Netherlands: Approaches in Dutch water governance to climate-adaptive urban planning. *International Journal of Water Resources Development*, *34*, 652–674. https://doi.org/10.1080/07900627.2017.1372273.

Debbage, N., & Marshall Shepherd, J. (2019). Urban influences on the spatiotemporal characteristics of runoff and precipitation during the 2009 Atlanta flood. *Journal of Hydrometeorology*, *20*, 3–21. https://doi.org/10.1175/JHM-D-18-0010.1.

Di Baldassarre, G., Viglione, A., Carr, G., Kuil, L., Salinas, J. L., & Blöschl, G. (2013). Socio-hydrology: Conceptualising human-flood interactions. *Hydrology and Earth System Sciences*, *17*, 3295–3303. https://doi.org/10.5194/hess-17-3295-2013.

Dong, S., Yu, T., Farahmand, H., & Mostafavi, A. (2020). Probabilistic modeling of cascading failure risk in interdependent channel and road networks in urban flooding. *Sustainable Cities and Society*, *62*. https://doi.org/10.1016/j.scs.2020.102398, 102398.

Elias, L. E. P., Walstra, D. J. R., Roelvink, J. A., Stive, M. J. F., & Klein, M. D. (2020). Hydrodynamic validation of Delft3D with field measurements at Egmond. In *Coast. Eng. 2000, Proceedings*. https://doi.org/10.1061/40549(276)212.

Elmore, A. J., & Kaushal, S. S. (2008). Disappearing headwaters: Patterns of stream burial due to urbanization. *Frontiers in Ecology and the Environment*, *6*, 308–312. https://doi.org/10.1890/070101.

Falconer, R. H., Cobby, D., Smyth, P., Astle, G., Dent, J., & Golding, B. (2009). Pluvial flooding: New approaches in flood warning, mapping and risk management. *Journal of Flood Risk Management*, *2*, 198–208. https://doi.org/10.1111/j.1753-318X.2009.01034.x.

Gaitan, S., & Ten Veldhuis, J. A. E. (2015). Opportunities for multivariate analysis of open spatial datasets to characterize urban flooding risks. *IAHS-AISH Proceeding Reports*, *370*, 9–14. https://doi.org/10.5194/piahs-370-9-2015.

Galloway, G. E., Reilly, A., Ryoo, S., Riley, A., Haslam, M., Brody, S. D., et al. (2018). *The growing threat of urban flooding: A national challenge*. College Park: A. James Clark School of Engineering.

Garcia, M., Juan, A., & Bedient, P. (2020). Integrating reservoir operations and flood modeling with HEC-RAS 2D. *Water*, *12*.

Gori, A., Blessing, R. B., Juan, A., Brody, S. D., & Bedient, P. B. (2019). Characterizing urbanization impacts on floodplain through integrated land use, hydrologic, and hydraulic modeling. *Journal of Hydrology*, *568*, 82–95.

Gori, A., Gidaris, I., Elliott, J. R., Padgett, J. E., Bedient, P. B., Panakkal, P., et al. (2020). Accessibility and recovery assessment of Houston's roadway network due to fluvial flooding during Hurricane Harvey. *Natural Hazards Review*, *21*, 04020005. https://doi.org/10.1061/(ASCE)NH.1527-6996.0000355.

Hammond, M. J., Chen, A. S., Djordjević, S., Butler, D., & Mark, O. (2015). Urban flood impact assessment: A state-of-the-art review. *Urban Water Journal*, *12*, 14–29. https://doi.org/10.1080/1573062X.2013.857421.

Johnson, J. M., Munasinghe, D., Eyelade, D., & Cohen, S. (2019). An integrated evaluation of the National Water Model (NWM)–Height Above Nearest Drainage (HAND) flood mapping methodology. *Natural Hazards and Earth System Sciences*, *19*, 2405–2420. https://doi.org/10.5194/nhess-19-2405-2019.

Juan, A., Gori, A., & Sebastian, A. (2020). Comparing floodplain evolution in channelized and un-channelized urban watersheds in Houston, Texas. *Journal of Flood Risk Management*, 1–19. https://doi.org/10.1111/jfr3.12604.

Knighton, J., Buchanan, B., Guzman, C., Elliott, R., White, E., & Rahm, B. (2020). Predicting flood insurance claims with hydrologic and socioeconomic demographics via machine learning: Exploring the roles of topography, minority populations, and political dissimilarity. *Journal of Environmental Management, 272*. https://doi.org/10.1016/j.jenvman.2020.111051, 111051.

Leandro, J., Schumann, A., & Pfister, A. (2016). A step towards considering the spatial heterogeneity of urban key features in urban hydrology flood modelling. *Journal of Hydrology, 535*, 356–365. https://doi.org/10.1016/j.jhydrol.2016.01.060.

Marelle, L., Myhre, G., Steensen, B. M., Hodnebrog, O., Alterskjaer, K., & Sillmann, J. (2020). Urbanization in megacities increases the frequency of extreme precipitation. *Environmental Research Letters*. https://doi.org/10.1088/1748-9326/abcc8f Manuscript.

Merz, B., Aerts, J., Arnbjerg-Nielsen, K., Baldi, M., Becker, A., Bichet, A., et al. (2014). Floods and climate: Emerging perspectives for flood risk assessment and management. *Natural Hazards and Earth System Sciences, 14*, 1921–1942. https://doi.org/10.5194/nhess-14-1921-2014.

Mobley, W., Sebastian, A., Highfield, W., & Brody, S. D. S. D. (2019). Estimating flood extent during Hurricane Harvey using maximum entropy to build a hazard distribution model. *Journal of Flood Risk Management, 12*, 1–16. https://doi.org/10.1111/jfr3.12549.

Murdock, H. J., de Bruijn, K. M., & Gersonius, B. (2018). Assessment of critical infrastructure resilience to flooding using a response curve approach. *Sustainability, 10*. https://doi.org/10.3390/su10103470.

NASEM. (2019). *Framing the challenge of urban flooding in the United States*. Washington, DC: The National Academies Press. https://doi.org/10.17226/25381.

Neal, J., Villanueva, I., Wright, N., Willis, T., Fewtrell, T., & Bates, P. (2012). How much physical complexity is needed to model flood inundation? *Hydrological Processes, 26*, 2264–2282. https://doi.org/10.1002/hyp.8339.

Niyogi, D., Lei, M., Kishtawal, C., Schmid, P., & Shepherd, M. (2017). Urbanization impacts on the summer heavy rainfall climatology over the eastern United States. *Earth Interactions, 21*. https://doi.org/10.1175/EI-D-15-0045.1.

Nobre, A. D., Cuartas, L. A., Hodnett, M., Rennó, C. D., Rodrigues, G., Silveira, A., et al. (2011). Height above the nearest drainage—A hydrologically relevant new terrain model. *Journal of Hydrology, 404*, 13–29. https://doi.org/10.1016/j.jhydrol.2011.03.051.

Pelletier, J. D., Murray, A. B., Pierce, J. L., Bierman, P. R., Breshears, D. D., Crosby, B. T., et al. (2015). Forecasting the response of Earth's surface to future climatic and land use changes: A review of methods and research needs Jon. *Earth's Future, 3*, 220–251. https://doi.org/10.1002/2014EF000290.Received.

Phillips, B. C., Yu, S., Thompson, G. R., & de Silva, N. (2005). 1D and 2D modelling of urban drainage systems using XP-SWMM and TUFLOW. In *10th Int. conf. urban drain*, 21–26 August.

Pregnolato, M., Galasso, C., & Parisi, F. (2015). *A compendium of existing vulnerability and fragility relationships for flood: Preliminary results* (pp. 1–8).

Rabori, A. M., & Ghazavi, R. (2018). Urban flood estimation and evaluation of the performance of an urban drainage system in a semi-arid urban area using SWMM. *Water Environment Research, 90*, 2075–2082. https://doi.org/10.2175/106143017X15131012188213.

Rosenzweig, B. R., McPhillips, L., Chang, H., Cheng, C., Welty, C., Matsler, M., et al. (2018). Pluvial flood risk and opportunities for resilience. *Wiley Interdisciplinary Reviews Water*. https://doi.org/10.1002/wat2.1302.

Rözer, V., Kreibich, H., Schröter, K., Müller, M., Sairam, N., Doss-Gollin, J., et al. (2019). Probabilistic models significantly reduce uncertainty in hurricane harvey pluvial flood loss estimates. *Earth's Future, 7*, 384–394. https://doi.org/10.1029/2018EF001074.

Rözer, V., Müller, M., Bubeck, P., Kienzler, S., Thieken, A., Pech, I., et al. (2016). Coping with pluvial floods by private households. *Water, 8*. https://doi.org/10.3390/W8070304.

Rubinato, M., Shucksmith, J., Saul, A. J., & Shepherd, W. (2013). Comparison between InfoWorks hydraulic results and a physical model of an urban drainage system. *Water Science and Technology, 68*, 372–379. https://doi.org/10.2166/wst.2013.254.

Sao, N. T. (2008). Storm surge predictions for Vietnam coast by Delft3D model using results from RAMS model. *Journal of Water Resources and Environmental Engineering, 06*, 39–47.

Schanze, J. (2006). Flood risk management—A basic framework. In *Vol. 67. Flood risk management—Hazards, vulnerability and mitigation measures* (pp. 1–20). https://doi.org/10.1007/978-1-4020-4598-1_1.

Sebastian, A., Gori, A., Blessing, R. B., van der Wiel, K., & Bass, B. (2019). Disentangling the impacts of human and environmental change on catchment response during Hurricane Harvey. *Environmental Research Letters*, 1–10. https://doi.org/10.1088/1748-9326/ab5234.

See, L. (2019). A review of citizen science and crowdsourcing in applications of pluvial flooding. *Frontiers in Earth Science, 7*, 1–7. https://doi.org/10.3389/feart.2019.00044.

Seyoum, D. S., Zoran, V., Price, R. K., & Sutat, W. (2012). Coupled 1D and noninertial 2D Flood inundation model for simulation of urban flooding. *Journal of Hydraulic Engineering, 138*, 23–34. https://doi.org/10.1061/(ASCE)HY.1943-7900.0000485.

Skougaard Kaspersen, P., Høegh Ravn, N., Arnbjerg-Nielsen, K., Madsen, H., Drews, M., Kaspersen, P. J., et al. (2015). Influence of urban land cover changes and climate change for the exposure of European cities to flooding during high-intensity precipitation. *IAHS-AISH Proceeding Reports, 370*, 21–27. https://doi.org/10.5194/piahs-370-21-2015.

Spekkers, M. H., Clemens, F. H. L. R., & Ten Veldhuis, J. A. E. (2015). On the occurrence of rainstorm damage based on home insurance and weather data. *Natural Hazards and Earth System Sciences, 15*, 261–272. https://doi.org/10.5194/nhess-15-261-2015.

Tarekegn, T. H., Haile, A. T., Rientjes, T., Reggiani, P., & Alkema, D. (2010). Assessment of an ASTER-generated DEM for 2D hydrodynamic flood modeling. *International Journal of Applied Earth Observation and Geoinformation, 12*, 457–465. https://doi.org/10.1016/j.jag.2010.05.007.

ten Veldhuis, J. A. E. (2011). How the choice of flood damage metrics influences urban flood risk assessment. *Journal of Flood Risk Management, 4*, 281–287. https://doi.org/10.1111/j.1753-318X.2011.01112.x.

Ten Veldhuis, J. A. E., & Clemens, F. H. L. R. (2010). Flood risk modelling based on tangible and intangible urban flood damage quantification. *Water Science and Technology, 62*, 189–195. https://doi.org/10.2166/wst.2010.243.

UNISDR. (2015). Sendai framework for disaster risk reduction 2015-2030. In *Third World Conf. disaster risk reduction, Sendai, Japan, 14-18 March 2015* (pp. 1–25). doi: A/CONF.224/CRP.1.

Van De Ven, F., Van Nieuwkerk, E., Stone, K., Veerbeek, W., Rijke, J., Van Herk, S., et al. (2011). *Building the Netherlands climate proof: Urban areas* (pp. 0–155).

Van Oldenborgh, G. J., Van Der Wiel, K., Sebastian, A., Singh, R., Arrighi, J., Otto, F., et al. (2017). Attribution of extreme rainfall from Hurricane Harvey, August 2017. *Environmental Research Letters, 12*. https://doi.org/10.1088/1748-9326/aa9ef2.

Villaret, C., Hervouet, J. M., Kopmann, R., Merkel, U., & Davies, A. G. (2013). Morphodynamic modeling using the Telemac finite-element system. *Computational Geosciences, 53*, 105–113. https://doi.org/10.1016/j.cageo.2011.10.004.

Wing, O. E. J., Bates, P. D., Smith, A. M., Sampson, C. C., Johnson, K. A., Fargione, J., et al. (2018). Estimates of present and future flood risk in the conterminous United States. *Environmental Research Letters, 13*. https://doi.org/10.1088/1748-9326/aaac65, 034023.

Wing, O. E. J., Pinter, N., Bates, P. D., & Kousky, C. (2020). New insights into US flood vulnerability revealed from flood insurance big data. *Nature Communications, 11*, 1–10. https://doi.org/10.1038/s41467-020-15264-2.

Xie, J., Chen, H., Liao, Z., Gu, X., Zhu, D., & Zhang, J. (2017). An integrated assessment of urban flooding mitigation strategies for robust decision making. *Environmental Modelling and Software, 95*, 143–155. https://doi.org/10.1016/j.envsoft.2017.06.027.

Ye, J., & McCorquodale, J. A. (1998). Simulation of curved open channel flows by 3D hydrodynamic model. *Journal of Hydraulic Engineering, 124*.

Zhang, W., Villarini, G., Vecchi, G. A., & Smith, J. A. (2018). Urbanization exacerbated the rainfall and flooding caused by hurricane Harvey in Houston. *Nature, 563*, 384–388. https://doi.org/10.1038/s41586-018-0676-z.

CHAPTER 6

Using machine learning to predict flood hazards based on historic damage

William Mobley and Russell Blessing
Institute for a Disaster Resilient Texas, Texas A&M University, College Station, TX, United States

The need for alternative flood hazard models and visualizations

Flooding in the United States negatively impacts millions of people each year. These impacts are reflected in increased costs to health, property, infrastructure, and the economy. Hurricane Harvey demonstrated the disparity and lack of existing information to convey where floodwaters would pool or move; how long it would take the water to recede; and how many people were obstructed by floodwaters and their exposure to flood risks. Having accurate information before, during, and after a natural disaster saves lives and protects critical infrastructure and property. Maps that link the natural and built environment to the locations of people, resources, vulnerabilities, and capabilities enable communities to prepare, respond, and recover more effectively from adverse events.

The creation of new visualizations and maps that better communicate the reality of flood risk is the starting point for improving community resilience. However, the United States continues to rely on the Federal Emergency Management Agency's (FEMA) 100-year floodplain as the primary communicator of flood risk even though this delineation has been shown to be an inadequate indicator of flood risk and a poor predictor of flood damage, especially in urban areas. To be fair, most 100-year floodplain maps were never intended to convey flood risk: the boundaries of the floodplain were drawn to set insurance rates. They do not and never were designed to convey information related to depth or duration of inundation, water flow, historical damage, or susceptibility to pluvial or compound flooding. In addition, the current flood maps do not inform investments in flood control and other infrastructure investments that positively or negatively influence urban flood risk. In short, decisions based off these maps are working on limited information often resulting in misguided efforts to increase flood adaptation and resilience. Despite this, the 100-year floodplain continues to provide the basis for community and household planning and mitigation decisions.

The effectiveness of the 100-year floodplain is undermined for multiple reasons that include but are not limited to: (1) outdated maps that have not kept pace with changes in development, (2) large amounts of time and money required to conduct conventional flood inundation models, and (3) modeling uncertainties. Floodplain maps can quickly become outdated, particularly in areas that experience rapid development and are sensitive to changing climate conditions (Sebastian, 2016) resulting in a disconnect between the regulatory floodplain boundaries and observed flood damages (Blessing, Sebastian, & Brody, 2017). This trend plays out across the nation where many floodplain maps are over 10-years old (Birkland, Burby, Conrad, Cortner, & Michener, 2003) and fail to capture much of the historic insured flood damage by as much as 40%–75% (Blessing et al., 2017; Brody, Highfield, & Blessing, 2015). Worse, many rural counties and resource-limited cities lack floodplain boundaries all together because the amount of time and money to conduct the hydraulic and hydrologic models necessary to estimate to the 100-year floodplain is prohibitive, costing anywhere between $3000 and $6000 per km of river reach (FEMA, 2005; Sangwan & Merwade, 2015). Finally, even areas with ample resources to run the models still must face the problem of uncertainty. For example, increasing urbanization, changes in flood infrastructure, rainfall nonstationarity, rainfall-runoff estimation, limited stream gauge and precipitation data, and model calibration (Gallien et al., 2018; Horritt & Bates, 2002; Renard, Kavetski, Kuczera, Thyer, & Franks, 2010; Reynolds, Halldin, Seibert, Xu, & Grabs, 2020; Winsemius, Schaefli, Montanari, & Savenije, 2009; Yilmaz & Perera, 2014) are all sources of uncertainty that can undermine traditional H&H models.

Machine learning as an alternative flood hazard model

The unreliability and, in some cases, absence of SFHA information for communities necessitates further exploration of alternative strategies that can help streamline the identification of flood hazards. The availability of high-resolution geospatial data with nationwide coverage, increasing access to greater computational resources, and state-of-the-art machine learning models has made it possible to estimate flood hazard at spatial scales and speeds that significantly reduce the costs and resources associated with conventional floodplain mapping.

Like other hazard-based research, water-resources research has relied on empirical models to understand systems when physical models are limited (Solomatine & Ostfeld, 2008). These empirical (i.e., data-driven) models are becoming increasingly popular for prediction and estimation of flooding in the hydrological sciences (Mosavi, Ozturk, & Chau, 2018). Data-driven approaches have also identified flood probability during a flood event (Mobley, Sebastian, Highfield, & Brody, 2019) or across longer time periods (Mobley, 2020). In areas with limited data availability for H&H models, this probabilistic approach can help better describe the hydrologic system (Rahmati & Pourghasemi, 2017).

In this chapter, our objective is to guide the reader through an alternative modeling framework for measuring flood hazard using a data-driven, machine learning approach. That is, we lay out some recommended steps and procedures to estimate flood hazard using historical flood damage data and underlying drivers of flood risk. In doing so, we hope to illustrate the advantages and promises of using nontraditional data-streams (e.g., flood insurance claims) and models (i.e., random forests) in identifying areas at risk of flooding. Although this chapter's case study is in Southeast Texas, we make a case for how this could be scaled to the nation and how similar datasets in the Netherlands could be leveraged to accomplish a similar task.

Establishing a machine learning workflow for flood hazard estimation

In this section, we outline the process of developing a flood hazard model. This will include an example of the workflow we used to predict flood hazard in Southeast Texas. The workflow is broken down into six sections: The first section "Case study: Southeast Texas" identifies the case study region used throughout the chapter; second section "Interpretation of flood hazard model outputs" explains what the model generates as an output and how this output should be interpreted; third section "Flood hazard concept measurement" describes concept measurement; this includes predicted variables and underlying drivers as well as potential sampling methods; fourth section "Overview of machine learning models for flood hazard estimation" describes several commonly used machine learning algorithms and includes benefits and limitations of each; fifth section "Model calibration" provides the process for model calibration; and sixth section "Interpreting the modeling results" provides the results from the case study. A discussion of the model implications and future potential follows this workflow. It is important to know that these techniques are flexible and will depend on what type of output the user wants to generate.

Case study: Southeast Texas

Southeast Texas is prone to large-scale damaging flood events. Flooding in the region has resulted in $16.8 billion in insured loss between 1976 and 2017 across 184,826 insurance claims. The clay soils and low topographic relief coupled with extreme precipitation results in wide and shallow floodplains. Regional flood events are driven by several dominant mechanisms, including mesoscale convective systems (MCS) (often referred to as thunderstorms) and tropical cyclones (Van Oldenborgh et al., 2017). Recent examples of MCS-driven events include the Memorial Day Flood (2015), Tax Day, and Louisiana–Texas Floods (2016). Several stalled tropical cyclones, including Tropical Storms Claudette (1979), Allison (2001), and Hurricane Harvey (2017), have also resulted in record-setting precipitation. Frequency estimates of tropical cyclone landfall range from once every 9 years in the eastern part of the study region to once every 19 years in the southwest part of the region (NHC, 2015). Historical storm surge ranges as high as 6 m along the coast. Notable surge events include the Galveston Hurricanes (1900, 1915), Hurricane Carla (1961), and Hurricane Ike (2008).

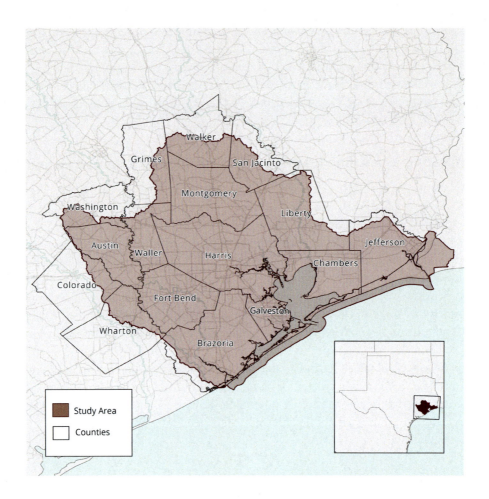

Interpretation of flood hazard model outputs

Before identifying the datasets used in the model, it is important to understand how to interpret the output of a flood hazard model. A flood hazard model provides a probability between 0 and 1 that a given location on the landscape flooded at some point during the sample time frame. This means that if you have locations of flood damage for a given flood event as inputs in a flood hazard model, then the model will generate a likelihood that any given location on the landscape would have flooded for that event. Or if the flood damage record spans 1 year, the model will generate a probability that a given location flooded during that year. Because precipitation events are spatially heterogeneous, the more events incorporated into the flood model the more likely the flood hazard will represent actual flood potential. In other words, the larger the temporal sample, the more representative the flood hazard output.

Flood hazard concept measurement

When developing a flood hazard model, the first step is to identify the dataset used for flooded locations. This requires deliberate selection of where it has and has not flooded. Three common examples of flood location datasets can include high water marks, flood notification and complaint databases maintained by municipalities, and geo-tagged flood observations posted on social media sites such as Twitter. Depending on the location, high water marks may be the most comprehensive dataset available as they are often recorded after each major event to help with response and recovery. For example, the US Geological Survey (USGS) has a network of stream gauges that collects data during flood events including; however, this data has limited spatial availability and requires site visits to measure the heights of high-water marks. High-water marks and stream gauge information are often combined to develop maps of inundated areas after a significant flood event. A second option could include flood-related reports and complaints collected by city governments and municipalities; in the United States, this is often through the 311 system (https://www.houstontx.gov/heatmaps/). Recently, some communities have started sharing this data as an API, which can be leveraged for a flood hazard in model. This dataset may have a shorter temporal resolution than high watermarks but will likely include data from smaller intensity storms that did not generate large-scale inundation. The third commonly used option for identifying flood-impacted areas are geo-tagged social media observations. For example, a group out of Vue-Amsterdam have developed an open-source database of tweets called the "Global Flood Monitor" that detects, in real-time, regions across the globe experiencing a flood event (https://www.globalfloodmonitor.org/).The downside using social media data is that little of it is geo-tagged (less than 2%), and requires a considerable amount of parsing, text disambiguation, and filtering to validate the location of posts that are geo-tagged.

Ideally, the model will be trained with the best available and most comprehensive dataset. For this chapter we used a fourth option, the National Flood Insurance Program (NFIP) flood loss database. Residential structures within the designated 100-year floodplain with a federally backed mortgage are required to purchase flood insurance through the NFIP. However, many homeowners voluntarily purchase flood insurance; this area, in particular, has a high degree of policy saturation with nearly 50% of the NFIP policies located outside of the 100-year floodplain (Highfield, Brody, & Blessing, 2014). Our NFIP flood loss dataset contains 184,826 claims that occurred from 1976 to 2017 and includes the dollar amount in structural damage paid per household.

All of the datasets mentioned have a common problem: they only provide a confirmation that an area has flooded. This means we are confident that the point location has flooded, but we are uncertain where flooding did not occur. This uncertainty can happen for various reasons: people may not complain to the city, tweet about flooding, or own flood insurance. Despite this uncertainty they can still flood. One work around is to generate pseudo-absences, or a random sample of locations we assume were dry (Barbet-Massin, Jiguet, Albert, & Thuiller, 2012). By generating pseudo-absences, the model will isolate variable characteristics for flooded locations against the statistical population. The model's discriminatory power can be further refined by selecting appropriate features from which to sample. For example, NFIP losses focus on the structural losses. These losses will not occur in areas where residential properties are not present such as streets or agricultural land. Selecting a random sample across the landscape may lead a biased model that distinguishes between the built and nonbuilt environment because flood loss can only occur where a structure is present. Using this sampling method, the model is overfit, and predicting characteristic that represent an area that has a structure and not an area that is flooding. Instead, creating a random sample of structure locations will ensure the model compares similar data. Sampling may differ depending on the chosen proxy for flooding. Municipal complaints may be dominated by road flooding; therefore, sampling roads would be more appropriate. For the model in this chapter, we randomly sampled structures from Microsoft's open source structures dataset (https://github.com/Microsoft/USBuildingFootprints). The number of samples required is dependent on the algorithm chosen. We talk about algorithm selection below; but, for this study we used a 1 to 1 sample of structures and claims.

The next step is to determine the underlying variables that influence an area's susceptibility to flood hazards. These will be used as independent variables within the modeling framework to spatially predict the likelihood of any given location's flood hazard probability by using the flood exposure dataset. These predictor variables should include topographic and location conditions. Topographic variables explain how water infiltrates and moves through the environment and are often significant predictors of flood hazard and often include variables such as elevation, imperviousness, and roughness. Location variables help the model identify where flooding has occurred in relation to hazardous locations such as streams and coastlines. For this chapter we use the variables found in Table 1.

Overview of machine learning models for flood hazard estimation

Selecting the appropriate statistical or machine learning algorithm is important to ensure accuracy and computational efficiency. Ideally, the algorithm should be as simple as possible while maintaining accuracy. This ensures that the model is interpretable, both in terms of output and for interpreting how variables affect the final model. In this section, we outline a few models that have been used to predict flood hazard, highlighting some of the pros and cons of each approach.

MaxEnt is a software used in ecological research that uses the Maximum Entropy algorithm to predict species' distribution (Phillips, Anderson, & Schapire, 2006). The

Table 1 Descriptions, spatial resolutions, and statistical ranges for the predictive features.

Name	Description	Initial resolution	Range
Accumulated flow	Hydrologic accumulation for contributing cells	30 m	$0–2.7 \times 10^6$ cells
Hydraulic conductivity (Ksat)[a]	Average soil water transmission for all contributing cells	30 m	0–140 μm/s
Manning's roughness coefficient[b]	Average landcover roughness for all contributing cells	30 m	0.001–0.39
Elevation[c]	Digital Elevation Model (DEM) from USGS	30 m	0–155 m
Distance to coast	Euclidean distance from the coast	10 m	0–155 km
Distance to stream	Euclidean distance from the nearest ordered stream	10 m	0–15 km
Height above nearest drainage (HAND)[d]	Relative vertical height compared with nearest stream, lake, or coastline	10 m	0–59 m
Imperviousness[d,e]	Percent impervious (NLCD)	30 m	0%–100%
Topographic wetness index (TWI)	Topographic Wetness Index adjusted to have no zero values	30 m	8.35–39.53

[a] USGS SSURGO Database website, https://www.nrcs.usda.gov/.
[b] Years: 2001, 2004, 2006, 2008, 2011, 2013, 2016.
[c] Landfire Database website, https://www.landfire.gov.
[d] NLCD Database website, https://www.mrlc.gov/data.
[e] Years: 2001, 2006, 2011, 2016.

ecological species distribution is a predecessor to flood hazard modeling. MaxEnt consistently provides accurate predictions of natural hazards (Bar Massada, Syphard, Stewart, & Radeloff, 2013; Mobley et al., 2019; Siahkamari, Haghizadeh, Zeinivand, Tahmasebipour, & Rahmati, 2017; Tehrany, Pradhan, & Jebur, 2013); because of this, it is ideal for estimating flood hazard. MaxEnt is a fully operational software that provides an out-of-the-box solution for flood hazard estimates, requiring only matching rasters for drivers, and flood location. The software can handle a wide range of data types, continuous, categorical, etc. The ecological literature finds that often MaxEnt is more accurate than other models (Bar Massada et al., 2013; Phillips et al., 2009). As its name suggests MaxEnt maximizes the independent variable entropy between the dependent variable, which can be represented nonlinearly.

An algorithm alternative to MaxEnt is the random forest classification algorithm. The idea behind random forest classification is to use a collection decision trees to democratize the classification prediction (Breiman, 2001). These series of trees have a subsample of data and driver variables, which tends to reduce overfitting while maintaining accuracy with minimal complexity. The consistent high degree of accuracy, minimal overfitting, and simple nature of the random forest classification make it a popular algorithm for machine learning and is provided in a series of libraries in Python and R. In instances where the flood variable is across multiple years, many of the drivers may change across that time. When possible, accounting for these dynamic changes will improve the model. Using a coding language such as Python allows for a flood hazard map to incorporate more dynamic landscapes such as changing roughness, and imperviousness over time.

Support vector machines (SVMs) have been shown to accurately represent flood hazard (Tehrany, Kumar, & Shabani, 2019). Like the random forest classification, SVMs are a highly generalizable algorithm for classification problems. SVMs identify hyperplanes that can distinguish between the two classifications, flood or not. If flood data is limited, SVMs are likely one of the best options for distinguishing flood hazard. However, computation time is quadratic with larger datasets, with the time required to train a model becomes quickly unfeasible (Li, Wang, & Hu, 2009) thereby limiting its application over large timeframes or areas.

The previous algorithms are nowhere near a comprehensive list; however, they represent models the authors have used in previous studies and represent algorithms commonly used for similar applications. In this chapter, the authors have used the random forest classification algorithm because of its simplicity, low computational requirements, and ability to account for dynamic landscapes. The flood hazard model developed from the region spanned 42 years and 360,000 data points. The random forest allowed for dynamic layers so that variables could be as representative as possible of the current state. In addition, the size of the dataset required a simpler approach that was not too computationally intensive.

Model calibration

The final stage before generating an output is to ensure that the model is properly calibrated and accurate. To ensure a properly calibrated model an independent training and testing dataset is needed to provide an unbiased evaluation of the model fit. Optimizing for hyperparameters and feature selection is done with the training dataset, and then validated on the test dataset. If the model is properly configured, then the accuracy metrics will be similar for both training and testing. If validation metrics are lower than training metrics, then the model is overfit. Iterating over samples of the training dataset helps with model configuration. Cross-validation uses subsamples of the training data while tuning parameters and variables. During cross-validation performance metrics are averaged, which gives a more accurate representation of how the model will perform on the test data. A 10-sample k-fold cross-validation sampling method can be used for this approach. A 10 k-folds cross-validation sampling method uses 90% sample of the training dataset to predict the other 10%. This is repeated 10 times with a new fold of the dataset. This sampling method helps to measure the robustness of the model, as variables are pruned and parameters are tweaked.

Researchers typically use performance metrics to ensure that a model is properly configured. We address two of these metrics, area under the curve (AUC) and accuracy.

AUC estimates that given two samples, a flooded and a nonflooded, the flooded sample will have a higher probability than the nonflooded sample (Metz, 1978). In practice, AUCs are robust for continuous probabilities and are not limited by the number of features in a model or a threshold used for classification. Accuracy is measured after the model has classified the sample. For the random forest, a sample is classified as flooded if half the trees predict that a sample flooded. The percentage of correctly identified flooded and nonflooded samples represents the accuracy.

Machine learning algorithms have various parameters that improve model performance. For the random forest there are two parameters: tree depth and number of trees. Increasing these parameters often increases accuracy but creating too large a tree will increase computation time. Beyond error rates, the number of trees helps with the final probability. Having too few trees reduces the final probability variation. For example, 100 trees will allow probabilities to vary by 1% while 25 trees would only provide a 4% variation.

Another way a model can improve performance is identifying the most important features for predicting and removing those with little explanatory power. As previously discussed, overfitting is where the trained model is fit too closely to the training data. When validated on the test data an overfit model performs poorly. Overfit models fail to account for other possibilities within the system. One cause of overfitting is using too many variables in the prediction. Different models have ways of identifying variables that cause overfitting. Random Forests provide an out-of-bag (OOB) error score, which isolates a subset of the training dataset used to measure error rates of each variable (Breiman, 1996) and generates feature importance. During the training with k-fold feature importance is measured and those with the lowest values can be removed. Alternatively, MaxEnt provides feature importance by removing variables and providing sensitivity. This provides an alternative approach for removing variables; the idea is that if the sensitivity is high when removing the variable then it should be removed from the final model.

For the example study, the NFIP flood claims dataset was split between the training and test dataset. The test dataset was based on 30% of the initial samples and the cross-validation assessment used a 10-sample stratified k-folds.

Interpreting the modeling results

This next section outlines how to interpret the flood hazard output through three different modeling stages: model tuning, accuracy metrics, and visualization. Tuning consists of variable selection, and parameters tweaking. During tuning, two variables were shown to have minimal importance to the final model: land cover and TWI. These two variables had low feature importance compared with the rest of the variables. After variable selection, hyperparameters were optimized. For this chapter, we used a tree depth of 90 and 200 decision trees. This allows the model to vary by 0.5% for the likelihood of flooding across the 42 years.

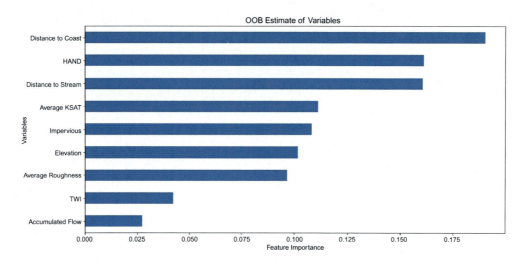

The cross-validation accuracy was 81.9%±0.00198, while the final test accuracy was 82.2%. Cross-validation sensitivities were 0.893±0.00184 for the k-folds while the final model produced a sensitivity of 0.895. This suggests the model performed similarly for training and test datasets. A good sign the model is not overfitting the training data.

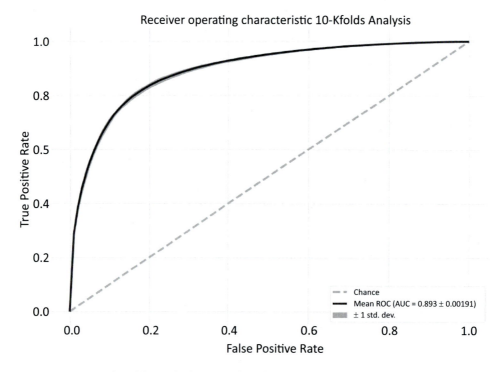

Once we are comfortable with the model calibration, we can predict the images. Since we know the number of years that NFIP was available, we can annualize the probability.

$$\text{Annual exceedence} = 1 - (1 - p_{\text{flood}})^{1/T}.$$

Notice, within the study area most high hazard areas follow along streams and the coast as expected. There are some areas where pluvial flooding cause higher flood hazard areas as well.

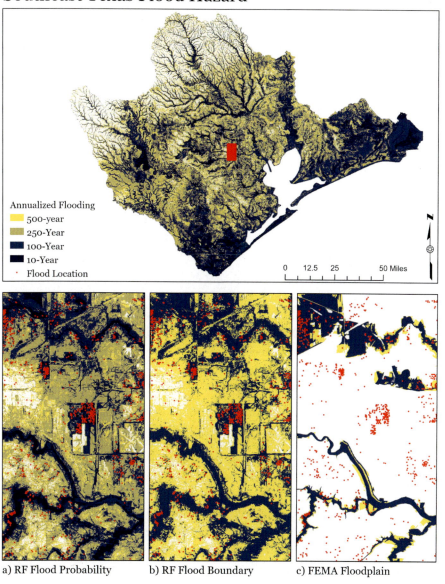

a) RF Flood Probability　　b) RF Flood Boundary　　c) FEMA Floodplain

A potential application for the Netherlands

Applying a similar flood hazard modeling method in the Netherlands would provide a unique case study primarily because of how the Dutch manage flood risk. As citizens of a country composed of deltas that are largely below sea level, the Dutch have historically focused on structural solutions to prevent flooding. This has resulted in a flood defense system with the highest safety standards in the world as developed by a centralized and independent group called the Delta Committee. Specifically, the Dutch employ a cost–benefit analysis to determine acceptable design and maintenance standards for their dams, dikes, and dunes. Based on this approach, they concluded that structural flood protection should be designed to withstand a 1-in-10,000-year storm in coastal areas and a 1-in-1250-year storm along rivers. Moreover, the Dutch also have not allowed development to occur in the floodplains since the 1980s and have even removed structures and entire neighborhoods as part of their "Room for the River" initiative, which restores the natural floodplain functions and morphology to increase the ability of the rivers to cope with higher discharges without flooding. This stands in stark contrast to the United States where many levees are only required to withstand the 1-in-200-year storm before overtopping and development can occur within the 100-year floodplain if the structures meet certain design criteria.

Despite the Netherlands extensive efforts to minimize flood risk, there will always remain some residual risk to flooding. One such risk that levees and dikes fail to prevent is pluvial flooding, which is emerging as a critical issue in water management worldwide and is expected to worsen as urban growth and climate change-induced extreme rainfall increases. While not perceived to be as serious as coastal and riverine floods, pluvial flooding is estimated to cost about $110 million (USD) per year and is expected to increase to as much as $245 million (USD) per year based on increased rainfall intensities and frequencies in the future. Areas that have been hit particularly hard by pluvial flooding in the Netherlands include the southwestern and northern areas (2004), Rotterdam (2009), Amsterdam (2010), Arnhem (2011, 2014, and 2016), and North Brabant and Limburg (2016 and 2017) (Pieters, 2016; Spekkers, Kok, Clemens, & Ten Veldhuis, 2013; Trell & van Geet, 2019; Van Riel, 2011).

Damages from this type of flooding are often not immediately noticeable or easy to detect leading some to refer to it as the "invisible hazard" (Houston et al., 2011) and in the Netherlands as *wateroverlast,* or 'water nuisance,' implying that it does not cause severe damages or disruption (Forrest, Trell, & Woltjer, 2021). Although causing relatively little damage compared to extreme flood events, the cumulative damage over time from these smaller flood events to urban drainage systems and homes can be considerable. Fortunately, a dataset has become available that can help identify where these types of flood damages are occurring in the Netherlands: a database of flood insurance claims from the Dutch Association of Insurers (Spekkers, Ten Veldhuis, Kok, & Clemens, 2011).

This database contains data on damages to both commercial and private buildings and content from flood events dating back to 1992 making it an ideal dataset for assessing flood hazards using the workflow discussed in this chapter. By doing so, the output of this model could help identify previously unmapped areas at risk of urban and pluvial flooding and help prioritize and spatially plan for and mitigate against future flood losses.

Summary and conclusions

This chapter outlines a novel, data-driven approach to estimate and map flood hazard using a machine learning algorithm. Not only is this approach scalable due to its computational efficiency, but it can also better capture the reality of flood risk by using a historic damage dataset. In doing so, this approach reduces the uncertainty and costs associated with conventional H&H models and opens opportunities to identify hazardous areas in previously unmapped regions. One of the key findings is that conventional floodplain boundaries consistently underestimate flood hazard, corroborating previous findings that have shown that large proportions of flood damage occur outside of the floodplain even for storms smaller than the 100-year event. Many of the areas that conventional floodplains fail to capture are hydrologically disconnected and subject to pluvial flooding (i.e., ponding) during moderate-to-intense rainfall events that can overwhelm the urban drainage system and cause damage to buildings. This is the type of flood risk that a machine learning algorithm trained on a database of structural flood damage accounts is tailored to capture and explains why its estimated flood boundaries are much more expansive than the regulatory floodplain.

There are many ways in which a data-driven flood hazard mapping approach can benefit cities and improve the resilience of built environments. One way is by enhancing how flood hazards are delineated, visualized, and ultimately communicated. Flood hazard identification should not stop with merely improving the model accuracy but should also include end-users and residents in the process of creating a flood hazard visualization that is understandable and interpretable. Doing so will improve decision-making and lessen the impacts of costly flood damages over the long-term. Another advantage of this approach is its flexibility. That is, it is not limited to identifying long-term flood hazard or insurance claims but can also be leveraged to identify areas impacted by specific storms or used to create a real-time flood exposure detection system. For example, damage reports from social media platforms such as Twitter or emergency requests from citizens can be pulled into a real-time data modeling pipeline to identify areas at risk of flooding as storms unfold or used in a post hoc fashion to determine the storm's inundation footprint and estimate total damages. Finally, as illustrated by the Netherlands example, this type of flood hazard identification is not limited to areas at risk of catastrophic or even chronic flooding but can be used to detect pluvial and nuisance flooding that have traditionally not been included within standard modeling practices.

References

Barbet-Massin, M., Jiguet, F., Albert, C. H., & Thuiller, W. (2012). Selecting pseudo-absences for species distribution models: How, where and how many? *Methods in Ecology and Evolution, 3*, 327–338.

Bar Massada, A., Syphard, A. D., Stewart, S. I., & Radeloff, V. C. (2013). Wildfire ignition-distribution modelling: a comparative study in the Huron–Manistee National Forest, Michigan, USA. *International Journal of Wildland Fire, 22*(2), 174–183.

Birkland, T. A., Burby, R. J., Conrad, D., Cortner, H., & Michener, W. K. (2003). River ecology and flood hazard mitigation. *Natural Hazards Review, 4*(1), 46–54.

Blessing, R., Sebastian, A., & Brody, S. D. (2017). Flood risk delineation in the United States: How much loss are we capturing? *Natural Hazards Review, 18*(3), 04017002.

Breiman, L. (1996). Out-of-bag estimation. *Technical report*. Department of Statistics, University of California.

Breiman, L. (2001). Random forests. *Machine Learning, 45*(1), 5–32.

Brody, S. D., Highfield, W. E., & Blessing, R. (2015). An analysis of the effects of land use and land cover on flood losses along the Gulf of Mexico coast from 1999 to 2009. *JAWRA Journal of the American Water Resources Association, 51*(6), 1556–1567.

FEMA. (2005). Floodplain management requirements: A study guide and desk reference for local officials. In *FEMA 480*.

Forrest, S. A., Trell, E.-M., & Woltjer, J. (2021). Emerging citizen contributions, roles and interactions with public authorities in Dutch pluvial flood risk management. *International Journal of Water Resources Development, 37*(1), 1–23.

Gallien, T. W., Kalligeris, N., Delisle, M.-P. C., Tang, B.-X., Lucey, J. T., & Winters, M. A. (2018). Coastal flood modeling challenges in defended urban backshores. *Geosciences, 8*(12), 450.

Highfield, W. E., Brody, S. D., & Blessing, R. (2014). Measuring the impact of mitigation activities on flood loss reduction at the parcel level: The case of the clear creek watershed on the upper Texas coast. *Natural Hazards, 74*(2), 687–704.

Horritt, M., & Bates, P. (2002). Evaluation of 1D and 2D numerical models for predicting river flood inundation. *Journal of Hydrology, 268*(1–4), 87–99.

Houston, D., Werrity, A., Bassett, D., Geddes, A., Hoolachan, A., & McMillan, M. (2011). *Pluvial (rain-related) flooding in urban areas: The invisible hazard*. Joseph Rowntree Foundation. Retrieved from http://www.jrf.org.uk/sites/files/jrf/urban-flood-risk-full.pdf.

Li, B., Wang, Q., & Hu, J. (2009). A fast SVM training method for very large datasets. In International joint conference on neural networks (IJCNN 2009) (pp. 1784–1789).

Metz, C. E. (1978). Basic principles of ROC analysis. Seminars in Nuclear Medicine, 8, 283–298. https://doi.org/10.1016/s0001-2998(78)80014-2.

Mobley, W. (2020). Flood hazard modeling output, V2, Texas data repository. https://doi.org/10.18738/T8/FVJFSW.

Mobley, W., Sebastian, A., Highfield, W., & Brody, S. D. (2019). Estimating flood extent during Hurricane Harvey using maximum entropy to build a hazard distribution model. Journal of Flood Risk Management, 12, e12549. https://doi.org/10.1111/jfr3.12549.

Mosavi, A., Ozturk, P., & Chau, K.-W. (2018). Flood prediction using machine learning models: Literature review. Water, 10, 1536. https://doi.org/10.3390/w10111536.

NHC. (2015). *Tropical cyclone climatology*. https://www.nhc.noaa.gov/climo/ (last accessed 4 March 2020).

Phillips, S. J., Anderson, R. P., & Schapire, R. E. (2006). Maximum entropy modeling of species geographic distributions. *Ecological Modelling, 190*(3), 231–259.

Phillips, S. J., Dudík, M., Elith, J., Graham, C. H., Lehmann, A., Leathwick, J., et al. (2009). Sample selection bias and presence-only distribution models: Implications for background and pseudo-absence data. *Ecological Applications, 19*(1), 181–197. https://doi.org/10.1890/07-2153.1.

Pieters, J. (2016). Thunder cause widespread flooding: More rain expected. Retrieved from https://nltimes.nl/2016/06/02/thunderstorms-cause-widespread-flooding-rain-expected.

Rahmati, O., & Pourghasemi, H. R. (2017). Identification of critical flood prone areas in data-scarce and ungauged regions: A comparison of three data mining models. Water Resources Management, 31, 1473–1487. https://doi.org/10.1007/s11269-017-1589-6.

Renard, B., Kavetski, D., Kuczera, G., Thyer, M., & Franks, S. W. (2010). Understanding predictive uncertainty in hydrologic modeling: The challenge of identifying input and structural errors. *Water Resources Research*, *46*(5), W05521.

Reynolds, J. E., Halldin, S., Seibert, J., Xu, C.-Y., & Grabs, T. (2020). Flood prediction using parameters calibrated on limited discharge data and uncertain rainfall scenarios. *Hydrological Sciences Journal*, *65*(9), 1512–1524.

Sangwan, N., & Merwade, V. (2015). A faster and economical approach to floodplain mapping using soil information. *JAWRA Journal of the American Water Resources Association*, *51*(5), 1286–1304.

Sebastian, A. G. (2016). *Quantifying flood risk and Hazard in highly urbanized coastal watersheds* (Dissertation). Rice University.

Siahkamari, S., Haghizadeh, A., Zeinivand, H., Tahmasebipour, N., & Rahmati, O. (2017). Spatial prediction of flood-susceptible areas using frequency ratio and maximum entropy models. Geocarto International, 33(9), 927–941.

Solomatine, D. P., & Ostfeld, A. (2008). Data-driven modelling: Some past experiences and new approaches. Journal of Hydroinformatics, 10, 3–22. https://doi.org/10.2166/hydro.2008.015.

Spekkers, M., Kok, M., Clemens, F., & Ten Veldhuis, J. (2013). A statistical analysis of insurance damage claims related to rainfall extremes. *Hydrology and Earth System Sciences*, *17*(3), 913–922.

Spekkers, M., Ten Veldhuis, J., Kok, M., & Clemens, F. (2011). Analysis of pluvial flood damage based on data from insurance companies in the Netherlands. In G. Zenz, & R. Hornich (Eds.), *Proceedings International Symposium Urban Flood Risk Management, UFRIM, 2011, September 21–23, Graz, Austria*.

Tehrany, M. S., Kumar, L., & Shabani, F. (2019). A novel GIS-based ensemble technique for flood susceptibility mapping using evidential belief function and support vector machine: Brisbane, Australia. PeerJ, 7, e7653. https://doi.org/10.7717/peerj.7653.

Tehrany, M. S., Pradhan, B., & Jebur, M. N. (2013). Spatial prediction of flood susceptible areas using rule based decision tree (DT) and a novel ensemble bivariate and multivariate statistical models in GIS. Journal of Hydrology, 504, 69–79.

Trell, E., & van Geet, M. (2019). The governance of local urban climate adaptation: Towards participation, collaboration and shared responsibilities. *Planning Theory & Practice*, *20*(3), 376–394.

Van Oldenborgh, G. J., Van Der Wiel, K., Sebastian, A., Singh, R., Arrighi, J., Otto, F., et al. (2017). Attribution of extreme rainfall from Hurricane Harvey, August 2017. Environmental Research Letters, 12. 124009. https://doi.org/10.1088/1748-9326/aaa343.

Van Riel, W. (2011). *Exploratory study of pluvial flood impacts in Dutch urban areas*. Delft, The Netherlands: Deltares.

Winsemius, H. C., Schaefli, B., Montanari, A., & Savenije, H. H. G. (2009). On the calibration of hydrological models in ungauged basins: A framework for integrating hard and soft hydrological information. Water Resources Research, 45, W12422. https://doi.org/10.1029/2009WR007706.

Yilmaz, A., & Perera, B. (2014). Extreme rainfall nonstationarity investigation and intensity-frequency-duration relationship. *Journal of Hydrologic Engineering*, *19*(6), 1160–1172.

CHAPTER 7

Compound flooding

Antonia Sebastian
Department of Earth, Marine and Environmental Sciences, The University of North Carolina at Chapel Hill, Chapel Hill, NC, United States

Introduction

Currently, more than half of the world's population lives within 100 km of a coastline, and this number is expected to grow dramatically during the 21st century. Future development will be concentrated in coastal megacities and port cities because of the strategic geographic connection that they provide to the hinterland, allowing for economic growth. However, these areas are also uniquely vulnerable to flooding caused by severe weather and high-tide events. The standard approach to flood hazard modeling typically addresses rainfall runoff, river discharge, and surge-induced flooding independently (FEMA, 2015; Moftakhari, Schubert, AghaKouchak, Matthew, & Sanders, 2019). However, multiple flood mechanisms can occur simultaneously in coastal regions, and their physical interactions can result in nonlinear increases in flood impacts (Bilskie & Hagen, 2018; De Ruiter et al., 2020). The occurrence of two or more flood hazards driven by a single weather event, or the occurrence of two or more independent events that produce flood hazards jointly in space and/or time, is known as a compound flood event. Research suggests that compound floods are likely to become more frequent and more severe in the future, exacerbated by upland development, sea-level rise, and changing storm characteristics as a result of climate change. Therefore, understanding compound flooding and the conditions under which it can occur is critical to coastal hazard assessment and mitigation.

Recognition of the threats posed by the interactions between multiple hazards and their drivers has led to several calls by national and international organizations to address them, including the World Climate Research Program (WCRP), European Union Sendai Framework, and the International Program on Climate Change (IPCC). As a result, there has been a dramatic increase in research focused on *compound events*—events that occur as a result of two or more hazard drivers and their interactions in space and/or time (Leonard et al., 2014; Zscheischler et al., 2018). Several working groups have also emerged, including DOMOCLES (European COST Action CA17109), Knowledge Action Network on Emergent Risks and Extreme Events (RISK KAN), and NOAA's Coastal Coupling Community of Practice, as well as numerous special issues in academic journals, colloquia, and organized meetings designed to collect and synthesize on-going research on this topic. The number of peer-reviewed articles that reference compound flooding has grown from less than one article per year prior to 2015 to more than 60 publications in 2020.

Compound floods are may be driven by a single weather system that produces multiple types of flooding. For example, tropical cyclones can generate both storm surges and extreme precipitation at landfall, which interact to exacerbate flooding. On the other hand, a high-tide event coinciding with heavy precipitation—otherwise independent events—can combine to generate moderate to severe flooding inland. In general, compound flood events are not well resolved in numerical models and, often, the statistical likelihood of their cooccurrence is unknown. This poses a challenge for coastal managers as these types of combined events need to be considered in the design of coastal infrastructure. For example, building a sea wall could inadvertently increase water levels inland because outflows are blocked, requiring the installation of pumps, or the design capacity of existing stormwater infrastructure may decrease over time because of rising sea levels, changing precipitation patterns, and their compound interactions.

The above examples illustrate the importance of understanding where compound hazard events could occur and the associated risks to human and natural systems. A complete review of the current literature is beyond the scope of this chapter. Instead, the following sections provide a general overview of the numerical and statistical methods for assessing compound flood risks in existing modeling frameworks, and their application to coastal areas. Several cases in Texas and the Netherlands are presented to demonstrate the ways in which different types of flood hazards (e.g., river flows, storm surge, tide) and their hydrometeorological drivers (e.g., tropical cyclones and extreme precipitation) can interact to exacerbate coastal flooding. The chapter ends with a brief discussion of the challenges of managing coastal hazards and the ways in which compound flooding should be considered in the future.

Modeling coastal flood hazards using numerical and statistical approaches

Efforts to mitigate coastal risks in communities have led to a growing need to establish standard approaches to incorporate compound flooding in risk assessment frameworks and hazard mitigation. The *transition zone* refers to the area where both inland and coastal processes contribute to total water levels and often interact to cause flooding that is more extreme than either river or coastal flooding alone (Bilskie & Hagen, 2018). In coastal watersheds, the backwater effects of elevated sea levels can extend far upstream and understanding the extent of the transition zone is important to understand where one transitions from surge-dominated to rainfall-dominated processes (Christian, Duenas-Osorio, Teague, Fang, & Bedient, 2013; Ray, Stepinski, Sebastian, & Bedient, 2011). To establish where compound flooding is likely to occur and map the associated hazard using numerical models, it is also important to determine the boundary conditions that lead to compound flooding. In the following section, some context for the numerical and statistical modeling approaches is provided.

Numerical modeling

Coastal flood hazards are difficult to model numerically due to the complexity of the physical processes that drive the hydrodynamic response in the coastal zone. Within inland river systems, gravitational forces driven by topography dominate the physical behavior of the flood wave. Hydrologic processes are therefore often simulated using simple rainfall-runoff models in which the physical characteristics of the system are averaged across hydrologic units and the resulting runoff volume is routed using empirical or reduced-physics approaches. The flood inundation extent is estimated using either stage-discharge relationships, simple raster-based approaches, or one-dimensional hydraulic wave models (see also Sebastian et al., Chapter 5, this volume). This comes with several limitations, particularly in urban areas and small coastal watersheds where the accumulation of water on the ground may occur at a rate faster than it can reach neighboring streams (see, e.g., Gori, Lin, & Xi, 2020). However, an advantage of simulating the inland hydrologic processes in this way is the low computational cost which enables the efficient simulation of flood dynamics across large spatial scales.

Within the coastal floodplain, flood-generating processes are often more complex than in inland regions. First, flat topography lends itself to wide and shallow floodplains and inundation can occur over much larger spatial areas than in inland river systems with well-defined floodplain areas, making the one-dimensional assumptions described above inadequate. Instead, two-dimensional hydrodynamic models which can resolve physics in both the along- and cross-channel direction are necessary to accurately simulate flooding. Because the vertical acceleration in the water column is negligible relative to the velocities in the planar directions, the two-dimensional assumption is typically sufficient for modeling flooding within the coastal plain and even for shallow estuaries and continental shelves. Second, coastal flooding may be driven by any combination of rainfall-runoff, river discharge, groundwater discharge, storm surge, and/or tides. The interactions between various flood drivers are difficult (and sometimes impossible) to replicate within existing model architectures. In fact, there is currently no single model that can fully resolve both wind- and ocean circulation-driven (e.g., tidal fluctuation, wave generation and dissipation, and storm surge) and precipitation-driven hydrologic processes (e.g., rainfall-runoff and infiltration) (Santiago-Collazo, Bilskie, & Hagen, 2019). Instead, most numerical modelers often choose to simulate the dominant flood hazards like storm surge or riverine flooding independently or using two or more models in which one model's output is used as a boundary condition for another model.

These multiple-model frameworks are generally classified as "loosely" or "tightly" coupled depending on the frequency and spatial resolution at which information is fed between the models. Loose coupling is straightforward as described above, but tight coupling requires that the models rest on similar architecture, enabling them to exchange information at various locations within the model domain and several points in time during the simulation. Ideally, tightly coupled models exchange information at each computational node and at every time step. This is difficult to accomplish because it requires that the model discretization be complementary so that both mass and momentum is conserved across their domains as data is exchanged. An example of a tightly coupled model is the ADCIRC+SWAN model in which waves are simulated simultaneously to tides and ocean currents. The two models operate on a single computational mesh, allowing for efficient simulation of complex solutions to the numerical equations that describe the physics of open water in response to wind, tides, and ocean circulation at each computational node (Dietrich et al., 2011). This type of elegant solution does not yet exist to describe the physical processes governing flooding within the coastal plain, which is characterized by physical processes occurring at various spatial and temporal scales.

Statistical modeling

Numerous statistical models have been applied to study compound events (Bensi, Mohammadi, Kao, & DeNeale, 2020; Leonard et al., 2014). Statistical models can be used to model the *joint probability* or the likelihood that two or more events occur simultaneously at a particular location. Prior studies have focused on identifying the probability of occurrence of conditions associated with compound flooding (e.g., high river flows and high sea levels) across multiple scales. Such models may be built using observed records at one or more gaged locations (Hendry et al., 2019; Wahl, Jain, Bender, Meyers, & Luther, 2015) or numerically simulated data (Paprotny, Vousdoukas, Morales-Nápoles, Jonkman, & Feyen, 2020; Santos, Casas-prat, Poschlod, Ragno, & Van Den Hurk, 2021). Statistical models can also be used to support numerical modeling, for example, by generating synthetic boundary conditions based on historical observed records (Couasnon, Sebastian, & Morales-Nápoles, 2018; Sebastian, Dupuits, & Morales-Nápoles, 2017), or in place of numerical models as is the case with surrogate models or emulators.

Under a joint probability framework, datasets of different flood drivers (i.e., precipitation, storm surge, wind speed, wave height) are treated as random variables. The random variables may represent observed or synthetic data and can be described by a univariate probability distribution known as a marginal distribution. Joint probability distributions of two or more random variables quantify the statistical dependence between variables and can be used to estimate the likelihood that a set of flood drivers will exceed a certain threshold. For the bivariate case, the joint probability of both variables can be represented by a cumulative distribution function $F_{XY}(x, y)$:

$$F_{XY}(x, y) = P(X \leq x, Y \leq y)$$

where X and Y are random variables, and the right-hand side of the equation represents the probability that the random variable X takes on a value less than or equal to x and the random variable Y takes on a value less than or equal to y. The primary limitation of this approach is that the random variables must be characterized by the same parametric family of univariate distributions (e.g., normal, log-normal, gamma) as their joint distribution.

Copula-based methods have emerged as a popular alternative to traditional joint probability distributions. They can be used to assess the likelihood of bivariate conditions at a single location or to support analyses of more complex networks of multivariate drivers (see e.g., Sebastian et al., 2017). According to Sklar (1959), there exists a cumulative distribution function $F_{XY}(x, y)$ defined as

$$F_{XY}(x, y) = C_{XY}(F_X(x), F_Y(y))$$

where C_{XY} is the bivariate copula that describes the dependence structure between the marginal distribution functions $F_X(x)$ and $F_Y(y)$ of the random variables X and Y. An advantage of copulas over traditional methods is the ability to model the dependence between a set of random variables independent of their marginal distributions (Genest & Favre, 2007). As such, copulas have been widely applied to estimate the statistical dependencies between flood hazard drivers at regional (Hendry et al., 2019; Klerk, Winsemius, Van Verseveld, Bakker, & Diermanse, 2015; Moftakhari, Salvadori, AghaKouchak, Sanders, & Matthew, 2017), continental (Paprotny, Vousdoukas, Morales-Nápoles, Jonkman, & Feyen, 2018; Wahl et al., 2015; Zheng, Westra, & Sisson, 2013), and global scales (Couasnon et al., 2020; Ikeuchi et al., 2017; Ward et al., 2018). They have also been applied to characterize changes in frequency and intensity of coinciding extrema under future climate conditions (see e.g., Ganguli, Paprotny, & Hasan, 2020). These studies provide important information about where compound hazards are likely to occur along coastlines, but higher spatial and temporal resolution information is needed to support flood risk reduction efforts.

Surrogate or machine learning models are emerging as an alternative to traditional numerical modeling to spatially map flood hazards. In general, surrogate modeling approaches leverage results from numerical models or observations of past flood conditions to train statistical models (e.g., artificial neural network, support vector machine, random forests) to predict flood hazards that have not yet been observed. Several recent studies suggest that these data-driven approaches hold promise for real- or near-real time flood hazard mapping, as well as coastal flood risk analysis. For example, in a study of compound flood events impacting the west side of Galveston Bay, researchers trained an artificial neural network on 223 simulated flood maps to generate a surrogate model that could predict the inundation extent based on forecasted tropical cyclone conditions (Bass & Bedient, 2018). In another study, researchers used modeled storm surge and remotely sensed imagery from Hurricane Matthew to train a surrogate model to generate estimates of compound flooding for the southeastern US Atlantic coastline (Muñoz, Muñoz, Moftakhari, & Moradkhani, 2021). While still an emerging area of research, it appears that surrogate models may be skilled at predicting compound flooding, potentially reducing the need for complex numerical solutions to model flood extent.

Managing coastal flood hazards

The consideration of multiple cooccurring or consecutive hazards in planning coastal flood protection is critical to developing resilient flood management systems. River levees, channels, sea walls, and quays have been go-to structural approaches to mitigate floods in both the United States and the Netherlands for more than a century. Some large-scale examples of coastal structures include the Delta Works in the Netherlands or the Mississippi River Gulf Outlet (MRGO) system in Louisiana or the proposed "Big U" in New York City. However, much of the existing flood control infrastructure was designed and built to mitigate one dominant hazard (e.g., storm surge). As a result, climate change, sea-level rise, and other anthropogenic processes may threaten the future performance of these coastal protection systems. Therefore, consideration of multiple hazards and their interactions, as well as how they may change under future conditions, is becoming increasingly important for the design of resilient coastlines.

While the types of events facing Texas and the Netherlands differ in magnitude and in terms of their hydrometeorological drivers, best practices and lessons learned from existing infrastructure systems can be incorporated in the transatlantic discussion of coastal protection. To illustrate the ways in which different hazards can combine to drive flooding in coastal areas, several geographic locations in Texas and the Netherlands are highlighted below.

Rotterdam

Rotterdam is located near the outlet of the Maas River and into the North Sea. It is home to the largest seaport in Europe and the second largest outside of Asia. Rotterdam is threatened by both high river levels and high sea levels. To mitigate the threat of coastal storm surge to the inland areas and complete the closure of the coastal barrier system (i.e., the Delta Works), a movable storm surge barrier, known as the Maeslant Barrier, or the Maeslantkering in Dutch, was built. The barrier consists of two 210 m steel truss gates that close off the Nieuwe Waterweg of the Rhine River Delta from the North Sea. Fig. 1 shows a map of the area protected by the Maeslant Barrier. An advantage of the moveable barrier is that it enables shipping traffic to transverse the river during nonextreme periods but will automatically close when sea levels are predicted to exceed 3 m at Rotterdam and/or 2.90 m at Dordrecht (Geerse, 2010). Because there is limited storage within the river system behind the barrier, high river flows must also be considered in the design and closure of the system to avoid backwater effects and flooding further upstream

Fig. 1 Satellite map of the Nieuwe Waterweg and the coastal barrier in Rotterdam. *(Source: Google Earth Engine.)*

(Zhong, van Overloop, van Gelder, & Rijcken, 2012). Currently, the barrier is expected to be closed once every 10 years on average, but under future sea-level rise, the barrier may be closed more often (Zhong et al., 2012).

The Maeslant Barrier example can provide a valuable blueprint for designing a movable barrier across a navigable waterway, including along the Texas coast. However, consideration of the optimal design and operation of the barrier for both high inland and coastal water levels is critically important, as these can exert considerable force on the system and could exacerbate conditions on the inland or upstream side of the flood barrier (Christian, Fang, Torres, Deitz, & Bedient, 2014; Torres et al., 2015, 2017).

Noorderzijlvest

Much of the Dutch landscape is managed through a series of polders: geographically contiguous hydrological units that are managed as a system. Groningen and the surrounding area are managed by the water board unit of Noorderzijlvest (1440 km^2) and are protected from coastal storm surge by sea dikes. Water levels inside the polder are managed using a series of pumps and canals. Excess water is pumped into the Lauwersmeer and then discharged via gravity into the North Sea during low tides. In January 2012, two consecutive low-pressure events, Cyclones Andrea and Ulli, caused 5 days of heavy rainfall (>70 mm) and high winds (>9 Beaufort) over the northern provinces of the Netherlands. Because of high sea levels, drainage to the North Sea could not occur over five tidal cycles, leading to high inland water levels along the canals. Saturated soils further exacerbated the inland conditions, leading to precautionary evacuations and managed flooding of designated overflow areas. In a postevent analysis using more than 800 years of simulated climate data, Van den Hurk, Van Meijgaard, de Valk, Van Heeringen, and Gooijer (2015) found that while neither the rain nor the surge event was extreme in itself, their cooccurrence led to extreme conditions. The researchers also concluded that the worst-case conditions do not necessarily occur during the peak surge (winter) or peak precipitation (summer) seasons, but rather occur when two moderate events driven by the same meteorological phenomenon coincide (typically in the fall), and that conditions could be further exacerbated by the occurrence of uncorrelated events, such as antecedent rainfall or neap tide. The Noorderzijlvest example highlights the importance of considering the dependence driven by seasonality in compound event analysis and the contributions of independent drivers (e.g., tide), as it is not always the most extreme events that generate the greatest impact.

West Side of Galveston Bay

Galveston Bay is the seventh largest estuary in the United States. It has a surface area of approximately 1554 km^2 and an average depth of 3 m. It is a shallow, wind-driven system separated from the Gulf of Mexico by two barrier islands: Bolivar and Galveston. Currently, more than 1.6 million people live within the Hurricane Evacuation Zones on the west side of Galveston Bay and it is projected that this number will reach 2.4 million by 2035. Recent tropical cyclones, like Hurricane Ike (2008), during which extreme coastal water levels coincided with heavy precipitation, and Hurricane Harvey (2017), during which more moderate coastal water levels restricted drainage from overland areas, have demonstrated the vulnerability of the west side of Galveston Bay to compound flooding. However, given the limited-relief topography, any built infrastructure can have significant impacts by redirecting flow into areas that water did not previously go or retaining water where it used to drain. In fact, upland development patterns are already driving dramatic increases in the extent of the floodplain (see, e.g., Gori, Blessing, Juan, Brody, & Bedient, 2019), and will likely exacerbate interactions between rainfall-runoff and storm surge by decreasing time to peak and increasing runoff volumes (Sebastian, 2016). This example demonstrates the importance of understanding how anthropogenic change will alter the frequency and severity of compound events over time, particularly for low-lying coastal cities (see also Brody, Chapter 15, this volume).

Galveston Island

Galveston Island and Bolivar Peninsula separate Galveston Bay from the Gulf of Mexico and act as the region's first line of defense against storm surges. In the wake of the 1900 Hurricane—the deadliest disaster in US history—a 5.2-m (17-ft) high sea wall was built on the gulf-side of the island to protect the downtown areas from storm surge. Portions of

Galveston Island were raised to meet the sea wall. The sea wall protects the island from the most damaging storm surges caused by tropical cyclones making landfall along the Texas coast; however, Galveston is still prone to moderate to severe flooding during extreme precipitation events. Flood conditions are further exacerbated when bay-side water levels (e.g., storm surge or high tides) prevent water from being conveyed off the island either overland or through the stormwater infrastructure network. As a result, inundation can occur along significant roadways (e.g., Broadway, 51st Street, and Harborside Drive) cutting off portions of the island from the mainland. Fig. 2 shows the combinations of daily coastal water levels and precipitation that have occurred since 1940. The figure demonstrates that the frequency of coincident high sea level and heavy precipitation days has increased in more recent years, a trend that is driven by rising sea levels. However, because most studies focus on modeling coincident extrema, floods which arise from less extreme conditions, or those which are classified as moderate or

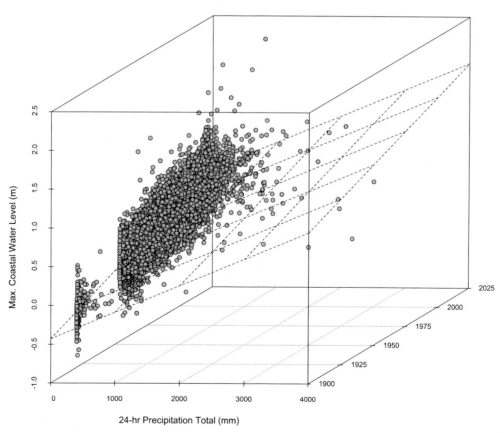

Fig. 2 Plot of maximum daily water levels (m) at Galveston Pier 21 and precipitation (mm) recorded on Galveston Island between 1940 and 2020.

nuisance are often overlooked despite that they pose significant safety concerns and can lead to local business disruptions (see, e.g., Hino, Belanger, Field, Davies, & Mach, 2019), and some researchers have suggested that over time, the cumulative cost of *nuisance flooding* could exceed the costs of larger, more extreme floods (Moftakhari, AghaKouchak, Sanders, & Matthew, 2017). Therefore, planning for small events is also critical to designing resilient flood risk reduction systems.

Conclusions

Coastlines are home to much of the global population and are uniquely threatened by flooding that can arise from multiple sources including heavy precipitation, extreme sea levels, high tide and river flows, and groundwater discharge. These hazard events may be extreme on their own but can also occur simultaneously or in succession. Both numerical and statistical modeling are important to determining where and to what extent flood hazard drivers could interact to exacerbate coastal flooding. While statistical models are useful across multiple spatial and temporal scales, in depth numerical modeling of compound coastal hazards and their potential impacts to communities is critically important for coastal management. The cases provided in the previous section highlight the importance of considering the correlation between multiple types of flood hazards, particularly for the design and operation of coastal infrastructure, as well as for land-use planning. They demonstrate the importance of understanding where and how frequently events coincide, as well as the ways in which coastal flooding will be exacerbated by a variety of natural and anthropogenic processes, including development of the coastal zone, rising sea levels, and increased storm frequency and intensity.

References

Bass, B., & Bedient, P. (2018). Surrogate modeling of joint flood risk across coastal watersheds. *Journal of Hydrology, 558*. https://doi.org/10.1016/j.jhydrol.2018.01.014.

Bensi, M., Mohammadi, S., Kao, S., & DeNeale, S. T. (2020). *Multi-mechanism flood hazard assessment: Critical review of current practice and approaches*. https://doi.org/10.2172/1637939.

Bilskie, M. V., & Hagen, S. C. (2018). Defining flood zone transitions in low-gradient coastal regions. *Geophysical Research Letters, 45*, 2761–2770. https://doi.org/10.1002/2018GL077524.

Christian, J., Duenas-Osorio, L., Teague, A., Fang, Z., & Bedient, P. B. (2013). Uncertainty in floodplain delineation: Expression of flood hazard and risk in a Gulf Coast watershed. *Hydrological Processes, 27*, 2774–2784. https://doi.org/10.1002/hyp.9360.

Christian, J., Fang, Z., Torres, J., Deitz, R., & Bedient, P. (2014). Modeling the hydraulic effectiveness of a proposed storm surge barrier system for the Houston ship channel during hurricane events. *Natural Hazards Review, 04014015*, 1–11. https://doi.org/10.1061/(ASCE)NH.1527-6996.0000150.

Couasnon, A., Eilander, D., Muis, S., Veldkamp, T. I. E., Haigh, I. D., Wahl, T., et al. (2020). Measuring compound flood potential from river discharge and storm surge extremes at the global scale and its implications for flood hazard. *Natural Hazards and Earth System Sciences Discussions*, 1–24. https://doi.org/10.5194/nhess-2019-205.

Couasnon, A., Sebastian, A., & Morales-Nápoles, O. (2018). A copula-based Bayesian network for modeling compound flood Hazard from riverine and coastal interactions at the catchment scale: An application to the Houston Ship Channel, Texas. *Water, 10*, 1190. https://doi.org/10.3390/w10091190.

De Ruiter, M. C., Couasnon, A., Homberg, M. J. C., Daniell, J. E., Gill, J. C., & Ward, P. J. (2020). Why we can no longer ignore consecutive disasters. *Earth's Future, 8*. https://doi.org/10.1029/2019ef001425.

Dietrich, J. C., Zijlema, M., Westerink, J. J., Holthuijsen, L. H., Dawson, C. N., Luettich, R. A., et al. (2011). Modeling hurricane waves and storm surge using integrally-coupled, scalable computations. *Coastal Engineering, 58*, 45–65. https://doi.org/10.1016/j.coastaleng.2010.08.001.

FEMA, 2015. Coastal flood risk study process[WWW document]. URL https://www.fema.gov/coastal-flood-risk-study-process.

Ganguli, P., Paprotny, D., & Hasan, M. (2020). Projected changes in compound flood Hazard from riverine and coastal floods in northwestern Europe. *Earth's Future*, 0–2. https://doi.org/10.1029/2020EF001752.

Geerse, C. P. M. (2010). Overzichtsdocument probabilistische modellen zoete wateren. In Hydra-VIJ, Hydra-B en Hydra-Zoet, C.P.M. Geerse (HKV), met medewerking van Herbert Berger en Robert Slomp (Waterdienst), HKV Lijn in Water, Lelystad, juli 2010, In opdracht van de Waterdienst.

Genest, C., & Favre, A.-C. (2007). Everything you always wanted to know about copula modeling but were afraid to ask. *Journal of Hydrologic Engineering, 12*, 347–368. https://doi.org/10.1061/(ASCE)1084-0699(2007)12:4(347).

Gori, A., Blessing, R. B., Juan, A., Brody, S. D., & Bedient, P. B. (2019). Characterizing urbanization impacts on floodplain through integrated land use, hydrologic, and hydraulic modeling. *Journal of Hydrology, 568*, 82–95.

Gori, A., Lin, N., & Xi, D. (2020). Tropical cyclone compound flood Hazard assessment: From investigating drivers to quantifying extreme water levels. *Earth's Future, 8*. https://doi.org/10.1029/2020EF001660.

Hendry, A., Haigh, I. D., Nicholls, R. J., Winter, H., Neal, R., Wahl, T., et al. (2019). Assessing the characteristics and drivers of compound flooding events around the UK coast. *Hydrology and Earth System Sciences, 23*, 3117–3139. https://doi.org/10.5194/hess-23-3117-2019.

Hino, M., Belanger, S. T., Field, C. B., Davies, A. R., & Mach, K. J. (2019). High-tide flooding disrupts local economic activity. *Science Advances, 5*, eaau2736. https://doi.org/10.1126/sciadv.aau2736.

Ikeuchi, H., Hirabayashi, Y., Yamazaki, D., Muis, S., Ward, P. J., Winsemius, H. C., et al. (2017). Compound simulation of fluvial floods and storm surges in a global coupled river-coast flood model: Model development and its application to 2007 cyclone Sidr in Bangladesh. *Journal of Advances in Modeling Earth Systems, 9*, 1847–1862. https://doi.org/10.1002/2017MS000943.

Klerk, W. J., Winsemius, H. C., Van Verseveld, W. J., Bakker, A. M. R., & Diermanse, F. L. M. (2015). The co-incidence of storm surges and extreme discharges within the Rhine-Meuse Delta. *Environmental Research Letters, 10*. https://doi.org/10.1088/1748-9326/10/3/035005.

Leonard, M., Westra, S., Phatak, A., Lambert, M., van den Hurk, B., Mcinnes, K., et al. (2014). A compound event framework for understanding extreme impacts. *Wiley Interdisciplinary Reviews: Climate Change, 5*, 113–128. https://doi.org/10.1002/wcc.252.

Moftakhari, H. R., AghaKouchak, A., Sanders, B. F., & Matthew, R. A. (2017). Cumulative hazard: The case of nuisance flooding. *Earth's Future, 5*, 214–223. https://doi.org/10.1002/2016EF000494.

Moftakhari, H. R., Salvadori, G., AghaKouchak, A., Sanders, B. F., & Matthew, R. A. (2017). Compounding effects of sea level rise and fluvial flooding. *Proceedings of the National Academy of Sciences, 114*, 9785–9790. https://doi.org/10.1073/pnas.1620325114.

Moftakhari, H. R., Schubert, J. E., AghaKouchak, A., Matthew, R. A., & Sanders, B. F. (2019). Linking statistical and hydrodynamic modeling for compound flood hazard assessment in tidal channels and estuaries. *Advances in Water Resources, 128*, 28–38. https://doi.org/10.1016/j.advwatres.2019.04.009.

Muñoz, D. F., Muñoz, P., Moftakhari, H., & Moradkhani, H. (2021). From local to regional compound flood mapping with deep learning and data fusion techniques. *Science of the Total Environment, 782*. https://doi.org/10.1016/j.scitotenv.2021.146927, 146927.

Paprotny, D., Vousdoukas, M. I., Morales-Nápoles, O., Jonkman, S. N., & Feyen, L. (2018). Compound flood potential in Europe. *Hydrology and Earth System Sciences Discussions*, 1–34. https://doi.org/10.5194/hess-2018-132.

Paprotny, D., Vousdoukas, M. I., Morales-Nápoles, O., Jonkman, S. N., & Feyen, L. (2020). Pan-European hydrodynamic models and their ability to identify compound floods. *Natural Hazards, 101*, 933–957. https://doi.org/10.1007/s11069-020-03902-3.

Ray, T., Stepinski, E., Sebastian, A., & Bedient, P. B. (2011). Dynamic modeling of storm surge and inland flooding in a Texas coastal floodplain. *Journal of Hydraulic Engineering*, *137*, 1103–1111. https://doi.org/10.1061/(ASCE)HY.1943-7900.0000398.

Santiago-Collazo, F. L., Bilskie, M. V. M. V., & Hagen, S. C. S. C. (2019). A comprehensive review of compound inundation models in low-gradient coastal watersheds. *Environmental Modelling and Software*, *119*, 166–181. https://doi.org/10.1016/j.envsoft.2019.06.002.

Santos, V. M., Casas-prat, M., Poschlod, B., Ragno, E., & Van Den Hurk, B. (2021). Statistical modelling and climate variability of compound surge and precipitation events in a managed water system : A case study in the Netherlands. *Hydrology and Earth System Sciences*, *25*, 3595–3615. 10.5194/hess-25-3595-2021.

Sebastian, A. (2016). *Quantifying flood Hazard and risk in highly urbanized coastal watersheds*. Rice University.

Sebastian, A., Dupuits, E. J. C., & Morales-Nápoles, O. (2017). Applying a Bayesian network based on Gaussian copulas to model the hydraulic boundary conditions for hurricane flood risk analysis in a coastal watershed. *Coastal Engineering*, *125*, 42–50. https://doi.org/10.1016/j.coastaleng.2017.03.008.

Sklar, M. (1959). *Fonctions de répartition à n dimensions et leurs marges* (p. 8). Université Paris.

Torres, J. M., Bass, B., Irza, N., Fang, Z., Proft, J., Dawson, C., et al. (2015). Characterizing the hydraulic interactions of hurricane storm surge and rainfall-runoff for the Houston-Galveston region. *Coastal Engineering*, *106*, 7–19. https://doi.org/10.1016/j.coastaleng.2015.09.004.

Torres, J. M., Bass, B., Irza, J. N., Proft, J., Sebastian, A., Dawson, C., et al. (2017). Modeling the hydrodynamic performance of a conceptual storm surge barrier system for the Galveston bay region. *Journal of Waterway, Port, Coastal, and Ocean Engineering*, *143*. https://doi.org/10.1061/(ASCE)WW.1943-5460.0000389.

Van den Hurk, B., Van Meijgaard, E., de Valk, P., Van Heeringen, K. J., & Gooijer, J. (2015). Analysis of a compounding surge and precipitation event in the Netherlands. *Environmental Research Letters*, *10*. https://doi.org/10.1088/1748-9326/10/3/035001.

Wahl, T., Jain, S., Bender, J., Meyers, S. D. S. D., & Luther, M. E. M. E. (2015). Increasing risk of compound flooding from storm surge and rainfall for major US cities. *Nature Climate Change*, *5*, 1–6. https://doi.org/10.1038/NCLIMATE2736.

Ward, P. J. P. J., Couasnon, A., Eilander, D., Haigh, I. D. I. D., Hendry, A., Muis, S., et al. (2018). Dependence between high sea-level and high river discharge increases flood hazard in global deltas and estuaries. *Environmental Research Letters*, *13*. https://doi.org/10.1088/1748-9326/aad400, 084012.

Zheng, F., Westra, S., & Sisson, S. A. (2013). Quantifying the dependence between extreme rainfall and storm surge in the coastal zone. *Journal of Hydrology*, *505*, 172–187. https://doi.org/10.1016/j.jhydrol.2013.09.054.

Zhong, H., van Overloop, P. J., van Gelder, P., & Rijcken, T. (2012). Influence of a storm surge Barrier's operation on the flood frequency in the Rhine delta area. *Water*, *4*, 474–493. https://doi.org/10.3390/w4020474.

Zscheischler, J., Westra, S., Van Den Hurk, B. J. J. M. B. J. J. M. B. J. J. M., Seneviratne, S. I. S. I., Ward, P. J. P. J., Pitman, A., et al. (2018). Future climate risk from compound events. *Nature Climate Change*, *8*, 469–477. https://doi.org/10.1038/s41558-018-0156-3.

SECTION II

Paying the price: Socioeconomic and political underpinnings of flood risk

CHAPTER 8

Cost-benefit analysis of a proposed coastal infrastructure for reducing storm surge-induced impact in the Upper Texas Coast

Meri Davlasheridze[a] and Kayode O. Atoba[b]
[a]Department of Marine and Coastal Environmental Science, Texas A&M University, Galveston Campus, Galveston, TX, Unites States
[b]Institute for a Disaster Resilient Texas, Texas A&M University, College Station, TX, United States

Introduction

The frequency and intensity of catastrophic events in recent years and future forecasts have been concerning scientists and policymakers (Allan & Soden, 2008; U.S. Climate Change Science Program, 2008; Emanuel, 2013; Grinsted, Moore, & Jevrejeva, 2013; Villarini & Vecchi, 2013; Walsh et al., 2016). According to the recent National Oceanic and Atmospheric Administration (NOAA) estimates, since 1980 the United States has experienced over 179 weather- and climate-induced "billion-dollar" disasters, each exceeding $1 billion dollars in economic losses (NOAA National Centers for Environmental Information (NCEI), 2020). The year 2020 has been historic in terms of setting a record number of such large-scale disasters such as hurricanes, floods, fires, and droughts. It has also set the record in terms of the largest number of the named storms that made landfall in the United States, which have also dominated the types of billion-dollar disasters (14 out of 16 events were related to hazards associated with severe storm events and cyclones). Alarmingly, the year 2020 was also the sixth consecutive year since 2015 in which the United States was impacted by 10 or more billion-dollar weather and climate disasters (NOAA National Centers for Environmental Information (NCEI), 2020). For all these reasons, proactive disaster management is a social imperative.

Unarguably, low-lying coastal areas, facing chronic and recurring risk of floods, storm surge and sea level rise, remain vastly vulnerable and are particularly challenged to find a solution to comprehensively address future risk (Wing, Pinter, Bates, & Kousky, 2020). It is also recognized that there may not be one solution that would solve this problem, but rather a suite of tangible mitigation options, ranging from gray (e.g., seawalls, dikes, levees, dams, etc.) and green infrastructure (open space protection, protected wetlands, etc.) to land-use planning, managed retreat, development restrictions, as well as more stringent building codes, resilient retrofits, and hazard insurance markets (Craig, 2019; de Vries & Fraser, 2012; Kousky, 2014; Marino, 2018; Zavar & Hagelman, 2016).

Comprehensively enumerating the benefits and costs of a mitigation option provides a solid and defensible means to inform and choose among policy alternatives, especially given the budgetary constraints facing many communities (Ganderton, 2005). Investment in a gray infrastructure is costly, and committing substantial public resources to address consequences of infrequent catastrophes in the future implies diverting resources from other alternative uses that may yield immediate benefits (Zerbe & Dively, 1994). In the United States, disaster risk management is largely decentralized and the federal government serves as an "insurer of a last resort" and steps in when the event overwhelms local resources to effectively respond to and recover from disasters (Bierbaum et al., 2013). Federal disaster aids allocated toward long-term mitigation projects, although limited, are also available postdisaster (Davlasheridze, Fisher-Vanden, & Klaiber, 2017). Increasing federal budget deficits often make disaster mitigation programs easy targets for budget cuts or postponement (DelRossi & Inman, 1999).

Notably, when there is no national guideline to address growing concerns of coastal inundation, there is a possibility of under- or oversupply of defense systems, depending on the funding structure. More specifically, if funded at a cost-share mechanism (because cost is shared and not entirely borne by local residents who accrue the most benefits of protection), there is a possibility that the amount of protection will be oversupplied. On the other hand, if funded primarily with local money, without recognizing that the benefits of structural mitigation spill over other jurisdictions and significant external benefits accrue to other localities, the size of defense will be suboptimal (Miao, Shi, & Davlasheridze, 2020).

This chapter provides an overview of cost-benefit analysis (CBA) of one type of structural defense proposed as the mitigation strategy for Galveston Bay communities in the state of Texas, USA. This defense structure is a coastal spine, which includes extension of an existing sea wall, a new gate barrier, and several miles of coastal dune system along the Gulf of Mexico to protect the Houston-Galveston region from the impact of hurricane-induced storm surge. The benefits are assessed in terms of loss avoidance for major residential assets and industrial operations afforded in part due to the coastal protection. The frameworks also assess indirect impacts on local, regional, and national economies. This chapter further draws inference about CBA practices in the Netherlands. The Netherlands represents a relevant reference case as the country has historically relied heavily on structural flood defense solutions to address chronic floods and rising sea levels. More importantly, the Netherlands is the only country in the world with a comprehensive and extensive storm suppression and defense system. Flood defense is a national priority and structures are regarded as a critical infrastructure, since 60% of the country lies below the sea level (Botzen, Aerts, & Van den Bergh, 2013; Hallegatte et al., 2011). Since 2000, performing CBAs is obligatory for major infrastructure projects and there exist formal, national guidelines on CBA (Eijgenraam et al., 2014).

Modeling approach

One approach to enumerating the benefits of surge protection barriers is to estimate the surge-induced losses without and with a coastal infrastructure in place. In order to achieve this objective, it requires creating an impact pathway and an understanding of (a) how storm surge-induced inundation impacts residential properties and industrial operations, and (b) how these direct impacts propagate through the other sectors and regions of the economy through intersectoral linkages, interactions of economic agents (e.g., households, suppliers, government) and trade. As illustrated in Fig. 1, hurricane events cause buildings to be inundated by floodwaters, thereby resulting in direct structural damages and indirect business disruptions, all of which affect the economy of the impacted area. The existence of a coastal infrastructure can significantly reduce both the direct and indirect economic impacts of hurricane events in coastal communities.

Research and modeling capabilities have advanced significantly toward assessing loss to residential property. For example, engineering-based models such as the Federal Emergency Management Agency's (FEMA) flood assessment software known as Hazards US Multi-Hazard (HAZUS-MH) have been used extensively to perform damage assessments and, by extension, CBAs (Ding, White, Ullman, & Fashokun, 2008; FEMA, 2006; Scawthorn et al., 2006; Scawthorn et al., 2006; Tate, Muñoz, & Suchan, 2014). These losses have been validated in the literature and have been used with confidence in predicting future losses under various mitigation, structure retrofit scenarios (e.g., see Atoba, Brody, Highfield, & Merrell, 2018; Kousky & Walls, 2014).

On the other hand, there remains a gap in quantifying impacts on industrial operations (e.g., refineries, petrochemical manufacturing, etc.). Historical loss data are almost nonexistent. Therefore, scientific understanding about hazard-loss relation is at best based on assumptions that cannot be validated (Burleson, Rifai, Proft, Dawson, & Bedient, 2015; Kirgiz, Burtis, & Lunin, 2009). In a handful of instances, companies publish reports indicating sustained losses to structures and operations. Relying on these reports and expert opinions, it appears that structurally (e.g., capital, property) industrial operations are fairly resistant to damaging surge events. For example, a report published by Phyllips 66's Bayway, one of the largest refineries operating in New York and New Jersey after Superstorm Sandy, indicated that its capital loss made up approximately 8% of total losses and the other 92% of losses were output losses associated with a prolonged power outage (Hydrocarbon Publishing Company, 2016). While capital losses may be negligible, these operations may suffer sizable output losses due to plant shutdown. Shutdowns could happen for precautionary purposes and during an active hazard warning (e.g., mandatory evacuation, stay in house) period because workers cannot get to workstations, or they may happen due to power outages or failure of electrical equipment and control room (e.g., systems and operating) (Hydrocarbon Publishing Company, 2016). According to the US Department of Energy, severe weather conditions represent the single leading

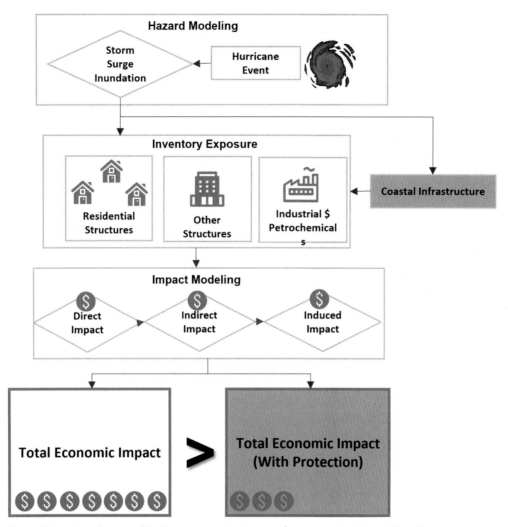

Fig. 1 Measuring direct and indirect economic impact of storm surge-induced flooding.

cause of power outages across the United States, costing billions in economic losses (US Department of Energy, 2013). For example, outages due to thunderstorms, hurricanes, and blizzards accounted for 58% of all outages observed since 2002 (US Department of Energy, 2013). The shutdown days depending on severity could last for a few months. For example, BP in Texas City was shut down for 77 days in response to hurricane Katrina according to the data published by the U.S. Department of Energy (2009). The same data showed shutdown and partial restarting days during Katrina lasted 18 days on average, which was approximately 33 days during 2008 hurricane Ike, and the overall average due to the two hurricanes was estimated at around 26 days. Using these average shutdown days and multiplying them by the daily value of production (i.e., sales) generated by these industries is one plausible proxy for plant-level output losses associated with production cessations.[a]

Meanwhile, estimating indirect impacts requires integrating direct impacts with the regional Input–Output (I—O) or other more advanced computable general equilibrium (CGE) models. I—O models are one of the widely used models for the impact and policy analysis (Bockarjova, Steenge, & van der Veen, 2004; Cochrane, 2004; Haimes & Jiang, 2001; Okuyama, 2004). However, this method has been criticized because fixed interindustry dependence may yield an overestimation of impacts across sectors. The variants of the so-called "adjusted" I—O models have been proposed in modeling economic impacts of disasters (e.g., Hallegatte, 2008; Rose, 2004; Rose & Blomberg, 2010; Rose, Oladosu, Lee, & Asay, 2009; Rose & Wei, 2013; Rose, Wei, & Wein, 2011). The CGE model is more realistic as it builds on general equilibrium theory and provides economic interactions across economic agents (e.g., consumers, producers, government, and the trade sector) through markets. The CGE models have been used widely to model policy changes at sectoral levels (Bergman, 1991; Böhringer, Rutherford, & Wiegard, 2003; Shoven & Whalley, 1992; Sue Wing, 2007a, 2007b), the effects of extreme weather phenomena (Rose et al., 2007; Rose & Guha, 2004; Rose & Liao, 2005; Rose, Oladosu, & Liao, 2007; Sue Wing, Rose, & Wein, 2016), sea level rise (Parrado et al., 2020), and climate change (Abler et al., 2009; Hsiang et al., 2017; Palatnik & Roson, 2012; Zhou, Hanasaki, Fujimori, Masaki, & Hijioka, 2018).

Both classes of models are based on the Social Accounting Matrix (SAM), depicting the interaction of a wide range of finer scale economic sectors. In our modeling approach, the impact pathways are built through a direct effect on a specific sector(s) output (e.g., residential housing, petroleum, chemical manufacturing), which then generates other sector-specific and regional ripple effects, and in case of the CGE also captures general equilibrium and multiplier effects. More details on the specificities of these models and how to best integrate them are provided in Davlasheridze et al. (2019) and Davlasheridze, Fan, Highfield, and Liang (2021).

---

[a] A more robust approach would establish a relationship between inundation levels and shutdown days via a simple regression in which plant-specific shutdown days are regressed on inundation levels at each point location. However, the shutdowns may happen not only because of the plant location-specific inundation, but rather because of the electric grid failure in servicing the end customers. Thereby, estimating how surge inundation impacts relevant electric grids, and how these impacts translate into a subsequent impact on end point customers (e.g., plants) would be an alternative, more defensible, and empirically testable approach. This approach will also offer a way to make predictions about the expected shutdown days for any levels of inundation.

In the following sections, we preview findings of model integrations in the context of a coastal spine protection as applied to the Houston-Galveston region.

Study area

The study area comprises Harris, Galveston, and Chambers counties. All these counties directly surround Galveston Bay which connects with the Gulf of Mexico along the upper southeastern coast of Texas. The region consists of 46 tidally and nontidally influenced watersheds that are highly prone to both precipitation-based flooding and hurricane-induced storm surge. The area is characterized by low topographic relief, large amounts of floodplains, and a long history of flood loss. Galveston Island and Bolivar peninsula specifically have experienced several large hurricanes, highlighted by the 1900 storm (the deadliest hurricane in US history) and Hurricane Ike in 2008. The impact of these storms was felt by areas further inland, such as portions of Houston, as storm surge ran up the creeks and bayous.

The long history of flooding and the exposure to storm surge events necessitate a need for both structural and nonstructural storm surge protection in the Houston-Galveston region. This region houses the largest petrochemical complex in the United States. It is also a major hub for industrial and petrochemical complexes, with several cargo traffic in the Houston Ship Channel and the Port of Houston. The main storm surge infrastructure in the region is the existing Galveston Seawall, which protects eastern sections of Galveston Island. The other protective structure is the Texas City Dike, which protects petrochemical complexes. Several other sections of this region are still exposed to storm surge-induced flooding.

The need for additional flood risk reduction strategies in the region has resulted in several flood infrastructure proposals. Most recently, a feasibility report for the storm surge infrastructure options for the region has been completed by the US Army Corps of Engineers (USACE) (USACE, 2020). One of the most prominent proposals is a shoreline levee system along Galveston Island and Bolivar Peninsula (also known as the Ike Dike) that would protect communities along the bay, including Houston and its surrounding suburbs. Fig. 2 shows the composition of the proposed coastal spine system. It extends the existing seawall on Galveston Island northeast to High Island southwest to San Luis Pass. A movable gate system similar to the Maeslant Barrier in Rotterdam will also be placed to prevent rising waters and storm surge from coming into Galveston Bay while also allowing navigation into the Houston ship channel. The engineering costs of this coastal barrier system are estimated at about $8.9 billion (Jonkman et al., 2015).

Results: Direct and indirect impacts

To estimate the direct impact component (e.g., losses to residential property and industrial operation), a two-step methodology was used: (1) estimating surge heights and

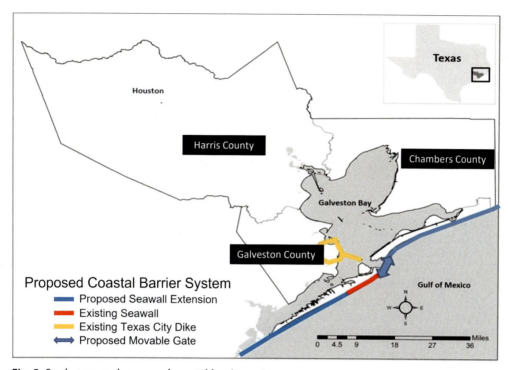

Fig. 2 Study area and proposed coastal barrier system.

inundation in the ADvance CIRCulation (ADCIRC) model and (2) estimating structural property losses using HAZUS-MH, and losses to petroleum and chemical manufacturing due to operation shutdowns using average production cessation days due to hurricanes Katrina and Ike. The ADCIRC model generated surge outputs for three synthetic storms differing in return probabilities (e.g., 10-year, 100-year, and 500-year) (Ebersole et al., 2015).[b] Scenarios were built around the storm intensity and impacts were quantified with and without a coastal spine, by spatially joining inundation depth [built using 1/9 arcsec (3 m) digital elevation models (DEM) for ground elevation] with a granular building inventory (e.g., updated for the first floor elevation, building, and foundation types) and petrochemical plant locations. Moreover, plant-specific employment data available through the Chemplant were combined with the zip-code level of the US Census of manufacturers data for the industry annual output value. Specifically, plant-level output value was calculated by multiplying the zip-code average output values per employee with the plant-specific employments. Subsequently, it was assumed that

---

[b] To validate model outputs, an additional "Ike-like" storm was recreated and surge levels were compared to observed surge heights reported for the 2008 hurricane Ike that devastated the region.

Table 1 Direct loss ($Millions) to residential building stock and production output loss to petrochemical complexes (without and with protection).

Loss estimates	500-yr	100-yr	10-yr
Baseline estimates without protection			
Residential building stock	8022.13	4351.74	527.71
Petrochemical complexes	7458.78	1894.94	52.63
Total	**15,480.91**	**6246.68**	**580.34**
Estimates with Coastal Spine			
Residential building stock	2331.46	1352.75	104.33
Petrochemical complexes	201.07	94.92	0
Total	**2532.53**	**1447.67**	**104.33**
Loss avoided (without-with)			
Residential building stock (%)	5690.67 (71%)	2998.99 (69%)	423.38 (80%)
Petrochemical complexes	2130.39 (91%)	1257.83 (93%)	104.33 (100%)
Total	**7821.06 (51%)**	**4256.82 (68%)**	**527.71 (91%)**

Notes: Petrochemical complexes correspond to petroleum, petrochemical and chemical manufacturing plants and output loss estimates are reported for the scenario when plants shut down for 33 days.
Source: Author calculations.

production was continuous around the year and the daily output was calculated from these annual estimates. Atoba et al. (2018) and Davlasheridze et al. (2019) provide more details about loss estimation for property and petrochemical complexes, respectively. The direct impacts associated with different storm surge scenarios with and without the spine are reported in Table 1.

The estimated losses were over $8 and $7.5 billion for residential properties and petrochemical complexes, respectively, from the most destructive storm surge considered (i.e., 500 year). With a coastal spine in place, the direct losses to residential properties were reduced by 71% and the output losses for the petrochemical industry were reduced by 91%. Other storm intensities also resulted in significant reduction in direct loss from storm surge-induced flooding. For example, the coastal spine reduces 100-year flood losses to residential structures by approximately 70% and the production output value is reduced by 93% with the proposed coastal infrastructure. Notably, while with the surge protection system, residential structures still experience some loss (although substantially smaller) with the least-intense simulated storm (i.e., 10-year), the coastal spine is expected to mitigate all the impacts on petrochemical operations.

Using direct loss estimates to sector outputs (e.g., residential and petrochemical manufacturing sectors) reported in Table 1, the IMPLAN model calibrated for the year 2012 (Minnesota IMPLAN Group (MIG), 2012) and built for the study region provided

Table 2 Indirect and induced impacts on output value for the HGA.

	Without spine	**With spine**
500-year storm		
Indirect	6383.00	504.2
Induced	1277.00	140.6
Total	7660.00	644.80
100-year storm		
Indirect	1926.00	280.8
Induced	426.3	77.78
Total	2352.30	358.58
10-year storm		
Indirect	116.3	0
Induced	28.94	0
Total	145.24	0.00

Notes: The table presents IMPLAN model results in terms of indirect and induced effects for the HGA region for the scenario when plants shut down for 33 days.

indirect and induced impact assessments on output values, income, employment, and value added in all other sectors.[c] In Table 2, results are reported in terms of production output (total sales value).[d]

As the final step, to factor in the storm-return probabilities, the average annualized losses are calculated as the weighted sum of total loss estimates from Tables 1 and 2, where the weights correspond to storm return probabilities (e.g., a 500-year storm has a 0.2% annual chance of occurrence, while probabilities for 100-year and 10-year storms correspond to 1% and 10%, respectively).

Comparing Avoided AAL with the annualized cost of a spine (discounted by an interest rate over a certain lifetime of a project) represents the Benefit to Cost ratio for this structure, and if greater than one would indicate that the expected benefits outweigh the cost of the project. In the context of the coastal spine for the HGA, the lower, average, and upper bounds of engineering costs estimated by Jonkman et al. (2015) were used, assuming a 5% discount rate and 100-year lifetime of the spine. The resultant Benefits to Costs ratio, as discussed in Davlasheridze et al. (2019), for a low-cost estimate was over 3, while for the medium- and high-cost projects the ratios were estimated at 1.8 and 1.3, respectively.

While in all the three cost cases the Benefits to Costs ratios were estimated to be greater than 1, thereby implying feasibility of the coastal spine, using fixed industry

---

[c] Indirect impacts capture intersectoral linkages through a supply chain, while the induced effects are changes in local spending of employees' wages and salaries of both directly and indirectly impacted sectors.

[d] The full set of model results is presented in Davlasheridze et al. (2019).

dependence of I—O models likely overestimated the benefits. However, we also note that several broader benefits and important costs were not accounted for in this exercise. Being a major hub for the largest petrochemical complexes in the world, the HGA supports energy and chemical related sectors in the United States Thereby, it is reasonable to expect storm surge-induced interruption of manufacturing processes to have wider repercussions across the United States as a whole. By estimating the benefits of a coastal spine to the HGA regional economy only and ignoring its connectedness to other markets, a substantial proportion of monetizable benefits are left unaccounted.

An important extension of the three-county HGA IMPLAN model was the development of a state-level dynamic recursive CGE model to integrate the spillover effects of surge impacts on the economies of other states and the US National Economy (see Davlasheridze et al., 2021). In addition to regional spillovers, the dynamic CGE model allowed modeling the long-term ramification of damaging surge events. To incorporate the direct output losses reported in Table 1 into relevant sectors in the CGE model, it was assumed that the surge would impact sector-specific total factor productivity (TFP). TFP broadly captures the efficacy and intensity of inputs utilized in the production process (Rausch & Rutherford, 2009; Sue Wing, 2007a, 2007b). The benefits in terms of loss avoidance were estimated as the difference between the scenario when the surge impacts the residential, petroleum, and chemical manufacturing sectors and the scenario that assumes uninterrupted (by storm surge) economy on a projected growth trajectory (referred to as Business As Usual, BAU).[e]

As presented in Davlasheridze et al. (2021), the CGE model results indicate that the Texas state Gross State Product (GSP) declines by an estimated 1.96% in the long term (50 years from the storm impact) without a coastal spine in response to a 500-year storm surge, which is mitigated fourfold with the coastal spine. The US GDP experiences a 0.3% decline corresponding to an approximately $32 billion in present value GDP loss without the coastal protection. Applying storm return probabilities to these impacts, the expected annualized impacts for the selected macroeconomic indicators are reported in Table 3. For example, the annual expected decline in Texas GSP is estimated at approximately 0.05% compared with that in the US GDP by 0.01% on average annually. The expected annual decline in net export value is estimated at 0.11% for Texas.

Dutch approach to coastal infrastructure BCA and lessons for the United States

Storm surge represents one of the most threatening aspects of tropical cyclones (Balaguru, Judi, & Leung, 2016). With asset losses estimated to exceed $990 billion by 2100 in the

---

[e] The BAU economic growth projections were based on the projections of the state-level population and key exogenous parameters such as annual growth rates of multifactor productivity and annual rate of improvement in labor quality (Davlasheridze et al. 2021).

Table 3 Average annualized impacts relative to BAU (% change).

	GDP	Welfare	Investment	Total consumption	Income	Government consumption	Net export
TX	−0.047	−0.049	0.036	−0.080	−0.040	0.008	−0.111
USA	−0.007	−0.006	0.006	−0.010	−0.005	0.000	−0.036

Notes: The table presents annualized (i.e., expected) impacts on selected macroeconomic variables measured in percentages, using the variable values relative to the BAU in 2066.

Source: Davlasheridze, M., Fan, Q., Highfield, W., & Liang, J. (2021). Economic impacts of storm surge events: examining state and national ripple effects. *Climatic Change*, *166*(1), 1–20.

United States (Neumann et al., 2015), there is substantial public interest in mitigating their effects. For the HGA communities, this interest has become particularly salient in response to the 2008 hurricane Ike. As demonstrated by our results, the magnitude of spatial and temporal effects of the storm surge could cross local boundaries and ripple to the economies of other states and the United States, as a whole, with long-lasting temporal adverse effects. Strategic assets (petrochemical manufacturing) located in the HGA underscore the important role the federal government may play in supporting both financially and technically the proposed coastal defense system (USACE, 2020). In a nutshell, drawing some parallels from the Netherlands may be especially informative.

The Netherlands is at the forefront of using storm surge barriers to reduce flood impacts, and it has traditionally used CBA to justify the construction of coastal infrastructure for several reasons. Approximately 60% of the country's land and more than half of its population lie and live below sea level. The country is increasingly vulnerable to storm surge impact and sea level rise and historically has relied heavily on flood defense structures as the risk management strategy (Botzen et al., 2013; Hallegatte et al., 2011). Because flood risk management represents a national priority, a specific flood control project is almost always placed in the context of national defense systems (Aerts & Botzen, 2013). Importantly, investment in flood control structures is guided by CBA, the performance of which has become obligatory for major infrastructure projects since 2000 in accordance with a formal, national guideline on CBA (Eijgenraam et al., 2014).

The Dutch history of CBAs for the flood control structures dates back to 1901 (e.g., enclosure of Zuiderzee and land reclamation).[f] It has become a more widely used tool for the central government to justify large public investment in flood control after the massive flooding event in 1950 and in particular to aid decision-making related to the Delta works (Eijgenraam et al., 2014; Aerts & Botzen, 2013). One notable change made in a simple benefit-cost analysis framework (Tinbergen, 1954) involved supplementing CBA with optimal dike heights analysis (van Dantzig, 1956). The latter led to the development of safety norms that have also accounted for the broader benefits of defense structures.

Specifically, Dantzig's formula for the economically optimal strength of dikes incorporated a probabilistic approach, and determined economically optimal safety levels, by comparing the costs of heightening dikes with its extra safety benefits. The extra benefits referred to loss of lives, other societal costs and damages to cultural assets often hard to enumerate. Initially, these extra safety benefits were calculated by multiplying the expected material damages by a factor 2 (Bos & Zwaneveld, 2017).

This model has improved markedly over the years by making it dynamic to account for the changing economic conditions and population growth. Furthermore, the analysis extended to incorporating linkages to other dikes and flood control systems, using a more realistic value of statistical life (estimated at 7 million euros) in benefits calculations

---

[f] Review of evolution of CBA for major defense barriers in the Netherlands is provided in Bos and Zwaneveld (2017).

afforded due to extra safety (Bos & Zwaneveld, 2017). It should be noted that environmental effects and assessment of environmental damages were not addressed in initial CBA. Nonmonetary impacts were treated with particular attention starting from the year 2000 with the Room for Rivers project, by providing cost estimates needed to restore or compensate for ecosystem service loss (Bos & Zwaneveld, 2017).

CBA represents an invaluable tool when choosing among competing alternatives (Brouwer & Kind, 2005). Initial Delta works analysis, for example, involved comparing the CBA of Delta works vs. raising and strengthening existing dikes along the waterways. Accounting for direct damages only, the latter alternative seemed the cheapest alternative; however, with supplemental benefits and costs, the Delta works proved to be the preferred alternative (Bos & Zwaneveld, 2017). Specifically, Tinbergen (1954) accounted for several important supplementary benefits of the new proposal including saving in traveling and transportation costs, benefits for fishing and agriculture due to alteration of water salinity levels, in addition to new opportunities for leisure activities. Other nonmonetized but discussed benefits included stimulus to hydraulic and engineering science, knowledge exploration and dissemination. However, various detrimental effects on ecosystems due to shortened waterways that were later realized were not accounted for.

One notable change in the CBA following the Delta Works was the official introduction of water safety norms, specifically determining standards to withstand a flood with the return probability of 1/10,000 years for densely populated and economically vibrant locations (e.g., Rotterdam). For less populated areas, safety standards of 1/1250 to 1/4000 years are required. These standards are the highest in the world (Brouwer & Kind, 2005; Kind, 2014).

The Dutch experience can be applicable to the upper Texas Coast. For example, the common practice of the USACE for engineering design corresponds to a flood risk with a return probability of 1/100 years. For densely populated areas, the exceedance standards of 1/500 years have been recommended (e.g., Galloway et al., 2006), but thus far they have not been implemented in practice. Furthermore, it is very rare for a CBA of a particular project to place the proposed project in the context of other defense structures or for broader economic benefits. More recently, recognizing the threats of sea level rise (SLR) and changing flood risk of a climate change, the Dutch strategy has shifted toward designing barriers that can be easily adaptable to uncertainty to SLR. Initial analysis indicates that if barriers allow, adaptability to SLR uncertainty makes structure upgrade to realized SLR less costly than building a brand-new barrier (Aerts & Botzen, 2013). Incorporating the effects of SLR is paramount in enumerating surge impacts, as recent studies have indicated that surge heights associated with 1/100-year storm events would become as frequent as one in every 4 years, simply because SLR creates higher launching points for them (Frumhoff, McCarthy, Melillo, Moser, & Wuebbles, 2007).

Building on a long history of the Dutch approach in CBA in the context of surge barriers and general flood defense structures, it seems reasonable to incorporate safety measures and uncertainty in SLR when deciding the barrier heights and strength. Extending the presented CBA for different scenarios of SLR will also provide defensible means for an easily adaptable barrier that will unarguably be more expensive. Importantly, enumerating the impacts of barriers on valuable ecosystems in the Galveston Bay is paramount to avoid irreparable environmental impacts caused by an embankment on one of the most diverse and complex ecosystems in the state of Texas and the United States (Lester, Gonzalez, Sage, & Gallaway, 2002). As the Dutch experience has shown, many of the environmental damages realized postconstruction were irreversible (Bos & Zwaneveld, 2017).

Furthermore, a surge may trigger a variety of cascading effects due to the spillage of dangerous toxins and chemicals posing human and ecosystem health threats (Burleson et al., 2015; Griggs, Lehren, Popovich, Singhvi, & Tabuchi, 2017; Krausmann & Cruz, 2013). Simulating potential cascading effects associated with the storms will inform the design standards for the protective structure. Last but not least, the framework presented in this chapter ignores surge impacts on human lives. According to the US Department of Transportation, the economic value of statistical life was estimated at approximately $9.6 million in 2016 (US Department of Transportation, 2016). Extending estimated benefits to account for avoided human loss due to storm surge, similar to the Dutch practice, will further support the feasibility of the proposed coastal spine, even if ignoring HGA significance for the US National Economy.

References

Abler, D., Fisher-Vanden, K., McDill, M., Ready, R., Shortle, J., Wing, I. S., et al. (2009). *Economic impacts of projected climate change in Pennsylvania: Report to the department of environmental protection*. Environment & Natural Resources Institute.

Aerts, J. C., & Botzen, W. J. (2013). Climate adaptation cost for flood risk management in the Netherlands. In *Storm surge barriers to protect New York City: Against the deluge* (pp. 99–113). https://doi.org/10.1061/9780784412527.007.

Allan, R. P., & Soden, B. J. (2008). Atmospheric warming and the amplification of precipitation extremes. *Science, 321*(5895), 1481–1484. https://doi.org/10.1126/science.1160787.

Atoba, K., Brody, S., Highfield, W., & Merrell, W. (2018). Estimating residential property loss reduction from a proposed coastal barrier system in the Houston-Galveston region. *Natural Hazards Review, 19*(3). https://doi.org/10.1061/(ASCE)NH.1527-6996.0000297.

Balaguru, K., Judi, D. R., & Leung, L. R. (2016). Future hurricane storm surge risk for the U.S. gulf and Florida coasts based on projections of thermodynamic potential intensity. *Climatic Change, 138*, 99–110. https://doi.org/10.1007/s10584-016-1728-8.

Bergman, L. (1991). General equilibrium effects of environmental policy: A CGE-modeling approach. *Environmental and Resource Economics, 1*(1), 43–61.

Bierbaum, R., Smith, J. B., Lee, A., Blair, M., Carter, L., Chapin, F. S., et al. (2013). A comprehensive review of climate adaptation in the United States: More than before, but less than needed. *Mitigation and Adaptation Strategies for Global Change, 18*(3), 361–406.

Bockarjova, M., Steenge, A. E., & van der Veen, A. (2004). On direct estimation of initial damage in the case of a major catastrophe: Derivation of the "basic equation". *Disaster Prevention and Management: An International Journal, 13*(4), 330–336.

Böhringer, C., Rutherford, T. F., & Wiegard, W. (2003). Computable general equilibrium analysis: Opening a black box *ZEW Discussion Papers 03-56*. Leibniz Centre for European Economic Research: ZEW.

Bos, F., & Zwaneveld, P. (2017). *Cost-benefit analysis for flood risk management and water governance in the Netherlands: An overview of one century*. CPB Background Document|August.

Botzen, W. J. W., Aerts, J. C. J. H., & Van den Bergh, J. C. J. M. (2013). Individual preferences for reducing flood risk to near zero through elevation. *Mitigation and Adaptation Strategies for Global Change, 18*(2), 229–244.

Brouwer, R., & Kind, J. (2005). Chapter 5: Cost-benefit analysis and flood control policy in the Netherlands. In R. Brouwer, & D. Pearce (Eds.), *Cost-benefit analysis and water resources management* (pp. 93–123). Cheltenham: Edgar Elgar.

Burleson, D. W., Rifai, H. S., Proft, J. K., Dawson, C. N., & Bedient, P. B. (2015). Vulnerability of an industrial corridor in Texas to storm surge. *Natural Hazards, 77*(2), 1183–1203.

Cochrane, H. (2004). Economic loss: Myth and measurement. *Disaster Prevention and Management: An International Journal, 13*(4), 290–296.

Craig, R. K. (2019). Coastal adaptation, government-subsidized insurance, and perverse incentive to stay. *Climatic Change, 152*, 215–226.

Davlasheridze, M., Atoba, K. O., Brody, S., Highfield, W., Merrell, W., Ebersole, B., et al. (2019). Economic impacts of storm surge and the cost-benefit analysis of a coastal spine as the surge mitigation strategy in Houston-Galveston area in the USA. *Mitigation and Adaptation Strategies for Global Change, 24*(3), 329–354.

Davlasheridze, M., Fan, Q., Highfield, W., & Liang, J. (2021). Economic impacts of storm surge events: examining state and national ripple effects. *Climatic Change, 166*(1), 1–20.

Davlasheridze, M., Fisher-Vanden, K., & Klaiber, H. A. (2017). The effects of adaptation measures on hurricane induced property losses: Which FEMA investments have the highest returns? *Journal of Environmental Economics and Management, 81*, 93–114.

de Vries, D. H., & Fraser, J. C. (2012). Citizenship rights and voluntary decision making in post-disaster U.S. floodplain buyout mitigation programs. *International Journal of Mass Emergencies and Disasters, 30*(1), 1–33.

DelRossi, A. F., & Inman, R. P. (1999). Changing the price of pork: The impact of local cost sharing on Legislators' demands for distributive public goods. *Journal of Public Economics, 71*(2), 247–273.

Ding, A., White, J. F., Ullman, P. W., & Fashokun, A. O. (2008). Evaluation of HAZUS-MH flood model with local data and other program. *Natural Hazards Review, 9*, 20–28.

Ebersole, B. A., Massey, T. C., Melby, J. A., Nadal-Caraballo, N. C., Hendon, D. L., Richardson, T. W., et al. (2015). *Interim report—Ike dike concept for reducing hurricane storm surge in the Houston-Galveston region*. Jackson State University. http://www.tamug.edu/ikedike/images_and_documents/Interim_ReportThe_Ike_Dike_Concept_for_Reducing_Hurricane_Storm_Surge_in_the_Houston-Galveston_Region.pdf. (Accessed 12 October 2020).

Eijgenraam, C. J. J., Kind, J., Bak, C., Brekelmans, R., den Hartog, D., Duits, M., et al. (2014). Economically efficient standards to protect the Netherlands against flooding. *Interfaces, 44*(1), 7–21.

Emanuel, K. A. (2013). Downscaling CMIP5 climate models shows increased tropical cyclone activity over the 21st century. *Proceedings of the National Academy of Sciences, 110*(30), 12219–12224. https://doi.org/10.1073/pnas.1301293110.

FEMA. (2006). *Hazus-MH flood technical manual*. Retrieved from https://www.fema.gov/media-library-data/20130726-1820-25045-8292/hzmh2_1_fl_tm.pdf.

Frumhoff, P. C., McCarthy, J. J., Melillo, J. M., Moser, S. C., & Wuebbles, D. J. (2007). Confronting climate change in the US Northeast. In *A report of the northeast climate impacts assessment* (pp. 47–61). Cambridge, Massachusetts: Union of Concerned Scientists.

Galloway, G. E., Baecher, G. B., Plasencia, D., Coulton, K. G., Louthain, J., Bagha, M., et al. (2006). *Assessing the adequacy of the national flood insurance program's 1 percent flood standard, Report*. Washington, DC: American Institutes for Research.

Ganderton, P. T. (2005). 'Benefit–cost analysis' of disaster mitigation: Application as a policy and decision-making tool. *Mitigation and Adaptation Strategies for Global Change, 10*(3), 445–465.

Griggs, T., Lehren, A. W., Popovich, N., Singhvi, A., & Tabuchi, H. (2017). More than 40 sites released hazardous pollutants because of hurricane Harvey. The New York Times. Retrieved from: https://www.nytimes.com/interactive/2017/09/08/us/houston-hurricane-harvey-harzardous-chemicals.html. Retrieved September 2017.

Grinsted, A., Moore, J. C., & Jevrejeva, S. (2013). Projected Atlantic hurricane surge threat from rising temperatures. *Proceedings of the National Academy of Sciences of the United States of America*, *110*(14), 5369–5373.

Haimes, Y. Y., & Jiang, P. (2001). Leontief-based model of risk in complex interconnected infrastructures. *Journal of Infrastructure Systems*, *7*(1), 1–12.

Hallegatte, S. (2008). An adaptive regional input-output model and its application to the assessment of the economic cost of Katrina. *Risk Analysis*, *28*(3), 779–799.

Hallegatte, S., Ranger, N., Mestre, O., Dumas, P., Corfee-Morlot, J., Herweijer, C., et al. (2011). Assessing climate change impacts, sea level rise and storm surge risk in port cities: A case study on Copenhagen. *Climatic Change*, *104*(1), 113–137.

Hsiang, S., Kopp, R., Jina, A., Rising, J., Delgado, M., Mohan, S., et al. (2017). Estimating economic damage from climate change in the United States. *Science*, *356*(6345), 1362–1369.

Hydrocarbon Publishing Company. (2016). *Power outage mitigation*. *Multi-client strategic reports*. Hydrocarbon Publishing Company.

Jonkman, S. N., Lendering, K. T., van Berchum, E. C., Nillesen, A., Mooyaart, L., de Vries, P., et al. (2015). *Coastal spine system—Interim design report* (Available from:) http://www.tamug.edu/ikedike/images_and_documents/20150620_Coastal_spine_system-interim_design_report_v06.pdf. (Accessed 15 December 2015).

Kind, J. J. (2014). Economically efficient flood protection standards for the Netherlands. *Journal of Flood Risk Management*, *7*(2), 103–117.

Kirgiz, K., Burtis, M., & Lunin, D. A. (2009). Petroleum-refining industry business interruption losses due to Hurricane Katrina. *Journal of Business Valuation and Economic Loss Analysis*, *4*(2), 1–13.

Kousky, C. (2014). Managing shoreline retreat: A US perspective. *Climatic Change*, *124*, 9–20.

Kousky, C., & Walls, M. (2014). Floodplain conservation as a flood mitigation strategy: Examining costs and benefits. *Ecological Economics*, *104*, 119–128.

Krausmann, E., & Cruz, A. M. (2013). Impact of the 11 March 2011, Great East Japan earthquake and tsunami on the chemical industry. *Natural Hazards*, *67*(2), 811–828.

Lester, J., Gonzalez, L. A., Sage, T., & Gallaway, A. (2002). *The state of the bay: A characterization of the Galveston Bay ecosystem*. Galveston Bay Estuary Program.

Marino, E. (2018). Adaptation privilege and voluntary buyouts: Perspectives on ethnocentrism in sea level rise relocation and retreat policies in the US. *Global Environmental Change*, *49*, 10–13.

Miao, Q., Shi, Y., & Davlasheridze, M. (2020). *Fiscal decentralization and natural disaster mitigation: Evidence from the United States*. Public Budgeting & Finance.

Minnesota IMPLAN Group (MIG). (2012). *Impact analysis for planning (IMPLAN) system*. Hudson, WI.

Neumann, J. E., Emanuel, K., Ravela, S., Ludwig, L., Kirshen, P., Bosma, K., et al. (2015). Joint effects of storm surge and sea-level rise on US Coasts: New economic estimates of impacts, adaptation, and benefits of mitigation policy. *Climatic Change*, *129*, 337–349. https://doi.org/10.1007/s10584-014-1304-7

NOAA National Centers for Environmental Information (NCEI). (2020). *U.S. billion-dollar weather and climate disasters*. https://www.ncdc.noaa.gov/billions/. 10.25921/stkw-7w73.

Okuyama, Y. (2004). Modeling spatial economic impacts of an earthquake: Input-output approaches. *Disaster Prevention and Management: An International Journal*, *13*(4), 297–306.

Palatnik, R. R., & Roson, R. (2012). Climate change and agriculture in computable general equilibrium models: Alternative modeling strategies and data needs. *Climatic Change*, *112*(3–4), 1085–1100.

Parrado, R., Bosello, F., Delpiazzo, E., Hinkel, J., Lincke, D., & Brown, S. (2020). Fiscal effects and the potential implications on economic growth of sea-level rise impacts and coastal zone protection. *Climatic Change*, 1–20.

Rausch, S., & Rutherford, T. F. (2009). *Tools for building national economic models using state-level implan social accounts* (Unpublished manuscript. Available online at:) http://www.cepe.ethz.ch/people/profs/srausch/IMPLAN2006inGAMS.pdf.

Rose, A. (2004). Defining and measuring economic resilience to disasters. *Disaster Prevention and Management: An International Journal*, *13*(4), 307–314.

Rose, A. Z., & Blomberg, S. B. (2010). Total economic consequences of terrorist attacks: Insights from 9/11. *Peace Economics, Peace Science and Public Policy*, *16*(1), 1–14.

Rose, A., & Guha, G. S. (2004). Computable general equilibrium modeling of electric utility lifeline losses from earthquakes. In *Modeling spatial and economic impacts of disasters* (pp. 119–141). Berlin Heidelberg: Springer. https://doi.org/10.1007/978-3-540-24787-6_7.

Rose, A., & Liao, S. Y. (2005). Modeling regional economic resilience to disasters: A computable general equilibrium analysis of water service disruptions*. *Journal of Regional Science, 45*(1), 75–112. https://doi.org/10.1111/j.0022-4146.2005.00365.x.

Rose, A. Z., Oladosu, G., Lee, B., & Asay, G. B. (2009). The economic impacts of the September 11 terrorist attacks: A computable general equilibrium analysis. *Peace Economics, Peace Science and Public Policy, 15*(2), 1–31.

Rose, A., Oladosu, G., & Liao, S. Y. (2007). Business interruption impacts of a terrorist attack on the electric power system of Los Angeles: Customer resilience to a total blackout. *Risk Analysis, 27*(3), 513–531. https://doi.org/10.1111/j.1539-6924.2007.00912.x.

Rose, A., Porter, K., Dash, N., Bouabid, J., Huyck, C., Whitehead, J., et al. (2007). Benefit-cost analysis of FEMA hazard mitigation grants. *Natural Hazards Review, 8*(4), 97–111.

Rose, A., & Wei, D. (2013). Estimating the economic consequences of a port shutdown: The special role of resilience. *Economic Systems Research, 25*(2), 212–232.

Rose, A., Wei, D., & Wein, A. (2011). Economic impacts of the ShakeOut scenario. *Earthquake Spectra, 27*(2), 539–557.

Scawthorn, C., Blais, N., Seligson, H., Tate, E., Mifflin, E., Thomas, W., et al. (2006). HAZUS-MH flood loss estimation methodology. I: Overview and flood hazard characterization. *Natural Hazards Review, 7*, 60–71.

Scawthorn, C., Flores, P., Blais, N., Seligson, H., Tate, E., Chang, S., et al. (2006). HAZUS-MH flood loss estimation methodology. II. Damage and loss assessment. *Natural Hazards Review, 7*(2), 72–81. https://doi.org/10.1061/(ASCE)1527-6988(2006)7:2(72).

Shoven, J. B., & Whalley, J. (1992). *Applying general equilibrium*. Cambridge University Press.

Sue Wing, I. (2007a). Computable general equilibrium models for the analysis of energy and climate policies. In *Prepared for the International Handbook of Energy Economics*.

Sue Wing, I. (2007b). *The regional impacts of US climate change policy: A general equilibrium analysis*. Manuscript, Department of Geography & Environment, Boston University and the Joint Program on the Science & Policy of Global Change, MIT.

Sue Wing, I., Rose, A. Z., & Wein, A. M. (2016). Economic consequence analysis of the ARkStorm scenario. *Natural Hazards Review, 17*, A4015002. https://doi.org/10.1061/(ASCE)NH.1527-6996.0000173.

Tate, E., Muñoz, C., & Suchan, J. (2014). Uncertainty and sensitivity analysis of the Hazus-MH flood model. *Natural Hazards Review, 16*(3), 04014030.

Tinbergen, J. (1954). De economische balans van het Deltaplan, bijlage bij het rapport van de Deltacommissie.

U.S. Army Corps of Engineers [USACE]. (2020). *Galveston district. Coastal texas protection and restoration feasibility study* (October 2020. Available online:) https://www.swg.usace.army.mil/Portals/26/docs/Planning/Public%20Notices-Civil%20Works/2020%20Coastal%20DIFR%20and%20dEIS/Coastal%20TX%20Executive%20Summary_20201019.pdf?ver=9fE_s4Hla4njYurhqiCYHQ%3d%3d.

U.S. Climate Change Science Program. (2008). *Weather and climate extremes in a changing climate*. Washington, DC: U.S. Climate Change Science Program.

U.S. Department of Energy. (2009). *Infrastructure security and energy restoration; Office of electricity delivery and energy reliability. Comparing the impacts of the 2005 and 2008 hurricanes on U.S. energy infrastructure.* (Report. February 2008).

U.S. Department of Energy. (2013). *Economic benefits of increasing electric grid resilience to weather outages* (Report. Available online:) https://www.energy.gov/sites/prod/files/2013/08/f2/Grid%20Resiliency%20Report_FINAL.pdf.

U.S. Department of Transportation. (2016). *Revised departmental guidance on valuation of a statistical life in economic analysis*. Available: https://www.transportation.gov/office-policy/transportation-policy/revised-departmental-guidance-on-valuation-of-a-statistical-life-in-economic-analysis.

van Dantzig, D. (1956). Economic decision problems for flood prevention. *Econometrica, 24*, 276–287.

Villarini, G., & Vecchi, G. A. (2013). Projected increases in North Atlantic tropical cyclone intensity from CMIP5 models. *Journal of Climate*, *26*(10). https://doi.org/10.1175/JCLI-D-12-00441.1.

Walsh, K. J. E., McBride, J. L., Klotzbach, P. J., Balachandran, S., Camargo, S. J., Holland, G., et al. (2016). Tropical cyclones and climate change. *WIREs Climate Change*, 7, 65–89. https://doi.org/10.1002/wcc.371.

Wing, O. E. J., Pinter, N., Bates, P. D., & Kousky, C. (2020). New insights into US flood vulnerability revealed from flood insurance big data. *Nature Communications*, *11*, 1444.

Zavar, E., & Hagelman, R. R. (2016). Land use change on U.S. floodplain buyout sites. 1990–2000. *Disaster Prevention and Management*, *25*(3), 360–374.

Zerbe, R. O., & Dively, D. D. (1994). *Benefit-cost analysis in theory and practice*. New York: Harper Collins College Publishers.

Zhou, Q., Hanasaki, N., Fujimori, S., Masaki, Y., & Hijioka, Y. (2018). Economic consequences of global climate change and mitigation on future hydropower generation. *Climatic Change*, *147*(1–2), 77–90.

CHAPTER 9

The role of insurance in facilitating economic recovery from floods

Samuel Brody[a,b], Wesley E. Highfield[a,b], and Russell Blessing[a]
[a]Institute for a Disaster Resilient Texas, Texas A&M University, College Station, TX, United States
[b]Department of Marine and Coastal Environmental Science, Texas A&M University, Galveston Campus, Galveston, TX, United States

Introduction

A central component of every country's flood management strategy is how to protect or recover from damaging events. Facilitating the economic recovery and well-being of households that experience flood losses has long been a priority, with different approaches being implemented over time. The foundation of flood risk management at the household level in the United States is the provision of federal flood insurance. As outlined in Chapter 2, this recovery-based approach stands in contrast to that of many other countries, such as the Netherlands, which takes a more protective stance for reducing flood impacts. This chapter describes the US National Flood Insurance Program (NFIP) as the cornerstone of flood risk management and household economic recovery. First, we trace the origins and founding logic of the program. Then we discuss its strengths, weaknesses, and the ongoing debate on how it can be improved. Finally, we briefly compare the NFIP to the noninsurance-based approach taken by the Netherlands, which has had its own successes and failures.

The US NFIP as the basis for household economic recovery

The United States formally adopted a recovery-based approach to flood mitigation for both fresh and saltwater inundation events with the adoption of the National Flood Insurance Program (NFIP) in 1968 (under the *National Flood Insurance Act*). Up to that point, the federal government was responsible for financially restoring communities and residents postflood disaster, a burden they wanted to shed. The logic at the time was that a government-based program could successfully offer insurance to homeowners and that the private industry could not because it could (a) pool risks, (b) spread losses over communities and events, (c) potentially borrow money from the federal government if there was a deficit for a given year, and, most importantly, (d) subsidize the true costs of the policies to ensure affordability (Michel-Kerjan & Kousky, 2010). This program was thought to be the most effective way to financially protect residents living in flood-prone areas without financially straining both households and federal funding entities. Policymakers embraced the concept of insurance policies against flood losses as an alternative to federal aid (Pasterick, 1998).

There are two major assumptions underlying the NFIP: (1) residents will be placed at risk of flooding and must have a fiscal mechanism to react and recover from an inundation event and (2) the primary marker of a regulatory unit for mitigating this risk is the 100-year floodplain. Floodplain maps are produced to identify the Special Flood Hazard Area (SFHA), or areas having a 1% probability of inundation each year (100-year floodplains). These areas serve as the spatial reference point for setting federal flood insurance requirements, flood insurance rates, and enforcing local floodplain management regulations. The *Flood Disaster Protection Act* of 1973 took a step further and required the mandatory purchase of flood insurance for property located in SFHAs for homeowners who have a federally backed mortgage.[a]

FEMA writes or underwrites flood insurance for participating NFIP communities in the United States. Individuals can purchase flood insurance directly through authorized FEMA representatives or through a traditional private insurer in what is known as the "Write Your Own" (WYO) program. Several characteristics distinguish the NFIP from the private insurance industry. First, flood insurance purchasers are held to a 30-day waiting period before the flood insurance coverage goes into effect. This policy effectively eliminates the possibility that a party can purchase flood insurance when there is an imminent risk of flooding. Second, the residential coverage amount is capped at $250,000 for buildings and $100,000 for contents. In many cases, this coverage ceiling is less than the total value of many structures, especially along the coast where storm surge-based flooding can result in catastrophic impacts. Finally, as mentioned above, there is a mandated requirement to purchase flood insurance for structures located within the 100-year floodplain that are being secured via bank loan. As a result of habitual noncompliance, this requirement has been more forcefully implemented through lenders and loan servicers, requiring them to determine and document whether a structure is in the 100-year floodplain and ensure that the mortgager maintains flood insurance throughout the life of the loan.

The NFIP has had several success in managing floods, including more widespread public identification of flood hazards and increased development standards in floodplain areas (Holway & Burby, 1990; U.S. Interagency Floodplain Management Review Committee, 1994). However, the program continues to suffer from a major flaw in that it focuses on economic recovery postflood event, rather than proactive mitigation to

---

[a] Includes secured mortgage loans from financial institutions, such as commercial lenders, savings and loan associations, savings banks, and credit unions, that are regulated, supervised, or insured by Federal agencies such as the Federal Deposit Insurance Corporation and the Office of Thrift Supervision. It also applies to all mortgage loans purchased by Fannie Mae or Freddie Mac in the secondary mortgage market.

reduce the risk in the first place. In fact, the vast amount of funding for coastal flood-related issues is provided by the federal government only after a disaster occurs, through emergency supplemental appropriations (NRC, 2014). A recovery-based approach predicated upon insurance payments is at its essence an acceptance of failure when it comes to avoiding adverse impacts from floods (Brody, Highfield, Merrell, & Lee, 2019). As the cornerstone of flood mitigation in the United States, the NFIP creates an expectation from the federal government down to the household level that residents will flood, incur damage, and need constant financial assistance to recover from their losses. This, in many ways, is a self-defeating strategy that continually strains the federal fiscal coffers (something the program was supposed to avoid) and has led to several unintended and undesirable consequences.

First, subsidized insurance has made it more affordable to purchase a home in a flood zone and has increased household exposure overall to flood risk over the long term. Subsidized insurance premiums create an often termed "perverse incentive" (Beatley, 2009) to locate in exposed areas because even if a home is flooded, the resident will receive financial recovery assistance. The incentive of subsidized insurance often overpowers local planning policies seeking to avoid development in the floodplains and other risk zones. The NFIP has unintentionally encouraged sprawling development patterns and associated adverse environmental impacts in sensitive coastal areas. Sprawling, low-density subdivisions in the urban fringe and in rural areas often have fewer resources that can be dedicated to drainage and flood protection infrastructure as compared to the more coordinated, municipal-level services provided by high-density urban areas. These residential subdivisions are more likely to have ad hoc or substandard local drainage and stormwater management systems that further exacerbate the flooding problem (Brody, Gunn, Highfield, & Peacock, 2011) and place more people and structures at risk when a major event takes place. Also, low-density development spiraling outward from urban areas requires larger amounts of impervious surfaces that increase surface runoff, change the spatial extent of floodplain boundaries faster than they can be officially mapped, putting downstream communities at greater risk (Brody, 2013). Thus, in an effort to protect residents against expected losses, the NFIP has actually made the flood problem worse.

Second, the NFIP forces homeowners and communities into a constant repetitive losses and disaster-recovery cycle. Once a structure is flooded, insurance payouts require the owner to repair or rebuild in the same way (unless there is a local regulation that mandates a structural change). This process can occur up to 3 times before a property buyout is considered. Because flooding tends to be chronic and spatially repetitive, homeowners in vulnerable areas are often trapped in a repetitive damage-rebuild process. These so-called "repetitive loss" (RL) properties are defined by FEMA as any insurable building for which two or more claims of more than $1000 were paid by the NFIP within any rolling 10-year period since 1978. Currently, there are over 122,000 RL properties nationwide, which over time create cumulative financial burdens (*http://www.fema.gov/txt/rebuild/repetitive*loss*faqs.txt, Accessed on September 10, 2021*). Texas leads the nation with over 7800 severe repetitive loss properties, meaning at least four separate flood insurance claims were filed for more than $5000 each or more claims filed where the payments exceeded the value of the property.

Third, increasing flood impacts to insured homes located in areas not considered at risk (e.g., outside of FEMA 100- and 500-year floodplains) has exacerbated the financial strain on the NFIP program. For example, a Houston, TX, resident in the FEMA X-zone (an area of minimal flood hazard or outside of the 500-year floodplain) may pay approximately $400 per year but be eligible for $350,000 to cover flood losses. During Hurricane Harvey in 2017, the majority of claims in Houston were outside of the SFHA. Payouts exceeding premiums at a large scale and on a continual basis put the NFIP on a shaky financial footing. This fiscal imbalance was initially supposed to be corrected by raising premiums and/or dropping coverage. However, a combination of payouts from large hazard events (Hurricane Ida in September 2021 being the latest at the time of writing) and the lack of actuarially sound rates has forced FEMA to regularly borrow money from the federal treasury to cover its deficit. The NFIP continues to accrue debt each year. For example, the Congressional Research Service (CRS) (2019) reported that a total of $20.5 billion is owed from the NFIP to the US Treasury as of December 2019. The US Congress attempted to address this fiscal dilemma by passing the *Flood Insurance Reform Act* in 2012, which focused on raising premiums based on actual risk in an effort to help the program become more actuarially sound. Since that time, however, many of the requirements have been relaxed or postponed, particularly for cases where increased premiums would place an undue financial burden on a property owner.

Lastly, the US federal government never entirely moved away from directly footing the bill for flood-disaster economic recovery at the household level. Noninsured residents are still able to obtain Individual Assistance (over $1.6 billion was distributed in Texas for Hurricane Harvey alone) if their dwelling was damaged by a flood. While the reimbursement amount is much lower than for federal insurance (around $2000 per claim), many homeowners are willing to forgo yearly premium charges knowing they will get some level of compensation in the low-probability event of a flood. There is also a myriad of other postflood federal support programs that unintentionally undermine the standing of the NFIP, such as Small Business Administration (SBA) loans, Housing and Urban Development (HUD) property buyouts, and US Department of Agriculture (USDA) nutritional supplements. Even nonhousehold assistance, such as FEMA Public Assistance grants, can erode the participation and effectiveness of the NFIP (Davlasheridze & Miao, 2019). In reality, most households rely on multiple sources of postdisaster financial assistance. Browne and Hoyt (2000), for example, found that insurance covered less than 10% of flood losses—an estimate consistent with Kunreuther's (1996) report that most risk area residents expect to finance recovery through their own resources (Lindell, Brody, & Highfield, 2017).

Record of historical loss through the NFIP

Although participation in the NFIP initially lagged, it has grown considerably since its inception and is now the primary financial recovery vehicle for flood loss at the household level in the United States. There are now more than 22,100 participating communities and more than 5.1 million NFIP policies in force, providing $1.25 trillion of content and building coverage. Since 1970, the program has received approximately 2.4 million insurance claims and paid out almost $70 billion. The number and amount of insurance claims has been steadily increasing over time (Bradt & Kousky, 2020). Based on FEMA data, insured losses increased, on average, $109 million per year (adjusted to 2018 dollars) from 1978 to 2018. In 2019, the average claim was $52,000 and the average annual insurance policy premium was $700 (https://www.fema.gov/data-visualization/historical-flood-risk-and-costs). The program overall collects approximately $4.6 billion in revenue each year from policyholders' premiums, fees, and surcharges (Congressional Research Service (CRS), 2021).

From a research and data analytics perspective, the NFIP provides a useful metric for understanding the location, extent, and drivers of flood loss, especially in areas that are heavily saturated with policies. Fig. 1 shows claim payouts from 1974 to 2014 by

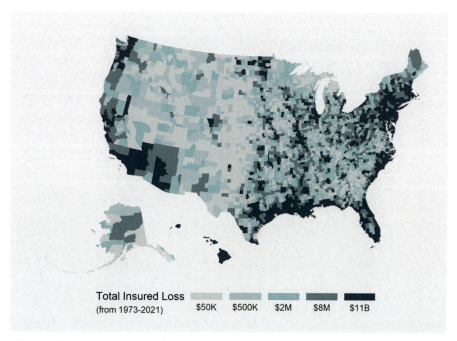

Fig. 1 Total NFIP losses, 1974–2021 by county. *Data source: FEMA. (2021). OpenFEMA Dataset: FIMA NFIP redacted claims – V1. Retrieved from https://www.fema.gov/api/open/v1/FimaNfipClaims on January 24. This product uses the FEMA OpenFEMA API, but is not endorsed by FEMA. The Federal Government or FEMA cannot vouch for the data or analyses derived from these data after the data have been retrieved from the Agency's website(s).*

county across the United States. The majority of losses occurred along the Gulf of Mexico and Eastern Seaboard.

Texas is second only to Florida when it comes to the number of NFIP policies and claims, with most losses concentrated in the Houston-Galveston region. Multiple catastrophic flood events have occurred in this region in recent decades, including Tropical Storm Allison (2001), Hurricane Ike (2008), and Hurricane Harvey (2017). The amount of losses and cost of recovery have increased with each successive storm. As shown in Fig. 2, the cumulative financial losses have mounted over time. Between 1978 and 2014, there were 316,336 claims made totaling $16,878,563,071 in insured flood loss (Galloway et al., 2018). After the most recent major storm, Hurricane Harvey, FEMA paid out over $8.8 billion in losses over approximately 91,000 claims spread out across 41 Texas counties. It is important to note that NFIP represents only a fraction of total losses communities incur during major flood events.

Analysis of NFIP data also reveals important trends that can provide guidance to decision makers. First, there are a large number of claims located outside of FEMA-defined floodplains and this trend has been increasing over time in Texas. For example, of the recorded 154,000 flooded homes in Houston during Hurricane Harvey, over 46% were located outside of the 500-year floodplain. Historically, over half of all claims have been

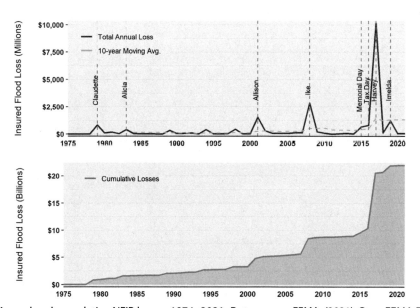

Fig. 2 Annual and cumulative NFIP losses, 1974–2021. *Data source: FEMA. (2021).* OpenFEMA Dataset: FIMA NFIP redacted claims – V1. *Retrieved from https://www.fema.gov/api/open/v1/FimaNfipClaims on January 24. This product uses the FEMA OpenFEMA API, but is not endorsed by FEMA. The Federal Government or FEMA cannot vouch for the data or analyses derived from these data after the data have been retrieved from the Agency's website(s).*

outside of the 100-year floodplain. This trend is in large part driven by features of the human built environment, such as the spread of impervious surfaces, fragmentation of natural hydrological processes, aging or inadequate drainage infrastructure, and blockage of overland flow (e.g., roads, sound walls, etc.) (see Chapter xx for more details). Second, research using NFIP data also shows which mitigation strategies are most effective in reducing flood loss, including open space (Atoba et al., 2021; Brody, Blessing, Sebastian, & Bedient, 2014; Brody & Highfield, 2013) and wetland protection (Brody, Highfield, & Blessing, 2015; Highfield, Brody, & Shepard, 2018), elevation of existing structures above base flood levels, locating development away from vulnerable areas (Blessing, Brody, & Highfield, W.E., 2019; Highfield, Brody, & Blessing, 2014), etc. These findings provide valuable guidance for local decision makers interested in improving their community's flood resilience over the long term.

Beyond the NFIP: Emergence of private insurance markets

Over the last couple of years, a small market for private residential flood insurance has emerged and is increasingly of interest to federal policymakers. The market thus far is limited to approximately 3.5%–4.5% of all primary flood policies and is concentrated mostly in Puerto Rico and Florida (Kousky, Kunreuther, Lingle, & Shabman, 2018). Private residential flood insurance may have advantages over the NFIP, such as an increased amount of overall coverage, more flexible flood policies, integrated coverage with homeowners insurance, or lower cost coverage for some consumers. Increased private coverage could reduce the overall financial risk to the NFIP and borrowing after major disasters (Horn & Webel, 2019). However, it is often noted that while NFIP will write a policy to anyone, private companies can only succeed if they can price lower than the NFIP or provide coverage when there is consumer demand. Private firms will only be able to provide policies in lower risk, higher income areas where the financials make sense, leaving the most vulnerable properties in the hands of the federal government (Kousky et al., 2018).

Flood risk reduction in the Netherlands: A counter approach

Given the issues mentioned above with the NFIP, decision makers in the United States are slowly realizing that an insurance-driven, recovery-based approach may not be the most effective way to reduce flood risk at the household level. Increasing investments in predisaster mitigation, large-scale flood barrier systems, property buyouts and open space protection, and growth of the FEMA Community Rating System (providing incentives for localities to pursue nonstructural mitigation activities) have all worked to shift emphasis toward a more proactive system of managing the increasing costs of flooding over the long term.

The slight shift in US strategy is partially inspired by the approach taken by the Netherlands, which is firmly rooted in a protective stance focusing on mitigating flood risk before an event takes place, or eliminating it altogether (Rijkswaterstaat, 2014). This more proactive and integrated approach to risk reduction assumes that residents should never bear the burden of flood loss, regardless of where they are located within coastal landscapes. Damage to property in this instance is considered a failure of the system rather than an expected consequence. The Dutch system relies on both system-based structural interventions and land use planning techniques that seek to remove or avoid structures from areas most vulnerable to flooding (Brody et al., 2019).

Since a catastrophic storm surge event in 1953, the Netherlands has become a world leader in constructing multifunctional, comprehensive, environmentally conscious flood barriers (see Chapters 2 and 22 for more details). Most notably, in 1998, the Delta Works organization completed the Maeslant Barrier (Maeslantkering) to protect the city of Rotterdam from storm surges off the North Sea. The structure consists of two immense movable gates that automatically close during a storm threat. At all other times, the gates are open to permit shipping activity, recreation, and natural water flow. Similarly, in 1986, the Eastern Scheldt storm surge barrier (Oosterscheldekering) was completed as a linear series of large sluice gates. This structure was placed across the mouth of the Eastern Scheldt Estuary to block damaging storm surges, but still permit tidal exchange, maintain fishery habitats, and keep the bay interior free from a disconnected scattering of mitigation structures. More recently, the Netherlands implemented its "Room for the River" program, which seeks to widen inland rivers to allow for natural water storage and reduce the probability of downstream flooding. These and other flood risk reduction initiatives across the Netherlands seek to protect residents so there is no need to rely on insurance or other financial recovery systems. They also represent a more systematic and integrated approach to flood risk reduction that seeks to generate multiple values as compared to the ad hoc, single value approach taken by the United States.

Interestingly, the Dutch have been considering shifting more toward a US-style insurance system to improve upon their existing programs. While private flood insurance is rarely used in the Netherlands, there is a postdisaster relief mechanism, established in 1998, that provides partial compensation by the government through the *Calamities and Compensation Act* (WTS). Aerts and Botzen (2011) noted that the WTS is problematic because households face uncertainty about the degree to which flood damage might be compensated and that the WTS is an ad hoc arrangement for which no reserves have been established. Because of these and other shortcomings, there has been debate whether the current dominant role of the Dutch government in providing flood damage relief is justified from an economic efficiency perspective (Botzen & Van Den Bergh, 2008). The idea of transferring the responsibility of flood disaster compensation to the private insurance market is attractive, but potentially complicated by accumulating risks, low capacity of insurers to pay for catastrophic damage, and problems with asymmetric information, among other issues (Aerts & Botzen, 2011; Jongejan & Barrieu, 2008). The issue of establishing an insurance market in the Netherlands is further investigated in Chapter 11 by Robinson and Botzen.

Conclusions

The beginning of this book presents two contrasting approaches to flood risk reduction: recovery-based in the United States and protective-oriented in the Netherlands. In this chapter, we play out this dichotomy through the use of insurance as a mechanism to reduce the financial impact of increasingly destructive flood events at the household level. Focusing on the NFIP in the United States, we describe the logic behind this system, as well as its various strengths and weaknesses. We then compare the United States approach to that of the Dutch, which has its own struggles with establishing sustainable financial mechanisms to protect its citizens from flood impacts. We find that each approach is deeply rooted in cultural and political norms, where in the United States responsibility to often placed on the individual, while in the Netherlands there is a strong history of federal government protection. Both countries have been pulled in the opposite direction when considering how to improve their programs and reduce the economic impacts of future flood events. As the potential impacts of storm events continue to increase, so will the need to provide fiscal protection to residents in both countries.

References

Aerts, J. C., & Botzen, W. W. (2011). Climate change impacts on pricing long-term flood insurance: A comprehensive study for the Netherlands. *Global Environmental Change, 21*(3), 1045–1060.

Atoba, K., Newman, G., Brody, S., Highfield, W., Kim, Y., & Juan, A. (2021). Buy them out before they are built: Evaluating the proactive acquisition of vacant land in flood-prone areas. *Environmental Conservation, 48*(2), 118–126.

Beatley, T. (2009). *Planning for coastal resilience: Best practices for calamitous times.* Washington, DC: Island Press.

Blessing, R., Brody, S. D., & Highfield, W. E. (2019). Valuing Floodplain protection and avoidance in a coastal watershed. *Disasters, 43*(4), 906–925.

Botzen, W. J., & Van Den Bergh, J. C. (2008). Insurance against climate change and flooding in the Netherlands: Present, future, and comparison with other countries. *Risk Analysis: An International Journal, 28*(2), 413–426.

Bradt, J., & Kousky, C. (2020). *Flood insurance in the US: Lessons from FEMA's recent data release (part I & II).* https://riskcenter.wharton.upenn.edu/lab-notes/lessonsfromfemadatapart1/.

Brody, S. D. (2013). The characteristics, causes, and consequences of sprawling development patterns in the United States. *Nature Education Knowledge, 4*(5), 2.

Brody, S. D., Blessing, R., Sebastian, A., & Bedient, P. (2014). Examining the impact of land use/land cover characteristics on flood losses. *Journal of Environmental Planning and Management, 57*(9), 1252–1265.

Brody, S. D., Gunn, J., Highfield, W. E., & Peacock, W. G. (2011). Examining the influence of development patterns on flood damages along the Gulf of Mexico. *Journal of Planning Education and Research, 31*(4), 438–448.

Brody, S. D., & Highfield, W. (2013). Open space protection and flood losses: A national study. *Land Use Policy, 32*, 89–95.

Brody, S. D., Highfield, W., & Blessing, R. (2015). An empirical analysis of the effects of land use/land cover on flood losses along the Gulf of Mexico coast from 1999 to 2009. *Journal of the American Water Resources Association*, *51*(6), 1556–1567.

Brody, S. D., Highfield, W. E., Merrell, W., & Lee, Y. (2019). Recovery versus protection-based approaches to flood risk reduction. In *The Routledge Handbook of Urban Disaster Resilience: Integrating Mitigation, Preparedness, and Recovery Planning*. Routledge.

Browne, M. J., & Hoyt, R. E. (2000). The demand for flood insurance: Empirical evidence. *Journal of Risk and Uncertainty*, *20*(3), 291–306.

Congressional Research Service (CRS). (2019). *Introduction to the National Flood Insurance Program (NFIP)*. https://fas.org/sgp/crs/homesec/R44593.pdf.

Congressional Research Service (CRS). (2021). *Introduction to the national flood insurance program*. CRS report R44593.

Davlasheridze, M., & Miao, Q. (2019). Does governmental assistance affect private decisions to insure? An empirical analysis of flood insurance purchases. *Land Economics*, *95*(1), 124–145.

Galloway, G. E., Reilly, A., Ryoo, S., Riley, A., Haslam, M., Brody, S., et al. (2018). *The growing threat of urban flooding: A national challenge*. College Park and Galveston: University of Maryland and Texas A&M University.

Highfield, W., Brody, S. D., & Blessing, R. (2014). Measuring the impact of mitigation activities on flood loss reduction at the parcel level: The case of the Clear Creek watershed along the upper Texas coast. *Natural Hazards*, *72*(2), 687–704. https://doi.org/10.1007/s11069-014-1209-1.

Highfield, W. E., Brody, S. D., & Shepard, C. (2018). The effects of estuarine wetlands on flood losses associated with storm surge. *Ocean and Coastal Management*, *157*, 50–55.

Holway, J., & Burby, R. (1990). The effects of floodplain development controls on residential land values. *Land Economics*, *66*(3), 259–271.

Horn, D. P., & Webel, B. (2019). *Private flood insurance and the national flood insurance program*. Congressional Research Service.

Jongejan, R., & Barrieu, P. (2008). Insuring large-scale floods in the Netherlands. *The Geneva Papers on Risk and Insurance-Issues and Practice*, *33*(2), 250–268.

Kousky, C., Kunreuther, H., Lingle, B., & Shabman, L. (2018). *The emerging private residential flood insurance market in the United States*. Wharton Risk Management and Decision Processes Center.

Kunreuther, H. (1996). Mitigating disaster losses through insurance. *Journal of Risk and Uncertainty*, *12*(2–3), 171–187.

Lindell, M. K., Brody, S. D., & Highfield, W. E. (2017). Financing housing recovery through hazard insurance: The case of the National Flood Insurance Program. In A. Sapat, & A.-M. Esnard (Eds.), *Coming home after disaster: Multiple dimensions of housing recovery* (pp. 50–65). Routledge.

Michel-Kerjan, E. O., & Kousky, C. (2010). Come rain or shine: Evidence on flood insurance purchases in Florida. *Journal of Risk and Insurance*, *77*(2), 369–397.

NRC. (2014). *Reducing coastal risk on the east and Gulf Coasts*. The National Academy Press.

Pasterick, E. T., Kunreuther, S. H., & Roth, R. J. (1998). The national flood insurance program. In *rc* (pp. 125–154). Washington DC: Joseph Henry Press.

Rijkswaterstaat. (2014). *The national flood risk analysis for the Netherlands*. Rijkswaterstaat VNK Project Office.

U.S. Interagency Floodplain Management Review Committee. Federal Interagency Floodplain Management Task Force. (1994). *Sharing the challenge: Floodplain management into the 21st century: Report of the interagency floodplain management review committee to the administration floodplain management task force*. Washington, DC: U.S. Government Printing Office.

CHAPTER 10

Behavioral insights into the causes of underinsurance against flood risks: Experimental evidence from the Netherlands

Peter J. Robinson[a] and W.J. Wouter Botzen[a,b,c]

[a]Department of Environmental Economics, Institute for Environmental Studies (IVM), VU University Amsterdam, Amsterdam, The Netherlands
[b]Utrecht University School of Economics (U.S.E.), Utrecht University, Utrecht, The Netherlands
[c]Risk Management and Decision Processes Center, The Wharton School, University of Pennsylvania, Philadelphia, PA, United States

Introduction

Flooding is the costliest natural disaster risk worldwide (Miller, Muir-Wood, & Boissonnade, 2008). This is especially the case for the United States, where over the course of the 20th-century floods were the number-one natural disaster in terms of property damage and loss of life (Perry, 2000). Several catastrophic flood events in coastal regions of Texas have occurred over the years. Examples of severe floods are Hurricane Harvey (2017) and Hurricane Ike (2008), which led to overall economic losses amounting to around $100 billion and $40 billion, respectively, due in large part to widespread coastal and inland flood inundation (Benfield, 2018; Ogashawara, Curtarelli, & Ferreira, 2013).

However, in the Netherlands, over half of the country is exposed to flood risk (Eijgenraam, Brekelmans, den Hertog, & Roos, 2017), the probability of flooding is low due to the country's large investments in dike infrastructure. Nevertheless, these defenses are incomplete protection, i.e., residual flood risk remains. For instance, in 1993 and 1995 the River Meuse overflowed its banks, leading to respective flood damage costs of 0.1 billion euro and 75 million euro, as well as the evacuation of 250,000 individuals in the latter case (van Stokkom, Smits, & Leuven, 2005; Wind, Nierop, Blois, & Kok, 1999). In 2003, dikes were breached in the village of Wilnis, causing around 16 million euros of flood damage (Aerts & Botzen, 2011). Furthermore, under climate change, flood probabilities will increase significantly over time, due to sea-level rise, intensified snowmelt, and increased winter precipitation, unless additional adaptation measures are undertaken (Middelkoop et al., 2001).

To raise protection standards to a point where flooding is no longer a threat is not economically viable (Gauderis, Kind, & van Duinen, 2013). Moreover, raising dikes increases potential flood damage amounts due to higher maximum flood water levels in the event of a dike failure (Botzen, Aerts, & van den Bergh, 2013). Raising dike levels also leaves individuals with an enhanced feeling of safety, which can result in more development in flood-prone areas (Vis, Klijn, de Bruijn, & van Buuren, 2003). Overall, it is arguable whether solely focusing on increasing flood protection standards is sufficient to manage increasing flood risks, or whether supplementary measures should be implemented at the homeowner level. Homeowners can take a number of actions to mitigate potential damages from flood events. This chapter will focus on flood insurance, which reduces the uncertainty of flood loss by spreading risk across a large number of insured individuals. However, there are some barriers to insuring flood risk in the Netherlands. That is, although flood risk is low-probability here, flood damage amounts can reach well into the billions of Euros (Paudel, Botzen, & Aerts, 2015), and low-probability/high-impact (LPHI) natural disaster events are especially difficult to insure on private markets.

Flooding was long considered to be uninsurable in the Netherlands (Jongejan & Barrieu, 2008). Despite this, in 2016 a flood insurance policy was introduced that can be purchased from private companies. However, to date few people have bought a separate flood insurance policy in the Netherlands (Suykens, Priest, van Doorn-Hoekveld, Thuillier, & van Rijswick, 2016), which is possibly due to a lack of awareness about flood risk and that flood insurance is available. Nevertheless, for flood insurance to be effective in the long run, there should be sufficient demand (Seifert, Botzen, Kreibich, & Aerts, 2013), i.e., the revenue generated by insurance companies should be high enough to reimburse claims following a flood. Another complexity is that some individuals demand low amounts of insurance against low-probability risks (Kunreuther & Pauly, 2004). For instance, in the United States, where flood insurance is relatively more commonplace, many homeowners in flood-prone areas tend to forgo purchasing flood insurance (Dixon, Clancy, Seabury, & Overton, 2006). In the aftermath of Hurricane Harvey, it turned out that less than 20% of homeowners in Harris County in Texas were holding an active National Flood Insurance Policy (Klotzbach, Bowen, Pielke Jr, & Bell, 2018).

There are various reasons why individuals may have low demand, related to systematic decision biases and difficulties with processing low probabilities. It has been proposed that individuals use mental shortcuts to deal with these limitations (Kahneman, 2003), which can lead to underestimation or dismissal of very low-probability risks (Meyer & Kunreuther, 2017). Based on a nationally representative sample of individuals residing in the United States, most assessed their likelihood of being killed by natural disasters such as hurricanes and floods as being below average (Viscusi & Zeckhauser, 2006). The same authors found that even in states with a high incidence of disasters, individuals greatly underestimate their risk of dying from these risks. Furthermore, in the Netherlands previous research has shown that a substantial proportion of individuals are not willing to pay positive monetary amounts for flood insurance (Botzen & van den Bergh, 2012), indicating that many individuals dismiss the probability of flooding altogether. These findings contrast rational theories of economic decision-making, such as expected utility theory

(EUT) which assumes that individuals can process probabilities well. However, there may also be rational reasons for low demand because individuals expect compensation for flood damage from the government (Coate, 1995).

At this point, it is useful to highlight some explanations of low demand for insurance by some against LPHI risk, on the basis of which we conducted several experimental studies. Some explanations of low demand can be accommodated within EUT. Whereas others are closer to bounded rationality, which implies that individuals make decisions in the presence of significant cognitive limitations with respect to both their knowledge about the decision at hand and their computational capacity (Simon, 1955, 1990).

Insufficient demand for flood insurance
Charity hazard

The government can provide compensation for uninsured flood damage via the Calamities and Compensation Act in the Netherlands and according to federal disaster assistance in the United States. However, the Netherlands has no established funds and no clear rules outlining under what circumstances flood damage will be compensated and by how much (Aerts & Botzen, 2011). Furthermore, in the United States it has been shown that federal disaster expenditures are subject to political will, and in particular depend upon whether there is an upcoming election (Garrett & Sobel, 2003). Therefore, it is currently uncertain whether households will receive compensation for flood damages in the Netherlands and regions prone to flooding in the United States, e.g., coastal Texas.

One rational reason for insufficient flood insurance demand is that individuals anticipate compensation from the government for damages suffered after a flood (charity hazard) (Raschky & Weck-Hannemann, 2007). Theory predicts a negative impact of unconditional transfers from the government on demand for insurance (Coate, 1995; Kelly & Kleffner, 2003; Lewis & Nickerson, 1989). However, it has been recognized that charity hazard is difficult to study empirically, based on actual insurance purchases, because of potential confounding variables such as objective levels of flood risk which makes it difficult to isolate the impact of government compensation on flood insurance demand (Kousky, Michel-Kerjan, & Raschky, 2018).

Economic experiments offer an ideal test bed when examining the influence of government compensation on demand for flood insurance because variables such as flood risk can be held constant. Economic experiments are especially suitable for studying flood insurance demand in the Netherlands, given market data on demand is not yet available here. We conducted an economic experiment in the Netherlands to investigate flood insurance demand under different conditions of government compensation (Robinson, Botzen, & Zhou, 2019). The experiment elicited insurance demand while also altering the loading factor (an index of the insurance premium which was set at either 30, 45, 60, or 240 euro experimental currency units) and the probability of loss (which was either 1 in 1000, 1 in 100 or 1 in 10) across 12 decisions that were displayed in a random order within the several government compensation conditions. The expected value of the flood risk (probability multiplied by possible damage) was held constant at 60 euro experimental currency units.

Overall, 200 students from a wide range of disciplines of VU University Amsterdam were recruited for the study. Subjects were randomly assigned to the various conditions of government compensation as follows: 52 faced the baseline (no government compensation) condition; a further 36 faced risky full government compensation first (where the compensation fully covered potential flood losses but was attached with a 1 in 2 probability of occurring), then certain half compensation (where in the event of an uninsured flood loss subjects were compensated for half of their losses); 39 subjects faced these schemes in the opposite order; moreover, 32 subjects were shown risky full government compensation first, then ambiguous full compensation (where the probability of being fully covered for potential losses was unknown). The remaining 41 subjects faced the latter schemes in the opposite order.

Subjects were paid a participation fee of €15 and were told that they could also earn money in the experiment itself based on the choices they made. According to various randomization devices, such as bingo cages and lottery ticket drawings, 48 subjects were paid based on their decisions with an exchange rate of 0.1% from the experimental currency used to euros, implying they could earn up to €60 on top of the participation fee. One subject was selected at random to be paid at an exchange rate of 1%, so they could earn up to €600. This method of incentivizing experimental decisions better aligns choices with those that would occur in practice, compared to purely hypothetical decisions (Robinson & Botzen, 2019b).

Systematic biases

Apart from charity hazards, systematic biases may shed light on why individuals underinsure disaster risks. Drawing on a large body of research, Meyer and Kunreuther (2017) conjectured that six biases cause low demand for insurance and underestimation of LPHI risks. Another two of our experiments focused on some of these systematic biases, namely, *simplification* and *inertia bias*.

Simplification

Simplification is the tendency to selectively consider a few of the relevant facts one might otherwise examine when making a decision involving risk. For instance, it has been proposed that individuals follow a threshold model heuristic (Slovic, Fischhoff, Lichtenstein, Corrigan, & Combs, 1977), which predicts that they dismiss low-probability flood risks if the probability is deemed to fall below their threshold level of concern. Moreover, it has been shown that biases in decision-making, which may include simplification, are associated with areas of the brain involved in emotion (de Martino, Kumaran, Seymour, & Dolan, 2006).

We conducted an online economic experiment with a large sample of 1041 Dutch homeowners to determine whether two emotions specific to flood risk, i.e., the expected regret individuals might feel about not purchasing flood insurance if a flood occurs and worry about flooding, influence individuals' threshold levels of concern for flood risk (Robinson & Botzen, 2018). We presented respondents with several increasing flood loss probabilities over time (between 1 in 10,000 and 19 in 20). The number of consecutive times individuals were willing to pay nothing to cover flood damage of 60,000 euro experimental currency units as the flood probabilities increased was used to proxy for the threshold (after those who were unwilling to pay for flood insurance across all decisions were omitted). Therefore, individuals who had a lower threshold level of concern were willing to pay for flood insurance under lower probabilities of flooding.

Although it is common for economic experiments to pay respondents based on the choices they make, in experiments involving low probabilities and high stakes, it was unclear whether payment mechanisms that randomly select a small number of individuals and choices can impact risk-taking attitudes. The experiment randomly selected an individual out of 624 individuals to be paid based on their decisions. The remaining 417 faced hypothetical decisions. To elicit incentivized flood insurance demand, the Becker, DeGroot, and Marschak (1964) procedure was used: The premium for which flood insurance was sold was selected at random between the upper and lower bounds of potential flood loss. The randomly selected subject was paid at an exchange rate of 1% from the experimental currency used to euros based on the outcome of the experiment. If she/he was willing to pay a value equal to or greater than the selected premium, then the subject had purchased insurance at the price of the premium, otherwise he/she faced flood risk uninsured. We supplemented the procedure with a visual aid (Fig. 1) and highlighted that it is in subjects' best interest to state their true willingness to pay.

Inertia bias

Inertia bias implies that individuals are reluctant to change from the status quo given the uncertainty as to the impact of a change, coupled with the time and attention required to modify one's behavior. It has also been suggested that individuals are anchored toward the status quo due to loss aversion (Samuelson & Zeckhauser, 1988), e.g., an individual may focus on the cost of insurance (which is a loss from the status quo of having no insurance coverage) relative to possible avoided future flood damages.

We experimentally studied the effectiveness of setting default options, i.e., altering the status quo, to raise flood insurance demand among 1187 homeowners within the Netherlands and the United Kingdom (UK) (Robinson, Botzen, Kunreuther, & Chaudhry, 2020). Two defaults were compared: opt-out (individuals had to remove flood insurance coverage from a prepurchased policy that included fire and burglary-related losses to cancel their coverage) and opt-in (individuals had to pay an extra insurance premium to include flood coverage in the policy). Note that we implemented

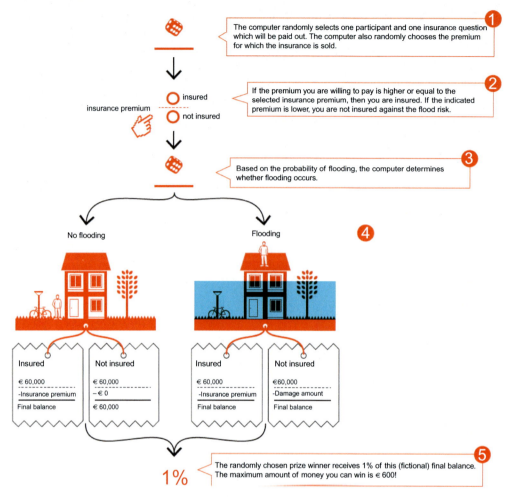

Fig. 1 Visual aid used to explain the Becker et al. (1964) method of incentivizing flood insurance demand.

a nonincentivized experiment, in contrast to the incentivized experiments outlined previously, where subjects were paid based on their choices. We believe that nonincentivized experiments involving simple choices are useful for supplying preliminary evidence on the effectiveness of certain policy interventions, especially since individuals have a low incentive to lie about their preferences.

There is some variation in actual flood risks faced by homeowners within both the Netherlands and the United Kingdom, but we chose one probability and damage amount to standardize the decision environment and to enable the comparison of default effects between the two countries. The probability of 1 in 1250 reflects the flood probability for homes situated in the river delta of the Netherlands, based on river-dike safety norms.

We used the average damage from flooding per residence for homes facing the 1 in 1250 probability that was calculated in Botzen and van den Bergh (2012), i.e., €80,000 in today's price levels. This amount was converted for the UK sample based on purchasing power parity (PPP). Moreover, the flood insurance premium was set at €64 (equal to probability multiplied by damage), which is actuarially fair. The insurance costs were provided with and without flood insurance, so it is unlikely that any effect of the default on decision-making is due to subjects' calculation capacities.

We studied to what extent the default effect can be attributed to loss aversion, as well as preconditions of the default effect (a positive difference in flood insurance purchase between opt-out and opt-in) like levels of experience related to the decision, namely flooding experience and experience purchasing flood insurance. That is, the cross-country focus of the study was motivated by differences in actual flooding experience and flood insurance purchasing. The flood insurance market in the Netherlands is new and penetration rates are currently low, moreover, recent flooding experiences have caused minor losses, whereas, in the United Kingdom, the opposite is true.

Experience may influence both susceptibility to default effects (List, 2003) and the direction of the impact of setting defaults on observed behavior (Reiter, McRee, Pepper, & Brewer, 2012). Low susceptibility in our context would follow from homeowners who have purchased flood insurance previously, or have been flooded in the past, knowing more about the benefits (or costs) of having insurance. This type of learning would drive stronger preferences toward flood insurance for experienced decision makers than individuals who have no experience purchasing flood insurance. Experience may increase decisiveness, leading to less inertia and lower default effects (Brown, Farrell, & Weisbenner, 2016; Sautua, 2017). Experienced individuals may also react to the setting of defaults, i.e., the default may backfire, due to a perceived threat to autonomy (Pavey & Sparks, 2009). This follows from evidence suggesting that individuals who have strong preferences for alternatives perceive opt-out policies as an infringement on their autonomy, and attempt to reinstate this by opting out, whereas this effect does not occur among individuals who are undecided on their preferences (Reiter et al., 2012).

Underweighting LPHI risk

Systematic decision biases are well suited to prospect theory (PT) (Tversky & Kahneman, 1992), which allows for boundedly rational behavior (Wakker, 2010). PT is an alternative theory of decision-making to the rational model of EUT. It posits that individuals evaluate choices with respect to a reference point and overweight losses relative to equal-sized gains due to loss aversion. Furthermore, in contrast to EUT which assumes that individuals process probabilities well, PT incorporates nonlinear weighting of probabilities. Nonlinear probability weighting is a fairly robust empirical phenomenon occurring within large general population samples in both the Netherlands and the United States

(Booij, van Praag, & van de Kuilen, 2010; Jaspersen, Ragin, & Sydnor, 2019). Concerning very low probabilities, it has been shown that these are often significantly underweighted in decision-making (Epper & Fehr-Duda, 2017; Fehr-Duda & Fehr, 2016; Hertwig, Barron, Weber, & Erev, 2004). This may extend to flood insurance decisions, whereby the unlikely event of flooding is underweighted resulting in a low willingness to pay for flood insurance.

Few studies have examined why individuals over/underweight LPHI risks (Barberis, 2013). We investigated the psychological mechanisms behind such behavior in the context of flood insurance demand (Robinson & Botzen, 2019a). We studied the impact of the anticipated regret one might feel about not purchasing flood insurance if a flood were to occur, as well as regret associated with paying for insurance if no flood were to occur; worry about flooding at one's home; incidental emotion, i.e., individuals' mood which is unrelated to the decision at hand; as well as the use of the threshold level of concern decision heuristic, on willingness to pay of zero for flood insurance. We used the latter as a proxy of probability neglect (rounding of very low probabilities to zero). In addition, we looked at whether the variables influence flood risk-taking attitudes in general based on the maximum amount individuals are willing to pay for flood insurance.

As well as the variables used to analyze probability neglect and maximum willingness to pay for flood insurance, we investigated the relation between probability weighting and locus of control (Robinson & Botzen, 2020),[a] which has been shown to influence protection decisions under high probability disaster risks (Sattler, Kaiser, & Hittner, 2000; Sims & Baumann, 1972). Locus of control refers to whether individuals judge that outcomes in their life are a consequence of their own actions or fate. These judgments prompt actions, e.g., protection against increasing flood risks. Flood risk is not static, and flood probabilities can rise at a given point in time due to increased river water levels.

Results

Charity hazard results

Robinson et al. (2019) showed that the availability of government compensation crowds out demand for flood insurance (Table 1). Whereas, certain half and risky full compensation, significantly negatively impacted flood insurance demand, compared to the baseline (no government compensation) in many comparisons, the impact of ambiguous full compensation is mostly insignificant. It seems that if individuals are compensated for certain in the case of uninsured losses, or know the probability that they will be compensated, they rely more on compensation than when the compensation probability is unknown. Perhaps individuals dislike the ambiguity of not knowing the probability of

---

[a] Both Robinson and Botzen (2019a) and Robinson and Botzen (2020) are based on the same data as Robinson and Botzen (2018).

Table 1 Significant percentage differences in the proportion of individuals purchasing insurance under the different conditions of government compensation relative to the baseline (no government compensation condition).

Insurance premium	Flooding probability	Certain half	Risky full	Ambiguous full
30	1 in 1000	–**	–*	–
	1 in 100	+	–	–
	1 in 10	–	–	+
45	1 in 1000	–	–	–
	1 in 100	–*	–***	–
	1 in 10	–**	–*	–
60	1 in 1000	–***	–***	–*
	1 in 100	–	–	–
	1 in 10	–	–	–
240	1 in 1000	–*	–**	–**
	1 in 100	–	–*	–
	1 in 10	–	–	–

Notes: *, ** and *** are significant differences in the proportion of individuals purchasing insurance between the government compensation conditions and the baseline condition (no government compensation) at the 10%, 5%, and 1% levels respectively. + and − indicate whether the difference is positive or negative.

compensation and insure flood losses more often because of this. This dislike of unknown probabilities relative to known ones is a common empirical phenomenon called ambiguity aversion (Ellsberg, 1961).

Threshold level of concern, probability neglect, willingness to pay for flood insurance, and probability weighting results

Recall that Robinson and Botzen (2018) examined two antecedents of use of the threshold level of concern heuristic, namely: whether individuals would feel regret about not purchasing flood insurance if a flood were to occur, as well as worry about flooding at one's home. Higher levels of regret and worry led to a lower threshold under which individuals were prepared to pay for insurance. In other words, these individuals were more likely to demand flood insurance even when flood probabilities are low.

We used the same data to investigate whether regret and worry influence willingness to pay for flood insurance and probability neglect of flood risk, as well as whether these influences depend on objective risk in Robinson and Botzen (2019a). Other variables included are a secondary regret variable, i.e., whether individuals would feel regret about paying for insurance if no flood were to occur; a survey measure of usage of the threshold level of concern heuristic; and individuals' mood.

Individuals who were more likely to anticipate regret for uninsured flood losses were less likely to neglect the flood probability, and the effect is most prominent for more likely flood risks, perhaps because uninsured losses and this type of regret are more probable. This regret is also related to higher willingness to pay for flood insurance, but the effect does not dependent on the flood probability.

There is also a positive level effect of anticipated regret about paying for insurance if no flood were to occur on probability neglect of flood risk. Whereas this regret negatively impacted willingness to pay for flood insurance only once the flood probability exceeded 1 in 10,000. Therefore, individuals who anticipate this regret may neglect very low-probability flood risks, but there are subgroups of individuals who either neglect the risk, or think about levels of insurance they are willing to pay under higher flood probabilities.

With respect to worry about flooding at one's home (which is an index of flood risk perception), more worry is related to a lower likelihood of probability neglect and to increased willingness to pay for flood insurance. The latter relation is higher in magnitude and significance under low flood probabilities. The increased significance of worry for low probabilities may be due to risk perceptions having more explanatory power for very low flood probabilities most of the sample face in reality.

In contrast to the risk-specific emotions regret and worry, mood had no impact on flood insurance demand, so emotions specific to flood risk have more explanatory power than emotions unrelated to the risk.

Our survey measure of reported use of the threshold level of concern heuristic by individuals is related to more probability neglect. However, the variable had no influence regarding levels of flood insurance demand. This indicates that individuals who have no concern for the flood probability round the probability to zero, consistent with the mechanisms underlying this heuristic.

To examine whether the psychological determinants of flood insurance demand are related to demand through probability weighting, in an exploratory analysis models of decision making were estimated based on PT in Robinson and Botzen (2020). It was found that utilization of the threshold level of concern decision heuristic (which is related to probability processing), is also associated with more underweighting/less overweighting of low probabilities. Whereas, anticipated regret for uninsured flood losses and internal locus of control types are more risk averse/less risk seeking across flood insurance decisions involving high flooding probabilities, meaning they are more likely to demand flood insurance, it was not possible to attribute these effects to probability weighting. An overview of the effects outlined in this section can be visualized in Fig. 2, which shows the relationships schematically (although not the dependence of these effects on the flood probability).

Improving individual flood preparedness results

Based on the results of Robinson et al. (2020), in the Netherlands, where demand for flood insurance and flooding experience is very low, setting an opt-out default resulted in significantly higher flood insurance demand (71% purchased actuarially fair insurance against a 1 in 1250 yearly likelihood of €80,000 flood damage), compared to opt-in (52% purchased insurance). In the United Kingdom, where demand for flood insurance and flooding experience is a lot higher there was no overall effect of the default (83% purchased insurance) (Fig. 3).

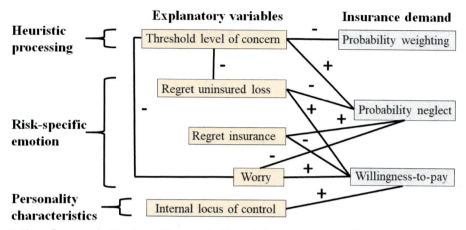

Fig. 2 The influence of utilization of the threshold level of concern decision heuristic, regret related to not purchasing insurance in the event of an uninsured flood loss (regret uninsured loss), regret related to purchasing insurance in the event of no flood loss (regret insurance), worry about flooding and internal locus of control, on components of flood insurance demand (probability weighting, probability neglect, and willingness to pay).

Using statistical control, we attributed some of the contrasting influence of the default on flood insurance demand between the Netherlands and the United Kingdom, to whether individuals reported to have been flooded in the past and to have purchased flood insurance. Experience with purchasing flood insurance and flooding led to more opting out of flood insurance, in line with the perceived threat to autonomy explanation (see "Inertia bias" section).

We examined the degree to which anticipated regret for uninsured flood losses and whether individuals perceive the flood insurance cost as expensive mediate the default effect found in the Netherlands. Mediation would imply that regret and perceived insurance cost explain to some extent the mechanisms underlying the default effect. These variables proxy for how individuals feel about losses from their reference point, which we assumed that individuals take as the status quo/default (opt-in or opt-out). That is, individuals assigned opt-out flood insurance policies may excessively focus on the regret they might feel if they were to remove flood coverage from their prepurchased policy and experience a flood loss, whereas individuals who were required to opt-in and pay an extra insurance premium to include flood coverage in the policy may instead focus on the cost of insurance. We found that together regret and perceived insurance cost mediated around one-third of the default effect in the Netherlands.

Levels of flood insurance demand

It was found that the payment procedure in Robinson and Botzen (2018, 2019a, 2020) did not impact whether individuals were willing to pay positive levels of flood insurance,

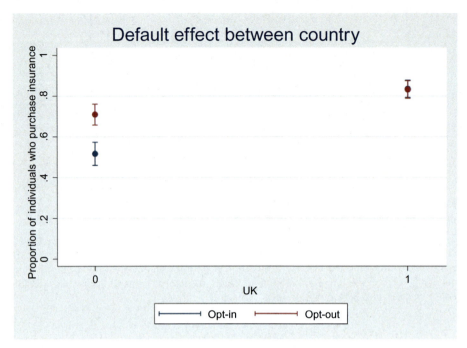

Fig. 3 Default effect between the Netherlands (UK=0) and the United Kingdom (UK=1).

whereas the procedure did influence levels of flood insurance individuals were willing to pay. Incentivized payments led to higher levels of risk aversion/lower levels of risk seeking, compared to nonincentivized payments.

Risk premiums are an indicator of individuals' levels of risk aversion, i.e., whether the amount of money that they are willing to give up in order to avoid a risk is greater than or less than the expected value of the risk (which in our case is the probability of flooding multiplied by possible flood loss). Therefore, an individual's maximum willingness to pay for flood insurance minus expected flood damage obtains their risk premium. The mean risk premiums for flood damage amount €60,000, across flood probabilities 1 in 10,000, 1 in 1000, 1 in 100, and 1 in 20, are €233, €471, €611, and €−704 for incentivized payments, and €167, €285, €164, and €−1325, for nonincentivized payments, respectively, after omitting those who are not willing to pay anything. Risk premiums greater than zero indicate risk aversion, equal to zero indicate risk neutrality, and less than zero indicate risk seeking. Note that the differences in risk premiums between incentivized and nonincentivized payments are always positive, indicating hypothetical bias.

Moreover, the subgroup of the sample who had zero willingness to pay for flood insurance across flood probabilities 1 in 10,000, 1 in 1000, 1 in 100, and 1 in 20 corresponds to 26%, 13%, 10%, and 8% respectively. Therefore, there is a subgroup of homeowners in the Netherlands for which demand for flood insurance is sufficient. The majority is willing to pay a positive amount for flood insurance and, on average, willingness to pay amounts exceed insurance premiums if the flood probability is 1 in 100 or lower. Nevertheless, there is also a sizable subgroup with insufficient demand, for which we provide several policy recommendations in the next section.

Policy recommendations

Based on our research, several recommendations can be made for Dutch and US policymakers interested in raising flood insurance penetration rates. These recommendations may have particular promise for the Netherlands and some risky areas of Texas given that penetration rates are low here (Klotzbach et al., 2018; Suykens et al., 2016). Increasing flood insurance demand may be seen as societally desirable in light of evidence that flood losses are predicted to rise in the future due to climate change. Firstly, bundling several risks into one insurance policy may overcome the threshold probability determining whether flood risk is devoted concern, among subgroups who utilize this heuristic. That is, the combined probability of at least one of the bundled risky events occurring may be more likely to exceed the threshold than each risk considered in isolation. For example, offering flood insurance together in a policy that covers fire risk is more likely to attract a high demand than offering flood coverage as a separate policy.

Another way to reduce the use of the threshold level of concern heuristic is by communicating the flood probability to homeowners over a longer time horizon. A flood probability communicated over the expected time homeowners typically reside at a given property (e.g., 1 in 4 probability of flooding at least once over 30 years) may attract more concern and result in higher flood insurance demand than the same flood probability communicated to individuals as a yearly likelihood (e.g., 1 in 100 probability of flooding per year). Bradt (2019) showed in a US sample, of which a large proportion of individuals had experienced flooding or natural disasters in the past, that extending the time horizon over which the flood probability is framed can raise levels of flood insurance demand. A similar pattern was also found in a study by Chaudhry, Hand, and Kunreuther (2018) in a sample of flood-prone individuals residing within states on the Gulf Coast, including Texas. This suggests that reframing the flood probability in this way might be effective for increasing insurance demand against flooding within communities where the risk of flooding is very salient, like the coastal regions of Texas.

Individuals may also be less likely to use heuristics if they better understand the components of risk. This can be facilitated through risk communication aids, education policies, or through information campaigns. Information campaigns which highlight to individuals that the "best return on an insurance policy is no return" may avoid that they anticipate regret for paying flood insurance premiums if no flood has occurred.

Lastly, our results reveal that providing flood insurance as a default option can be an effective way of increasing the insurance penetration rate among individuals who have never experienced flooding nor purchased flood insurance before, as is the case for many homeowners in the Netherlands. Whereas, in places where flood insurance is relatively commonplace or where flooding is a more frequent occurrence, like within coastal communities of Texas, default flood insurance policies may backfire, i.e., lead to reactance and lower penetration rates among those who need to opt-out to cancel their flood coverage, compared to those who need to opt-in to insure against flooding.

Glossary list

Concept	Proposed explanation	Source	Found in chapter by [author(s)]
Expected utility theory	A model of rational decision-making where individuals choose between risky or uncertain prospects by comparing their expected utilities, i.e., the sum of utility transformed outcomes multiplied by their respective probabilities of occurrence. It is the baseline theory of rational individual choice in economics.	Mongin, P., 1997. *Expected utility theory*. In: J. Davis, W. Hands, U. Maki (Eds.). Handbook of economic methodology. London, UK: Edward Elgar; 1997. pp. 342–350.	Robinson & Botzen
Calamities and compensation act	In the Netherlands, this act provides legal arrangements for disaster loss compensation. In cases of loss as a result of disasters such as floods, the state aims to compensate the loss using public resources, which derive from tax income and state loans. *(Full name in Dutch: Wet Tegemoetkoming Schade bij rampen en zware ongevallen)*	Keskitalo, E.C.H., Vulturius, G. and Scholten, P., 2014. Adaptation to climate change in the insurance sector: Examples from the UK, Germany and the Netherlands. *Natural Hazards*, 71(1), pp. 315–334.	Robinson & Botzen

Continued

Concept	Proposed explanation	Source	Found in chapter by [author(s)]
Prospect theory	Prospect Theory is an alternative theory of decision-making that deviates from Expected Utility Theory by positing that individuals evaluate choices with respect to a reference point. They tend to overweight losses compared to equal sized gains and process probabilities in a nonlinear way.	Tversky, A. and Kahneman, D., 1992. Advances in prospect theory: Cumulative representation of uncertainty. *Journal of Risk and Uncertainty*, 5(4), pp. 297–323.	Robinson & Botzen

Abbreviations list

Abbreviation	In full	Found in chapter by [author(s)]
LPHI	Low-probability/high-impact	Robinson & Botzen
EUT	Expected utility theory	Robinson & Botzen
PT	Prospect theory	Robinson & Botzen
PPP	Purchasing power parity	Robinson & Botzen

References

Aerts, J. C., & Botzen, W. J. W. (2011). Climate change impacts on pricing long-term flood insurance: A comprehensive study for the Netherlands. *Global Environmental Change*, 21(3), 1045–1060.

Barberis, N. C. (2013). The psychology of tail events: Progress and challenges. *The American Economic Review*, 103(3), 611–616.

Becker, G. M., DeGroot, M. H., & Marschak, J. (1964). Measuring utility by a single-response sequential method. *Behavioral Science*, 9(3), 226–232.

Benfield, A. (2018). *Weather, climate and catastrophe insight: 2017 annual report*. Available from: http://thoughtleadership.aonbenfield.com/Documents/20180124-ab-if-annual-report-weather-climate-2017.pdf.

Booij, A. S., van Praag, B. M., & van de Kuilen, G. (2010). A parametric analysis of prospect theory's functionals for the general population. *Theory and Decision*, 68(1–2), 115–148.

Botzen, W. J. W., Aerts, J. C., & van den Bergh, J. C. (2013). Individual preferences for reducing flood risk to near zero through elevation. *Mitigation and Adaptation Strategies for Global Change*, 18(2), 229–244.

Botzen, W. J. W., & van den Bergh, J. C. (2012). Risk attitudes to low-probability climate change risks: WTP for flood insurance. *Journal of Economic Behavior & Organization*, 82(1), 151–166.

Bradt, J. (2019). Comparing the effects of behaviorally informed interventions on flood insurance demand: An experimental analysis of 'boosts' and 'nudges'. *Behavioural Public Policy*. https://doi.org/10.1017/bpp.2019.31.

Brown, J. R., Farrell, A. M., & Weisbenner, S. J. (2016). Decision-making approaches and the propensity to default: Evidence and implications. *Journal of Financial Economics, 121*(3), 477–495.

Chaudhry, S. J., Hand, M., & Kunreuther, H. (2018). *Extending the time horizon: Elevating concern for rare events by communicating losses over a longer period of time.* The Wharton School, University of Pennsylvania Working Paper.

Coate, S. (1995). Altruism, the Samaritan's dilemma, and government transfer policy. *The American Economic Review, 85*(1), 46–57.

de Martino, B., Kumaran, D., Seymour, B., & Dolan, R. J. (2006). Frames, biases, and rational decision-making in the human brain. *Science, 313*(5787), 684–687.

Dixon, L., Clancy, N., Seabury, S. A., & Overton, A. (2006). *The national flood insurance program's market penetration rate: Estimates and policy implications.* Santa Monica, CA: RAND Corporation.

Eijgenraam, C., Brekelmans, R., den Hertog, D., & Roos, K. (2017). Optimal strategies for flood prevention. *Management Science, 63*(5), 1644–1656.

Ellsberg, D. (1961). Risk, ambiguity, and the savage axioms. *The Quarterly Journal of Economics, 75*, 643–669.

Epper, T., & Fehr-Duda, H. (2017). *A tale of two tails: On the coexistence of overweighting and underweighting of rare extreme events.* University of St. Gallen Working Paper.

Fehr-Duda, H., & Fehr, E. (2016). Game human nature. *Nature, 530*(7591), 413–415.

Garrett, T. A., & Sobel, R. S. (2003). The political economy of FEMA disaster payments. *Economic Inquiry, 41*(3), 496–509.

Gauderis, J., Kind, J., & van Duinen, R. (2013). Robustness of economically efficient flood protection standards: Monte Carlo analysis on top of cost-benefit analysis. In F. Klijn, & T. Schweckendiek (Eds.), *Comprehensive flood risk management. Research for policy and practice. Proceedings of the 2nd European conference on flood risk management* CRC Press.

Hertwig, R., Barron, G., Weber, E. U., & Erev, I. (2004). Decisions from experience and the effect of rare events in risky choice. *Psychological Science, 15*(8), 534–539.

Jaspersen, J. G., Ragin, M. A., & Sydnor, J. R. (2019). *Predicting insurance demand from risk attitudes.* National Bureau of Economic Research. No. w26508.

Jongejan, R., & Barrieu, P. (2008). Insuring large-scale floods in the Netherlands. *The Geneva Papers on Risk and Insurance-Issues and Practice, 33*(2), 250–268.

Kahneman, D. (2003). Maps of bounded rationality: Psychology for behavioral economics. *The American Economic Review, 93*(5), 1449–1475.

Kelly, M., & Kleffner, A. E. (2003). Optimal loss mitigation and contract design. *The Journal of Risk and Insurance, 70*(1), 53–72.

Klotzbach, P. J., Bowen, S. G., Pielke, R., Jr., & Bell, M. (2018). Continental US hurricane landfall frequency and associated damage: Observations and future risks. *Bulletin of the American Meteorological Society, 99*(7), 1359–1376.

Kousky, C., Michel-Kerjan, E. O., & Raschky, P. A. (2018). Does federal disaster assistance crowd out flood insurance? *Journal of Environmental Economics and Management, 87*, 150–164.

Kunreuther, H., & Pauly, M. (2004). Neglecting disaster: Why don't people insure against large losses? *Journal of Risk and Uncertainty, 28*(1), 5–21.

Lewis, T., & Nickerson, D. (1989). Self-insurance against natural disasters. *Journal of Environmental Economics and Management, 16*(3), 209–223.

List, J. A. (2003). Does market experience eliminate market anomalies? *The Quarterly Journal of Economics, 118*(1), 41–71.

Meyer, R., & Kunreuther, H. (2017). *The ostrich paradox: Why we underprepare for disasters.* Philadelphia, PA: Wharton Digital Press.

Middelkoop, H., Daamen, K., Gellens, D., Grabs, W., Kwadijk, J. C., Lang, H., et al. (2001). Impact of climate change on hydrological regimes and water resources management in the Rhine basin. *Climatic Change, 49*(1–2), 105–128.

Miller, S., Muir-Wood, R., & Boissonnade, A. (2008). An exploration of trends in normalized weather-related catastrophe losses. In H. F. Diaz, & R. J. Murnane (Eds.), *Climate extremes and society* Cambridge University Press.

Ogashawara, I., Curtarelli, M. P., & Ferreira, C. M. (2013). The use of optical remote sensing for mapping flooded areas. *International Journal of Engineering Research and Applications*, *3*(5), 1–5.

Paudel, Y., Botzen, W. J. W., & Aerts, J. C. (2015). Influence of climate change and socio-economic development on catastrophe insurance: A case study of flood risk scenarios in the Netherlands. *Regional Environmental Change*, *15*(8), 1717–1729.

Pavey, L., & Sparks, P. (2009). Reactance, autonomy and paths to persuasion: Examining perceptions of threats to freedom and informational value. *Motivation and Emotion*, *33*(3), 277–290.

Perry, C. A. (2000). *Significant floods in the United States during the 20th century: USGS measures a century of floods*. US Department of the Interior, US Geological Survey.

Raschky, P. A., & Weck-Hannemann, H. (2007). Charity hazard—A real hazard to natural disaster insurance? *Environmental Hazards*, *7*(4), 321–329.

Reiter, P. L., McRee, A. L., Pepper, J. K., & Brewer, N. T. (2012). Default policies and parents' consent for school-located HPV vaccination. *Journal of Behavioral Medicine*, *35*(6), 651–657.

Robinson, P. J., & Botzen, W. J. W. (2018). The impact of regret and worry on the threshold level of concern for flood insurance demand: Evidence from Dutch homeowners. *Judgment and Decision making*, *13*(3), 237–245.

Robinson, P. J., & Botzen, W. J. W. (2019a). Determinants of probability neglect and risk attitudes for disaster risk: An online experimental study of flood insurance demand among homeowners. *Risk Analysis*, *39*(11), 2514–2527.

Robinson, P. J., & Botzen, W. J. W. (2019b). Economic experiments, hypothetical surveys and market data studies of insurance demand against low-probability/high-impact risks: A systematic review of designs, theoretical insights and determinants of demand. *Journal of Economic Surveys*, *33*(5), 1493–1530.

Robinson, P. J., & Botzen, W. J. W. (2020). Flood insurance demand and probability weighting: The influences of regret, worry, locus of control and the threshold of concern heuristic. *Water Resources and Economics*, *30*, 100144.

Robinson, P. J., Botzen, W. J. W., Kunreuther, H., & Chaudhry, S. J. (2020). *Default options and insurance demand*. National Bureau of Economic Research. No. w27381.

Robinson, P. J., Botzen, W. J. W., & Zhou, F. (2019). *An experimental study of charity hazard: The effect of risky and ambiguous government compensation on flood insurance demand*. Utrecht School of Economics (USE). Working Paper series, 19 (19).

Samuelson, W., & Zeckhauser, R. (1988). Status quo bias in decision making. *Journal of Risk and Uncertainty*, *1*(1), 7–59.

Sattler, D. N., Kaiser, C. F., & Hittner, J. B. (2000). Disaster preparedness: Relationships among prior experience, personal characteristics, and distress. *Journal of Applied Social Psychology*, *30*(7), 1396–1420.

Sautua, S. I. (2017). Does uncertainty cause inertia in decision making? An experimental study of the role of regret aversion and indecisiveness. *Journal of Economic Behavior & Organization*, *136*, 1–14.

Seifert, I., Botzen, W. J. W., Kreibich, H., & Aerts, J. C. (2013). Influence of flood risk characteristics on flood insurance demand: A comparison between Germany and the Netherlands. *Natural Hazards and Earth System Sciences*, *13*(7), 1691–1705.

Simon, H. A. (1955). A behavioral model of rational choice. *The Quarterly Journal of Economics*, *69*(1), 99–118.

Simon, H. A. (1990). Bounded rationality. In *Utility and probability*. London: Palgrave Macmillan.

Sims, J. H., & Baumann, D. D. (1972). The tornado threat: Coping styles of the north and south. *Science*, *176*(4042), 1386–1392.

Slovic, P., Fischhoff, B., Lichtenstein, S., Corrigan, B., & Combs, B. (1977). Preference for insuring against probable small losses: Insurance implications. *The Journal of Risk and Insurance*, *44*(2), 237–258.

Suykens, C., Priest, S. J., van Doorn-Hoekveld, W. J., Thuillier, T., & van Rijswick, M. (2016). Dealing with flood damages: Will prevention, mitigation, and ex post compensation provide for a resilient triangle? *Ecology and Society*, *21*(4).

Tversky, A., & Kahneman, D. (1992). Advances in prospect theory: Cumulative representation of uncertainty. *Journal of Risk and Uncertainty*, *5*(4), 297–323.

van Stokkom, H. T., Smits, A. J., & Leuven, R. S. (2005). Flood defense in the Netherlands: A new era, a new approach. *Water International*, *30*(1), 76–87.

Vis, M., Klijn, F., de Bruijn, K. M., & van Buuren, M. (2003). Resilience strategies for flood risk management in the Netherlands. *International Journal of River Basin Management*, *1*(1), 33–40.

Viscusi, W. K., & Zeckhauser, R. J. (2006). National survey evidence on disasters and relief: Risk beliefs, self-interest, and compassion. *Journal of Risk and Uncertainty*, *33*(1–2), 13–36.

Wakker, P. P. (2010). *Prospect theory: For risk and ambiguity*. Cambridge University Press.

Wind, H. G., Nierop, T. M., Blois, C. D., & Kok, J. D. (1999). Analysis of flood damages from the 1993 and 1995 Meuse floods. *Water Resources Research*, *35*(11), 3459–3465.

CHAPTER 11

Assessing economic risk, safety standards, and decision-making

Matthijs Kok
Department of Hydraulic Engineering, Faculty of Civil Engineering and Geosciences, Delft University of Technology, Delft, The Netherlands

Introduction

Flood risk can be perceived and analyzed in many ways. It can be seen as a natural hazard, but others consider it as a by-product of our technological society. And indeed, without advanced flood defenses, the risk would be much different in many areas. However, because of the benefits of living in flood-prone areas and the protection of flood defenses against natural hazards, the majority of humankind lives in Delta regions. In all these regions, the social benefits of protection have been considered higher than the social costs of protection.

In this chapter, we follow the probabilistic risk analysis approach as advocated in Bedford and Cooke (2001), which aims to quantify the risk caused by an uncertain event in situations where classical statistical analysis is difficult or impossible. In this approach, the risk has to do with the probability of the uncertain event (that may be the extreme weather conditions or the flood) and the consequences of the event (economic damages, loss of life, environmental losses, etc.).

We will discuss flood risk approaches in three countries: the United States of America, the United Kingdom, and the Netherlands. We will see that these approaches have much in common, especially if we look at the way economic risk is defined. However, if we look more specifically at the safety standards in the three countries, we see many differences.

Risk approaches

Risk is often defined as a combination of probability and consequences of unwanted events. The probability has a strong relationship with the hydraulic loads (discharges, water levels, waves) and the strength of the flood defense system (levees, hydraulic structures, storm surge barriers). Consequences depend heavily on the hydrodynamic characteristics (e.g., water depth, duration of the flood), the land use (in urban and industrial areas the damage is much higher compared with rural areas), and the evacuation strategy (preventive evacuation can lower the number of casualties dramatically).

There are many common elements in the flood risk approach all over the world. In general, the probability that a flood defense structure will fail is determined by the probability of a particular load and the probability that the structure will not be able to withstand this load (CIRIA, 2013). A flood can occur in an almost endless variety of ways, depending on factors such as the conditions in which it occurs, the location of levee breaches, and the stability of linear elements in the landscape such as raised roads and railway lines. The impact of a flood depends on the vulnerability of the area affected and the decisions taken by members of the public and the authorities as the threat of flooding increases. The success of any preventive evacuation depends to a great extent on the time available and the conditions in which the evacuation must take place. Evacuation can reduce the number of victims, but traffic chaos in a low-lying polder could in fact cause many casualties in a flood. All these factors are uncertain, and we can consider them for example in terms of probabilities of scenarios. Combining all the possible effects (consequences) with their probabilities gives us a complete picture of the flood risk.

In Kok, Jongejan, Nieuwjaar, and Tanczos (2017), five steps have been proposed to assess the risk:

(a) *Load*: assessing probability distributions of hydraulic loads.
(b) *Flooding probability*: for possible loads, assess the strength of flood defenses to assess the flooding (or failure) probability.
(c) *Flood scenario*: for each failure, assess flood scenarios, that is the likely progress of flooding that might occur, and the probability of the flood scenarios.
(d) *Consequences*: for each scenario, assess the economic and social consequences (e.g., loss of life), taking uncertainties in hydraulic characteristics, land use data, and evacuation planning into account.
(e) *Risk*: combine the probabilities of flooding with the consequences to obtain a representation of the risk.

In Fig. 1, an example of a flood scenario in the Netherlands is given. On the left-hand side, an overview of all possible flood scenarios is given. On the right-hand side, one possible scenario is shown, with the *red dot* showing the breach location along the river Lek (which is one of the Rhine branches). There is a 10 m elevation difference between the river and the lake (10 vs 0 m above sea level, which is indicated by NAP).

An important issue in flood risk is the interpretation of risk. It seems that sometimes the failure probability is seen as a property of the flood defense. And, that seems quite obvious: a strong flood defense has a smaller failure probability compared with a fragile flood defense, assuming that these flood defenses are attacked with the same hydraulic loads. However, in the risk analysis literature, it is also widely known that the failure probability also depends on the knowledge of the flood defense itself. Compare for example two identical levees: one levee where the soil below the levee is known, and another levee where the soil is not known. The uncertainties of the second levee are much larger, and hence it might be expected that the failure probability of the second levee is larger

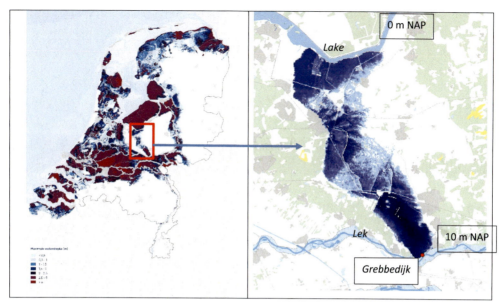

Fig. 1 All possible flood scenarios in the Netherlands *(left)* and example of flood scenario: Gelderse Vallei flooded by a breach of a flood defense along the river Lek *(right)*.

than the levee which properties are perfectly known. This interpretation is called the Bayesian interpretation: the probability of flooding is a measure of the likelihood that a flood will occur, given the knowledge at our disposal. The difference between inherent and knowledge uncertainty is irrelevant in the Bayesian interpretation, according to which the probability that a flood will occur is not uncertain; the probability is a measure of uncertainty. The probability is no longer, therefore, a physical property but a subjective "degree of belief." According to the Bayesian interpretation, a person can give only one answer to the question of whether the probability of flooding is lower than the standard. However, the probability estimates of different people can differ. In practice, such differences can be overcome by exchanging data, second opinions, and the establishment of best practices.

In all risk approaches, much attention is given to the economic and individual risk (Kok et al., 2017). *Economic risk* concerns the cost of risk bearing, expressed in euros (or US dollars), or in euros (US dollars) per year. In cost-benefit analyses, economic risk is often equated with the annual expected value of the damage, the product of probability and damage. The idea behind this is that the government can efficiently spread the cost of any damage among all residents of the Netherlands. If this does not happen and everyone has to bear the cost of the damage they themselves sustain, the cost of risk bearing will often exceed the annual expected value of losses. For example, if the yearly economic risk is privately insured, the insurance costs will often be (much) higher than the expected yearly damage, which can also be seen as a "risk averse" premium (see also van Erp, 2017).

Economic risk relates to risk in the flood-prone area. It does not provide any insight into the risks for individual persons. The local individual risk (LIR) is a measure of risk that expresses the probability that a person who is permanently present at a particular location will die as a result of flooding, taking into account the potential for evacuation. Using local individual risk provides everyone in the Netherlands regions protected by levees with a basic level of protection.

Economic optimization

An important point in the flood risk approach is the economic optimization: what would be the best decision if we only look at the economic costs and benefits, where also non-monetary values (such as loss of life) are taken into account? Such a question can only be answered if we make assumptions about the future, and that we have knowledge about failure probabilities and consequences of a flood. This knowledge is always uncertain, but this is not a special property of floods: every risk analysis of technical installation (like missiles) or hazard (like earthquakes) has these issues. What is important is that these assumptions are "most likely," since these assumptions are not demonstrable.

The classic way of economic optimization is well explained in Kok et al. (2017). More investment in the reliability of flood defenses reduces the flood risk. The investments and the risk are the total costs to society. Minimizing the total costs allows the optimum reliability of the flood defenses to be identified. This principle was first put into practice by the original Dutch Delta Committee in 1953 (van Dantzig, 1956), and is schematically represented in Fig. 2.

The optimum investment strategy is also associated with a certain progression in the probability of flooding over time. This takes the form of a sawtooth wave because the probability of flooding reduces immediately when a levee is reinforced, then gradually rises due to subsidence, increasing river discharge rates and sea-level rise. The scale of reinforcement (and thus the reduction in the probability of flooding after reinforcement) and the time until the next round of reinforcements are strongly influenced by the relationship between the *fixed* and *variable* costs of the intervention. The fixed costs of the intervention are not related to the size of the intervention measure, like for example a storm surge barrier. If the fixed costs are relatively high, it is economically advisable to postpone a new intervention for as long as possible. If the fixed costs are relatively low, as they are along the sandy coastline, it makes more economic sense to make small interventions more frequently.

Safety standards

Standards are the result of a political process based on the results of risk calculations and a cost-benefit analysis. In many cases, the results have been adopted, with due consideration of the uncertainty associated with the input.

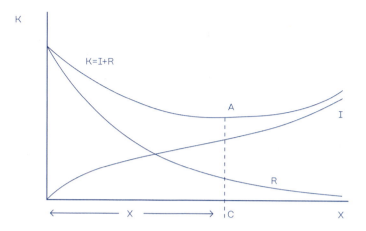

K. Total costs
I. Investment costs
R. Risk costs

Fig. 2 The basic principle of economic optimization. The total costs (K) are equal to the investment costs (I) associated with improving reliability (here: heightening levees) plus the present value of the risk (R). The optimum lies at the point where the total costs (I+R) are lowest. (Source: Kok, M., R. Jongejan, M. Nieuwjaar and I. Tanczos, 2017. Fundamentals of flood protection. Ministry of Infrastructure and the Environment and Expertise Network for Flood Protection. https://www.enwinfo.nl/publish/pages/183541/grondslagenen-lowresspread3-v_3.pdf.)

Safety standards are not a goal in itself but have to serve the framework of "acceptable risk." In the concept of acceptable risk, two questions are important:
(a) How safe is it?
(b) Is it safe enough?
From a theoretical point of view, it can be remarked that the same risk framework should be applied to all kinds of natural hazards (earthquakes, floods, hurricanes, etc.) and technical artifacts (ferries, airports, traffic, plants, dikes, powerplants, missiles, etc.), because risk levels can be compared with each other. However, we see in reality that such an approach seems not achievable. Moreover, safety is not a separate discipline, but part of the design of technical artifacts, so the risk framework is (or should be) part of the technical guidelines.

Having a risk framework is not a goal but needs to serve more underlying objectives. In Vrijling (2008), the following aims are mentioned:
(a) a predictable decision process
(b) a basis for technical design
(c) to avoid changes after each disaster
(d) discharge of the engineering profession

Especially the last mentioned is important from an engineering point of view. Without a framework, it cannot be detected whether the engineer made a mistake for which he is responsible, or whether he or she was acting in a socially responsible way.

There are many ways to relate the risk framework to safety standards. We will discuss three of them, and in the next paragraph, we will discuss the pros and cons of these arrangements. It is important to notice that there is, from a scientific point of view, no best way to set up an arrangement.

(a) Exceedance probability of design water level

In this approach, the flood defenses are designed and maintained to withstand a hydraulic load with a fixed exceedance probability, for example a water level which is on average exceeded in 100 years (so an annual exceedance probability of 1/100). As such the standard seem to focus only on the hydraulic load. Though the strength of the flood defenses played a major role, it was not explicitly reflected in the numerical standard.

(b) Flooding probability

In this approach, the probability of flooding is the key parameter of the safety standard (VNK, 2015). The main reason for defining the safety standard to the probability of flooding is that it properly reflects the degree of protection from flooding. The probability of flooding depends both on the hydraulic load (water levels and wave action) and on the strength of the defenses (height, width, type of material etc.). In a safety standard, the food probability norm can be based on the risk of flooding. Risk refers to both the probability and the consequences of flooding (see above). The possible consequences focus often on fatalities and victims. For example, the loss-of-life risk can play an explicit role in the updating of standards for flood defenses. An example is the risk limit in the protected areas in the Netherlands where the government has decided that the probability of loss of life due to flooding may not exceed 1/100,00 per year.

(c) Costs < benefits

In this arrangement, the safety standard is not a number, but a decision rule: if costs of measures are lower than the risk reduction by these measures, these measures should be applied.

It can be seen directly that the second and third alternatives are directly related, because in order to assess the costs and benefits, the flooding probability has to be assessed. There are many methods to assess this probability. The concept of probabilities is directly related to uncertainties. These uncertainties can arise from, for example, data constraints and lack of knowledge. This means that data collection and further investigation can lead to a change in the probability of flooding. If the probability of flooding were regarded as a property of a flood defense system, this would of course be impossible. The probability of flooding is not an easily definable property of a flood defense structure, like its height, but a judgment based on knowledge of the structure. The probability of flooding is therefore also a measure of our uncertainty, as the probability depends on the knowledge and information available to us. For instance, our uncertainty about the flood defense capacity of a new dam, for example, declines dramatically once the reservoir has been filled. After it has been filled, we judge the probability of a breach to be considerably smaller as it is now successfully retaining a large body of water, though the properties of the dam have not changed at all. This also means that a flood probability is not the same as the probability that a particular water level will be exceeded, leading to flooding. This would only be the case if we were able to know exactly what water level the flood defense will breach. This is however uncertain in practice, because of our lack of knowledge about the subsurface, for example. As a result of this uncertainty, there is a chance that the flood defenses will breach even at a relatively low water level, but it is also possible that this will not happen until the water level is relatively high (see Kok et al., 2017).

The flooding probability as a safety standard has been recently introduced in the Netherlands. This probability standard relates to flood defense sections of about 20–30 km is between 1/100 and 1/1,000,000, and this depends on the consequences (economic losses and loss of life) of a breach in the flood defense. An overview of the standards is given in Fig. 3.

The *flooding probability* as a safety standard has been adapted in the Netherlands in 2017. Until now, the new approach has still some challenges in practice. For example, the assessment carried out by the water authorities shows some remarkable results (ENW, 2019). Some dike sections have been judged with a failure probability of 1/10, which is quite remarkable since these dike sections have not seen failures in the past 100 years. Recommendations are given in ENW (2019) on how to improve the method and the application of the method (assessments of the flood defenses), for example, to have independent reviews of the assessments.

The *exceedance probability of designing water level* as a safety standard can be seen in the United States. Often, a standard of 100 years or 500 years is used as the exceedance probability (which is the inverse of the average return period). The exceedance probability has a relation with the flooding probability, but this is not a one-one relation. This can be easily explained, because the flooding probability also depends on the strength of the flood defenses. For the design of the flood defenses, guidelines and manuals are available (for example, the International Levee handbook). Also, the design and the deterioration of these flood defenses have a large influence on the flooding probabilities.

The *Cost < Benefits* approach is being applied in the United Kingdom. There are many small rivers in the United Kingdom with relatively small damages if there is a flood, so it seems that this approach is more suited for these small-scale damages and relatively high costs to prevent these damages. Also, appropriate spatial planning seems important to reduce flood risk in the future, where more extreme rainfall events might be expected.

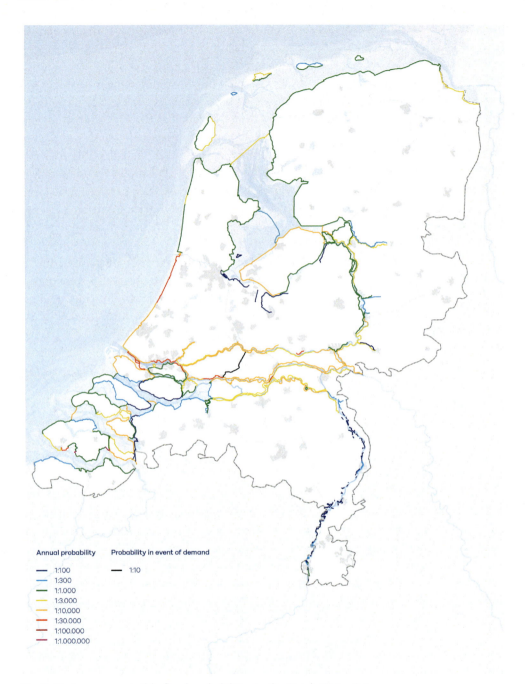

Fig. 3 Maximum permissible flood probabilities in the Dutch Water Act.

Examples

When illustrating the different risk frameworks, it must always be kept in mind that the outcomes of the three frameworks depend heavily on the location and the available information. We use the example of a river stretch of 10 km long and catastrophic damage of 10 billion euro to illustrate two of the frameworks. This example was inspired by the Grebbedijk in the Netherlands, which is located along the river Lek, and shown in Fig. 2.

Exceedance probability framework

The exceedance probability framework is a way to define the design load, for example, the design water level (or for rivers the design discharge, which needs to be transferred to local water levels). See Fig. 4 for an example for the river Rhine in the Netherlands, in this figure the exceedance frequency line discharge of the river at the border with Germany is shown.

An example of this approach: for the Grebbedijk a safety standard of 1/1250 was chosen in the recent past, this means that before January 1, 2017 the flood defenses had to be designed in such a way that these flood defenses can withstand the design water level with an exceedance frequency of 1/1250 (1/year).

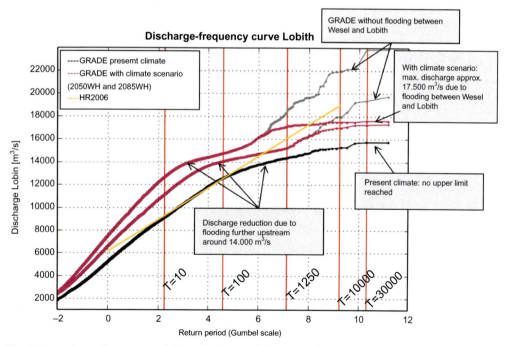

Fig. 4 Exceedance frequency of river discharge of the river Rhine at Lobith for current situation, upstream flooding, and climate change scenarios (Hegnauer, Beersma, van den Boogaard, Buishand, & Passchier, 2014).

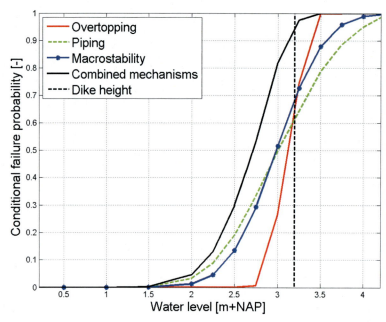

Fig. 5 Example of a fragility curve in the Netherlands. *(Source: Wojciechowski, K., G. Pleijter, M. Zethof, F. J. Havinga, D.H. van Haaren, W.L.A. ter Horst, 2017. Application of fragility curves in operational flood risk assessment, Geotechnical safety and risk, T. Schweckendiek et al. (Eds.) https://doi.org/10.3233/978-1-61499-580-7-528.)*

In the United States, the National Flood Insurance Program uses the (minimum) standard of flood protection of a 100-year return period, often extended with a safety margin of 0.3 m. Although this standard was derived as a criterion for entering the insurance program, it has become a safety standard in practice. Often, improving levees up to stricter standards than a 100-year return period was not considered until recently. Nowadays, new federal flood protection standards are investigated, making the nation more resilient.

Flooding probability framework

In order to assess the flooding probability, not only the hydraulic load (such as water levels) is taken into account, but also the strength of a levee. One way to do so is by using fragility, which shows the conditional probability of failure, given the hydraulic load. An example of a fragility curve is given in Fig. 5, where the conditional probability of failure is given as a function of the water level, and different failure mechanisms of the flood defense are shown.

The flooding probability is then obtained by combining the fragility curve with the probability density function of the hydraulic load (often the water level).

Costs-benefits framework

In the appraisal guide in the United Kingdom (Environment Agency, 2010), no indication of safety standards is provided, leaving the appropriate standards to be determined on a risk basis. This means that the benefit-cost ratio becomes relatively more important. In the 2008–2009 to 2010–2011 investment program, every £1 of capital investment in flood and coastal erosion risk management in the United Kingdom provided an average long-term benefit in reduced damage of approximately £8, a ratio of 8. And in more recent years, this ratio went up even further to 9.5 (Jonkman, Jorissen, Schweckendieck, & van den Bos, 2017).

Comparison and discussion

What are the advantages and disadvantages of the three different arrangements?

Exceedance probability of design water level

Advantage	Disadvantage
It looks rather simple: there is a hydraulic load (for example the peak water level) and all decisions about the strength of the flood defense are based on this load	The hydraulic load can have more than one uncertain variable. For example, waves can come is, and of course, water levels and waves are not independent. Also, the duration of load can come into play for some failure mechanisms of the flood defense. Last, but not least, if you want to perform a cost-benefit analysis within this approach

Flooding probability

Advantage	Disadvantage
It indicates a level of protection against floods	The methods to assess the flooding probability is rather complex, since it involves the assessment of failure mechanisms of flood defenses (like for example piping, overflow, etc.) with its associated length effects (that is that the failure probability of a longer dike section is higher than a shorter dike section, with the assumption that all dike sections are all identical)

Costs < benefits

Advantage	Disadvantage
Investments are efficient because these investments are only done if these are cost-efficient	There is no standard, so the need for investments cannot be justified because "it does not fulfill the safety standard" Assessment of flooding probabilities needs to be done, and also the consequences

It can be seen that the three arrangements also have a lot in common because they all are based on the risk approach.

Concluding remarks

In this chapter, we have presented the steps in the risk approach and the way safety standards can be defined. Three different arrangements have been shown, which are used in three different countries. From the overview of advantages and disadvantages, it can be concluded that there is no one way to define safety standards that are superior to other ways.

References

Bedford, T., & Cooke, R. (2001). *Probabilistic risk analysis. Foundations and methods.* Cambridge University Press.

CIRIA. (2013). *The International Levee Handbook (C731).* CIRIA, Ministry of Ecology, USACE, ISBN: 978-0-86017-734-0. Number of pages: 1348.

Environment Agency. (2010). *Flood and coastal erosion risk management appraisal guidance.*

ENW. (2019). *Naar geloofwaardige overstromingskansen (Towards credible flooding probabilities).* https://www.enwinfo.nl/adviezen/advies-overstromingskansen/.

Hegnauer, M., Beersma, J. J., van den Boogaard, H. F. P., Buishand, T. A., & Passchier, R. H. (2014). *Generator of rainfall and discharge extremes (GRADE) for the Rhine and Meuse basins.* Final report of GRADE 2.0. © Deltares.

Jonkman, S. N., Jorissen, R. E., Schweckendieck, T., & van den Bos, J. P. (2017). *Flood defences, Lecture notes CIE5314* (2nd ed.). Delft University of Technology.

Kok, M., Jongejan, R., Nieuwjaar, M., & Tanczos, I. (2017). *Fundamentals of flood protection.* Ministry of Infrastructure and the Environment and Expertise Network for Flood Protection. https://www.enwinfo.nl/publish/pages/183541/grondslagenen-lowresspread3-v_3.pdf.

van Dantzig, D. (1956). Economic decision problems for flood prevention. *Econometrica, 24,* 276–287.

van Erp, N. (2017). *A Bayesian framework for risk perception.* https://doi.org/10.4233/uuid:1ff6ae46-c2bd-4375-aeb1-a4a9313ec560.

VNK. (2015). *The national flood risk analysis for the Netherlands.* Final report Rijkswaterstaat VNK Project Office. https://www.helpdeskwater.nl/onderwerpen/waterveiligheid/programma-projecten/veiligheid-nederland/english/flood-risk-the/.

Vrijling, J. K. (2008). *Flood prevention in the Netherlands and the Safety chain.* Presentation at the English Embassy.

SECTION III

Place-based design and the built environment

CHAPTER 12

Infrastructure impacts and vulnerability to coastal flood events

Jamie E. Padgett, Pranavesh Panakkal, and Catalina González-Dueñas
Department of Civil and Environmental Engineering, Rice University, Houston, TX, United States

Introduction

Coastal infrastructure systems are a vital part of urban and rural development, coastal socio-demographic dynamics, and the global economy. For instance, ports and industrial facilities act as a link in marine and land transportation of goods acting as a major source of employment and an economic catalyst (Dwarakish & Salim, 2015). However, their strategic geographical location also makes them vulnerable to both chronic and punctuated flood-related hazards such as sea-level rise or hurricanes events that threaten infrastructure performance now and in the future. Moreover, while flood inundation alone carries significant implications for damage or loss of functionality of various infrastructure (e.g., housing, power systems, transportation), the multihazard and compound nature of severe storms along the coast further hampers infrastructure performance. For example, multihazards from hurricanes or tropical cyclones, including wind, rain, storm surge and waves, produce complex loading conditions that induce significant damage to structures and infrastructure systems with loss of functionality and other cascading consequences. Flood-related damages to coastal infrastructure can result in threats to public safety and quality of life, particularly given risks to housing and transportation systems used in emergency response (Stearns & Padgett, 2012; Testa, Furtado, & Alipour, 2015); health and environmental impacts, given potential coastal industrial failures leading to spills of hazardous materials such as oil (Bernier, Elliott, Padgett, Kellerman, & Bedient, 2017; Cruz & Krausmann, 2009); and far-reaching economic implications due to disruption to business operations or infrastructure services, such as intermodal transport of goods (Becker et al., 2013; Nair, Avetisyan, & Miller-Hooks, 2010).

Given the importance and potential vulnerability of structures and infrastructure systems to coastal flood events, coastal risk and resilience assessment frameworks have received growing attention in the literature (Kameshwar et al., 2019; Kammouh, Gardoni, & Cimellaro, 2020). These frameworks often rely on inventory models for the built environment along with the understanding of exposure to scenario-based or probabilistic hazards.

The effects of these hazards on infrastructure performance are often assessed through fragility, or vulnerability, models that may vary in the fidelity with respect to uncertainty treatment, performance metric of interest, consideration of single or multihazard loads, or incorporation of cascading effects like debris, to name a few. Depending on the aim of the analysis, risk models may move beyond infrastructure damage or functionality quantification to include such consequences as economic losses or environmental impacts (Bernier et al., 2017; Bernier & Padgett, 2019a). Resilience frameworks increasingly emphasize the value of assessing not only immediate postevent infrastructure performance (vital to emergency response or inspection deployment), but also the long-term functionality and recovery over time (with implications for planning and resilience enhancement interventions) (Balomenos, Hu, Padgett, & Shelton, 2019). Furthermore, the role of infrastructure systems in supporting broader community resilience (Kameshwar et al., 2019; Koliou et al., 2020) and coupled modeling of natural-built-human systems along the coast has received heightened attention in recent years (Ellingwood et al., 2016; Fereshtehnejad et al., 2020).

In the next section of this chapter, international case studies of coastal flood impacts on infrastructure are posed to highlight key considerations in risk assessment, leverage potential comparative analyses, and showcase the results from a series of PIRE place-based research studies. The Port of Rotterdam and the Houston-Ship Channel—two of the most important petrochemical complexes and port regions in the world—are adopted as case studies to analyze the effects of coastal hazards on infrastructure systems. Given the vital role of industrial and transportation infrastructure in supporting broader community resilience in such regions, along with the significant consequences of damage or functionality loss, select industrial and transportation infrastructures are considered for the case studies. Furthermore, this chapter will subsequently highlight future opportunities for philosophical shifts in infrastructure design and management in flood-prone regions. In particular, concepts "smart resilience" and performance-based coastal engineering are explored as promising paradigms.

International case studies of coastal flood impacts on infrastructure

Storage tanks in coastal port and industrial complexes: The Netherlands and The Gulf Coast

Aboveground storage tanks (ASTs) are prevalent in port and industrial complexes often used to store bulk chemicals such as oil and gas, and are among the most vulnerable components responsible for spillage of hazardous materials during coastal flood events. As the largest port in Europe with 127 km^2 of the port area (Port of Rotterdam, 2019), the Port of Rotterdam has over 3000 ASTs located in the port regions of Maasvlakte, Europort, and Botlek. Bernier (2018) explored the influence of multihazard conditions on the AST infrastructure performance, considering the vulnerability of ASTs to flood (as detailed in

Chapter 29) as well as the potential for debris impact. The Botlek region is located inside the area protected by the Maeslant barrier and dikes and includes ASTs with elevations ranging between 0.9 and 4 m above the mean sea level. This relatively low elevation makes the ASTs vulnerable to flood and debris when considering the event of failure of the barrier or overtopping of the dike system (Bernier, 2018). Bernier (2018) leveraged the 10,000-year probabilistic flood maps developed by Deltares (2015) for the years 2050, and 2100 considering sea-level rise to evaluate the vulnerability of the ASTs to debris impact and flotation failure from flooding. Under these conditions, flood levels and velocities between 0.5–2.0 m and 0.5–2.0 m/s, respectively, are expected for the area. Fig. 1 depicts the probability of failure of ASTs (using the parameterized fragility models proposed by Kameshwar and Padgett, 2018) for the years 2050 and 2100, under the 10,000-year design event. Results show that the probability of failure does not surpass 30% due to the relatively low flood-elevations levels (i.e., no more than 2.0 m). Using aerial imagery for the Port of Rotterdam, Bernier (2018) identified cars and shipping containers as the principal sources of debris, and verified its flotation potential (Korswagen, 2016), to evaluate the debris impact risk for the ASTs in the Botlek area. However, preliminary analyses suggested that cars did not inflect significant damage to the ASTs and were neglected in the subsequent assessments. Finite-element models considering imperfection in the tank shell, variation in material properties, internal liquid and external surge loads, as well as hydrodynamic effects, were then used to develop an artificial neural network (ANN) regression model covering different types of geometries and debris properties. The ANN model developed by Bernier (2018) and Bernier and Padgett (2019b) shows a 94% accuracy and was implemented to evaluate the conditional probability of damage of ASTs under shipping container impact. Results show that 146 ASTs and 434 ASTs for the years 2050 and 2100, respectively, are vulnerable to debris impact damage (probability of failure over 50%) in the area. The results of the case study in the Port of Rotterdam underscore the significant impact of cascading failures on infrastructure performance in coastal regions as well as the influence of a changing climate (e.g., flood estimates future time horizons) on infrastructure risks.

By way of comparison, the Houston-Ship Channel (HSC) is the United States Gulf Coast's largest container port (Port Houston, 2019) and the second-largest petrochemical complex in the world (Ebad Sichani et al., 2020). More than 4500 ASTs located in this region, as opposed to Rotterdam, are susceptible to additional multihazard loads due to the occurrence of seasonal hurricane events. Therefore, the vulnerability analysis of ASTs must consider the effects of the concurrent wave, surge, and wind loads, as well as the potential of cascading effects such as debris impact, to accurately estimate its risk. Bernier and Padgett (2019a) developed parameterized buckling and dislocation (for both anchored and unanchored ASTs) fragility models for ASTs under multihazard storm conditions. Buckling of ASTs is of particular importance when high wind or water pressures are expected to occur, while dislocation from the ground is usually driven by storm surge

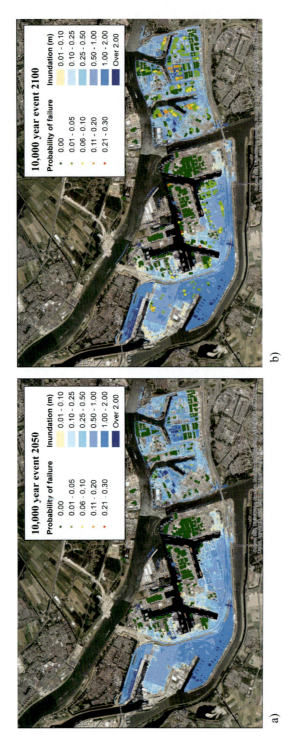

Fig. 1 Probability of failure of above storage tanks (ASTs) in the Port of Rotterdam for the years (A) 2050 and (B) 2100 under the 10,000-year design event (Kameshwar, 2018).

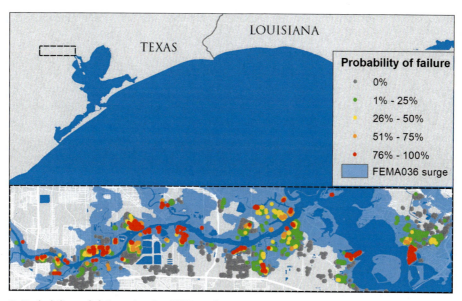

Fig. 2 Probability of failure in the HSC under storm FEMA036 (500-year return period storm) considering multihazard effects on dislocation of unanchored ASTs (Bernier & Padgett, 2019a).

effects and can occur under uplift, sliding, or overturning mechanisms (Bernier & Padgett, 2019a). Both failure modes can lead to a spill of the ASTs content, the former by the breakage of the tank shell caused by large deformations and the latter by the rupture of connecting pipes. As an illustration, Fig. 2 (Bernier & Padgett, 2019a) presents the probabilities of failure for a 500-year return period storm (FEMA036 (Ebersole et al., 2015)) in the HSC considering the unanchored dislocation parameterized fragility models. The ranges of storm parameters (i.e., surge and wave height, wave period, current and wind velocities) can be found in Bernier and Padgett (2019a). The study revealed that while inundation from coastal surge was the primary driver of AST damage, neglecting multihazard loading conditions, particularly associated with hydrodynamic forces and wave load effects during storms, could significantly underestimate the damage and spill risks. Furthermore, it is evident that the failure probability of ASTs is significantly higher for the Houston Ship Channel case study than that for the ASTs in the Rotterdam Port, even when considering a much lower return period event (e.g., 500 vs 10,000 years). This can be attributed primarily to the hazard characteristics of the regions and nature of flood events considered, but also in part to the siting and relative elevations of vulnerable infrastructure.

To explore the effects of waterborne debris impacts on ASTs, Bernier and Padgett (2020) proposed a probabilistic framework to evaluate the vulnerability of ASTs under the impact of shipping containers. Finite-element models were used to develop parameterized fragility models using logistic regression considering the damage to the AST shell and sliding. These fragility models were then used to assess the risk of the impact of a case study storage terminal in the HSC. Results showed that disregarding the effect of debris impact significantly underestimates the probability of damage. Moreover, recent work highlights the increased vulnerability of petrochemical infrastructure to a changing climate by analyzing the effects of sea-level rise and shifts in hurricane forward velocity in the expected economic losses in the HSC (Ebad Sichani et al., 2020). Given the expected changes in climate in the coming years, risk and resilience assessments of ASTs in such regions as the HSC should consider both present and future climate conditions in port and industrial complexes, alongside multihazard loading effects.

Transportation infrastructure: The Netherlands and The Gulf Coast

Flood impact on transportation infrastructure can be either short-term or long-term. In the short-term, inundated roadways and overtopped bridges could cripple emergency response and isolate several regions from access to critical facilities such as fire stations. These isolated areas are at an elevated risk of potential cascading consequences. For example, flood or storm surge damage to industrial facilities could result in fire or release of hazardous materials, such as the AST spill risk concerns from "Storage tanks in coastal port and industrial complexes: The Netherlands and The Gulf Coast" section. If timely access to the failure location is not available for supporting repair, containment, or cleanup efforts, the ensuing cascading consequences could include severe environmental and economic losses. This section presents select case studies on the impact of coastal flooding on transportation infrastructure performance, again leveraging the Port of Rotterdam and Gulf Coast regions.

Panakkal (2019) investigated the flood impact on road transportation accessibility to at-risk industrial facilities, in particular the ASTs located in the Botlek region. This study coupled probabilistic flood maps from Deltares (2015) with road network models to identify flooded roads and connectivity between fire stations and ASTs. The probabilistic flood hazard maps from Deltares considered potential climate change, sea-level rise, and reliability of flood protection systems for return periods ranging from 100 to 30,000 years for base years 2015, 2050, and 2100. Results from Panakkal (2019) (Fig. 3) show a significant increase in the percentage of flooded roads (in terms of road length) due to potential sea-level rise and climate change. For example, for a 10,000-years return period flood event, while only 12% of roads are flooded in 2015, about 40% of roads will be flooded in 2100 due to potential climate change and sea-level rise. The failure of 40% of roads would significantly limit emergency response access to the vulnerable ASTs in the Botlek area identified by Kameshwar (2018) and Bernier (2018). Especially, adjacent residential communities in Rozenburg could be at risk of potential cascading impacts. In addition, considering a possible future increase in flood risk, even a 100-year return period event could cause a failure of 12% of roads in 2100. This study highlights the importance of

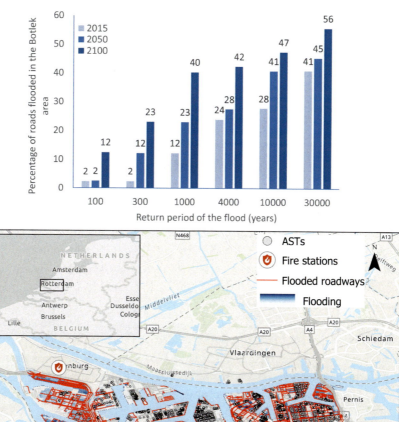

Fig. 3 (A) Percentage of roads flooded in the Botlek area, the Port of Rotterdam, for different return period scenarios and points in time; (B) location of inundated roadways for 10,000-year return period scenario in 2100 (Panakkal, 2019).

considering factors such as climate change, sea-level rise, and land-use patterns while estimating flood risk and flood impact on transportation systems.

In addition to short-term impacts, flood and storm surge could also result in long-term consequences due to structural failure of critical infrastructure components, such as roads and bridges. A majority of bridge failures in the United States are due to hydraulic actions (Cook, Barr, & Halling, 2015). Several studies have also highlighted the vulnerability of bridges to coastal hazards including surges and waves during hurricanes (Padgett et al., 2008; Stearns & Padgett, 2012). Balomenos et al. (2019) proposed a framework to examine the impact of coastal flood events on residents' ability to access health services considering both long-term and short-term impacts. In this study, storm surge models were used to estimate surge conditions at bridges and road links. Bridge fragility functions from Ataei and Padgett (2013) were then used to identify bridge damage in addition to temporary road closures due to the overtopping of bridges and roads. While inundation could result in short-term transportation impact, a bridge failure could significantly impact connectivity for a longer period. Finally, network analyses can estimate both the short-term and long-term impact of flood hazards on residents' access to health care facilities. Balomenos et al. (2019) presented a case study application of this framework for the Harris County region, Houston, Texas, using two synthetic storm scenarios. They discovered that infrastructure vulnerability could significantly impact network performance and spatial accessibility. Further, by coupling spatial accessibility with sociodemographic indicators, Balomenos et al. (2019) noted that vulnerable populations, such as low-income groups or people over the age of 65, are more likely to have limited access to healthcare facilities even after a low-level storm.

Similarly, Bernier, Gidaris, Balomenos, and Padgett (2019) presented a scenario-based framework to assess the accessibility of petrochemical facilities by emergency responders and workers during or after a storm surge event. The framework coupled surge models with fragility models of ASTs to identify impacted locations and the likely time period of potential failures. Storm surge models are then used with bridge fragilities and road network models to determine the short-term and long-term impacts of storm surges. Finally, probabilistic network analysis can estimate the accessibility of impacted facilities by workers or emergency responders. Case study results from Bernier et al. (2019) corroborate Balomenos et al. (2019) findings and highlight the importance of considering the structural vulnerability of critical transportation infrastructure during flooding or storm surge events. Considering only the short-term impact of storm surges could underestimate the extent and duration of network disruptions. Further, Bernier et al. (2019) highlight the need to consider and propagate uncertainties associated with network analysis and infrastructure models to facilitate risk-informed decision-making; performing only deterministic network analysis could result in overly conservative accessibility results or bias decision-making.

Envisioning the future of coastal infrastructure design and management
Advancing performance-based coastal engineering
As highlighted in the abovementioned case studies, coastal settings are susceptible to concurrent and individual hazards that pose major challenges to the design and planning of

infrastructure systems. These challenges are expected to grow when considering the changes in climate, land-use, property value, and demographic shifts. Therefore, strategies to cope with the complexity of coastal settings while reducing the risk associated with existing and future infrastructure exposure to coastal flooding are needed to create truly resilient coastal cities.

Since first proposed, performance-based engineering frameworks have established a comprehensive methodology to evaluate the performance of structures during their service life and provided means to estimate the consequences (usually measured in incurred losses yet not limited to them) of particular designs or retrofitting strategies under specified hazards. This is done by computing the probability of exceedance of a decision variable DV (e.g., economic loss, casualties) using the joint probability distribution of the random variables of the problem at hand which are defined based on six basic steps: (1) performance objectives, (2) hazard analysis, (3) structural characterization, (4) structural analysis, (5) damage analysis, and (6) performance analysis. However, depending on the specific hazard under consideration, more analysis components can be added to the basic methodology. For coastal regions, the performance-based engineering frameworks for wind (Ciampoli, Petrini, & Augusti, 2011), hurricane (Barbato, Petrini, Unnikrishnan, & Ciampoli, 2013), and tsunami hazards (Attary, Unnikrishnan, van de Lindt, Cox, & Barbosa, 2017; Attary, Van De Lindt, Barbosa, Cox, & Unnikrishnan, 2019) posed key advancements in coastal infrastructure planning by introducing the consideration of multihazard and successive analyses, as well as the interaction between the structure and its proximal environment (e.g., fluid-structure interaction, soil-structure interaction). However, the intrinsic dynamic nature of coastal areas, the upcoming climate changes, and the interdependencies between systems necessitates the incorporation of a performance analysis that takes into consideration time-varying factors such as structural degradation over time, shifts in frequency and intensity of hazards, as well as potential cascading effects.

To pave a path for future design or risk management of coastal infrastructure, Gonzalez-Duenas and Padgett (2021) recently proposed a performance-based coastal engineering (PBCE) framework that incorporates both time-varying factors and cascading effects in the performance assessment of individual structures and systems, while still accounting for multihazard scenarios and environment interactions. Moreover, the methodology is constructed based on a dynamic Bayesian network approach, which allows the incorporation of evidence in the model to update the joint probability distribution of the decision variable DV. Such a framework provides flexibility in improving performance and risk estimates as new information becomes available. Furthermore, given that a dynamic Bayesian network is a probabilistic graphical model, its construction not only unveils correlations and interdependencies between factors and systems but also helps to disseminate information to stakeholders and bring together experts from different fields, which is key to tackling modern world problems from an integral point of view. Thus, future work should address the use of the PBCE framework to support resilience assessments of infrastructure systems and coastal communities, its use in adaptation engineering to address future challenges, and the effective use of information to enhance our existing probabilistic risk estimates of coastal systems.

Smart resilience

Beyond PBCE frameworks, which are particularly poised for supporting future coastal infrastructure design, upgrade, and adaptation, additional opportunities exist to harness the data revolution for improving coastal infrastructure performance. Many communities are becoming increasingly smart and interconnected. New campaigns and data collection efforts in modern cities are expected to yield an unprecedented amount of information on features ranging from infrastructure conditions to urban climate to system demands. This poses both challenges and opportunities for enhancing infrastructure resilience and supporting decision-making by organizations affected by or responsible for managing flood risk. A new paradigm of "smart resilience" can pave a path to the design and management of infrastructure in coastal regions in which data from smart systems or technologies is leveraged to enhance the system's ability to adapt, respond, and recover from stressors.

Several technologies can be leveraged to enhance the resilience of critical infrastructure systems in the face of flood events. For example, the Internet of things (IoT) is a collection of interconnected sensors integrating the physical world into the internet (Van Ackere et al., 2019). Physical sensors such as cameras (Lo, Wu, Lin, & Hsu, 2015), water level gauges (Perumal, Sulaiman, & Leong, 2015), smart sewage (Granlund & Brännström, 2012), accelerometers (Chiang, 2018) could provide real-time information on hazard conditions. Further, structural health monitoring devices (Abdelgawad & Yelamarthi, 2016; Mahmud, Bates, Wood, Abdelgawad, & Yelamarthi, 2018; Tokognon, Gao, Tian, & Yan, 2017) could provide real-time data on infrastructure state and performance. In addition to IoT platforms, real-time situational awareness models based on physics-based simulations (Fang, Bedient, & Buzcu-Guven, 2011; Panakkal, Juan, Garcia, Padgett, & Bedient, 2019; SSPEED Center, 2020; University LS, n.d.), as well as social-media analysis (de Bruijn, de Moel, Jongman, Wagemaker, & Aerts, 2018; De Longueville, Smith, & Luraschi, 2009; Jongman, Wagemaker, Romero, & De Perez, 2015) and crowdsourcing data (Liu, Shieh, Ke, & Wang, 2018; See, 2019; Wang, Mao, Wang, Rae, & Shaw, 2018) could provide crucial and timely information on natural hazards, infrastructure state, as well as the community response to disasters. The plethora of data generated by IoT and other sources could vary in temporal and spatial resolution and reliability. These factors necessitate complex data processing and analysis workflows using big data and data fusion methods such as Kalman filters (Chen, 2003). Past studies have shown the successful application of big data (Pollard, Spencer, & Jude, 2018) for assessing and managing flood risk (Sood, Sandhu, Singla, & Chang, 2018), for flood detection (Anbarasan et al., 2020; Cian, Marconcini, & Ceccato, 2018), and for facilitating emergency response (Smith, Liang, James, & Lin, 2017). Furthermore, surrogate models including the use of machine learning to model flood hazards and their interaction with built infrastructure are gaining interest (e.g., Bass & Bedient, 2018; Mobley et al., 2020; Mosavi, Ozturk, & Chau, 2018) and can afford computational efficiency for practical applications of the smart resilience paradigm. Future studies are required to formalize and validate the smart resilience paradigm in various contexts, including but not limited to, infrastructure performance monitoring and risk management, emergency preparedness, disaster response, and recovery.

Conclusions

Coastal infrastructure plays an important role in the safety, economic vitality, and resilience of a community; yet repeated events have highlighted the vulnerability of various systems to coastal storms and flood-related hazards. Both punctuated (e.g., hurricanes) and chronic stressors (e.g., sea-level rise) can hinder short- and long-term performance of systems, such as transportation infrastructure, given inundation as well as physical damage induced by multihazard loading. The case study examples from Houston, Texas, United States as well as Rotterdam, The Netherlands highlight the flood risks to aboveground storage tanks common in port and industrial facilities as well as the risks to transportation infrastructure affecting mobility and access around the regions. The results highlight that concurrent multihazard effects, such as combined wind, surge, and wave, may be particularly important in some regions, such as Houston, which has significant tropical cyclone or hurricane hazards contributing to its flood risk. In both regions, cascading hazards, such as debris effects, were shown to influence infrastructure risk estimates. In fact, such phenomena have received relatively little attention in coastal infrastructure risk and resilience studies and should be more rigorously addressed in future studies. These international case studies reveal that the risk to diverse infrastructure in both regions may be significantly exacerbated by projected sea-level rises in future climate conditions. Opportunities exist to more broadly consider other temporally evolving parameters associated with climate, infrastructure aging and deterioration, or demand shifts.

Practices (e.g., design level events) and policies (e.g., regulations or restrictions on siting of infrastructure in hazard-prone regions) in the two regions differ significantly, with The Netherlands tending toward a practice that places heavy weight on risk mitigation and avoidance, and the United States that balances mitigation with response preparedness. Future paradigms in infrastructure design and management may provide opportunities to build on these philosophies. For example, PBCE frameworks, such as the recent dynamic Bayesian network formulation, can enable risk-based design or mitigation where time-varying factors, cascading effects, multihazard conditions are considered along with possibilities for incorporating new information or evidence in the model. The smart resilience paradigm can promote infrastructure management and adaptation that leverages data sources emerging from smart systems or technologies, such as sensor data, authoritative sources, camera data, or model output. Moving forward, smart technologies offer significant potential for enhancing situational awareness, reducing uncertainty, and supporting decision-making surrounding the resilience of infrastructure exposed to coastal flooding. Simultaneously, they also engender challenges in ensuring privacy (Van Ackere et al., 2019), securing against malicious attacks (Alaba, Othman, Hashem, & Alotaibi, 2017; Zhu, Leung, Shu, & Ngai, 2015), and handling complexity (Jovanović & Vollmer, 2017). Thus, with new promising paradigm shifts, new challenges will continually emerge.

References

Abdelgawad, A., & Yelamarthi, K. (2016). Structural health monitoring: Internet of things application. In *2016 IEEE 59th int. midwest symp. circuits syst., IEEE* (pp. 1–4).

Alaba, F. A., Othman, M., Hashem, I. A. T., & Alotaibi, F. (2017). Internet of things security: A survey. *Journal of Network and Computer Applications, 88*, 10–28.

Anbarasan, M., Muthu, B., Sivaparthipan, C. B., Sundarasekar, R., Kadry, S., Krishnamoorthy, S., et al. (2020). Detection of flood disaster system based on IoT, big data and convolutional deep neural network. *Computer Communications, 150*, 150–157.

Ataei, N., & Padgett, J. E. (2013). Probabilistic modeling of bridge deck unseating during hurricane events. *Journal of Bridge Engineering, 18*, 275–286.

Attary, N., Unnikrishnan, V. U., van de Lindt, J. W., Cox, D. T., & Barbosa, A. R. (2017). Performance-based tsunami engineering methodology for risk assessment of structures. *Engineering Structures, 141*, 676–686. https://doi.org/10.1016/j.engstruct.2017.03.071.

Attary, N., Van De Lindt, J. W., Barbosa, A. R., Cox, D. T., & Unnikrishnan, V. U. (2019). Performance-based tsunami engineering for risk assessment of structures subjected to multi-hazards: Tsunami following earthquake. *Journal of Earthquake Engineering*, 1–20. https://doi.org/10.1080/13632469.2019.1616335.

Balomenos, G. P., Hu, Y., Padgett, J. E., & Shelton, K. (2019). Impact of coastal hazards on residents' spatial accessibility to health services. *Journal of Infrastructure Systems, 25*, 04019028. https://doi.org/10.1061/(ASCE)IS.1943-555X.0000509.

Barbato, M., Petrini, F., Unnikrishnan, V. U., & Ciampoli, M. (2013). Performance-based hurricane engineering (PBHE) framework. *Structural Safety, 45*, 24–35. https://doi.org/10.1016/j.strusafe.2013.07.002.

Bass, B., & Bedient, P. (2018). Surrogate modeling of joint flood risk across coastal watersheds. *Journal of Hydrology, 558*, 159–173.

Becker, A. H., Acciaro, M., Asariotis, R., Cabrera, E., Cretegny, L., Crist, P., et al. (2013). A note on climate change adaptation for seaports: a challenge for global ports, a challenge for global society. *Climatic Change, 120*, 683–695.

Bernier, C. (2018). Impacts of water driven debris impacts on above storage tanks. In B. B. Kothuis, Y. Lee, & S. Brody (Eds.), *NSF-PIRE coast. flood risk reduct. program. Authentic learn. transform. educ. Vol. I—2015-2017* (p. 248).

Bernier, C., Elliott, J. R., Padgett, J. E., Kellerman, F., & Bedient, P. B. (2017). Evolution of social vulnerability and risks of chemical spills during storm surge along the Houston ship channel. *Natural Hazards Review, 18*, 1–14. https://doi.org/10.1061/(ASCE)NH.1527-6996.0000252.

Bernier, C., Gidaris, I., Balomenos, G. P., & Padgett, J. E. (2019). Assessing the accessibility of petrochemical facilities during storm surge events. *Reliability Engineering and System Safety, 188*, 155–167.

Bernier, C., & Padgett, J. E. (2019a). Fragility and risk assessment of aboveground storage tanks subjected to concurrent surge, wave, and wind loads. *Reliability Engineering and System Safety, 191*. https://doi.org/10.1016/j.ress.2019.106571, 106571.

Bernier, C., & Padgett, J. E. (2019b). *Neural networks for estimating storm surge loads on storage tanks*.

Bernier, C., & Padgett, J. E. (2020). Probabilistic assessment of storage tanks subjected to waterborne debris impacts during storm events. J Waterw Port, Coastal. *Ocean Engineering, 146*, 4020003.

Chen, Z. (2003). Bayesian filtering: From Kalman filters to particle filters, and beyond. *Statistics (Ber), 182*, 1–69.

Chiang, C.-T. (2018). Design of a CMOS MEMS accelerometer used in IoT devices for seismic detection. *IEEE Journal on Emerging and Selected Topics in Circuits and Systems, 8*, 566–577.

Ciampoli, M., Petrini, F., & Augusti, G. (2011). Performance-based wind engineering: Towards a general procedure. *Structural Safety, 33*, 367–378. https://doi.org/10.1016/j.strusafe.2011.07.001.

Cian, F., Marconcini, M., & Ceccato, P. (2018). Normalized difference flood index for rapid flood mapping: Taking advantage of EO big data. *Remote Sensing of Environment, 209*, 712–730.

Cook, W., Barr, P. J., & Halling, M. W. (2015). Bridge failure rate. *Journal of Performance of Constructed Facilities, 29*, 4014080.

Cruz, A. M., & Krausmann, E. (2009). Hazardous-materials releases from offshore oil and gas facilities and emergency response following Hurricanes Katrina and Rita. *Journal of Loss Prevention in the Process Industries, 22*, 59–65.

de Bruijn, J. A., de Moel, H., Jongman, B., Wagemaker, J., & Aerts, J. C. J. H. (2018). TAGGS: grouping tweets to improve global geoparsing for disaster response. *Journal of Geovisualization and Spatial Analysis, 2*, 2.

De Longueville, B., Smith, R. S., & Luraschi, G. (2009). "OMG, from here, I can see the flames!" a use case of mining location based social networks to acquire spatio-temporal data on forest fires. In *Proc. 2009 int. work. locat. based soc. networks* (pp. 73–80).

Deltares. (2015). *Potentiële inundatie botlek*.

Dwarakish, G. S., & Salim, A. M. (2015). Review on the role of ports in the development of a nation. *Aquatic Procedia, 4*, 295–301. https://doi.org/10.1016/j.aqpro.2015.02.040.

Ebad Sichani, M., Anarde, K. A., Capshaw, K. M., Padgett, J. E., Meidl, R. A., Hassanzadeh, P., et al. (2020). Hurricane risk assessment of petroleum infrastructure in a changing climate. Front. *Built Environment, 6*, 104.

Ebersole, B. A., Massey, T. C., Melby, J. A., Nadal-Caraballo, N. C., Hendon, D. L., Richardson, T. W., et al. (2015). *Interim report—Ike Dike concept for reducing hurricane storm surge in the Houston-Galveston region*. Jackson, MS: Jackson State Univ.

Ellingwood, B. R., Cutler, H., Gardoni, P., Peacock, W. G., van de Lindt, J. W., & Wang, N. (2016). The centerville virtual community: A fully integrated decision model of interacting physical and social infrastructure systems. *Sustainable and Resilient Infrastructure, 1*, 95–107.

Fang, Z., Bedient, P. B., & Buzcu-Guven, B. (2011). Long-term performance of a flood alert system and upgrade to FAS3: A Houston, Texas, case study. *Journal of Hydrologic Engineering, 16*, 818–828.

Fereshtehnejad, E., Gidaris, I., Rosenheim, N., Tomiczek, T., Padgett, J. E., Cox, D. T., et al. (2020). *Probabilistic risk assessment of coupled natural-physical-social systems: the cascading impact of hurricane-induced damages to civil infrastructure in Galveston, Texas*. In Review.

González-Dueñas, C., & Padgett, J. E. (2021). Performance-based coastal engineering framework. *Frontiers in Built Environment*, 100.

Granlund, D., & Brännström, R. (2012). Smart city: the smart sewerage. In *37th annu. IEEE conf. local comput. networks-workshops, IEEE* (pp. 856–859).

Jongman, B., Wagemaker, J., Romero, B. R., & De Perez, E. C. (2015). Early flood detection for rapid humanitarian response: harnessing near real-time satellite and Twitter signals. *ISPRS International Journal of Geo-Information, 4*, 2246–2266.

Jovanović, A., & Vollmer, M. (2017). Do smart technologies improve resilience of critical infrastructures? *Critical Infrastructure Protection Review, 37*.

Kameshwar, S. (2018). Understanding vulnerability and acceptable flood risk to storage tanks. In B. B. Kothuis, Y. Lee, & S. Brody (Eds.), *NSF-PIRE coast. flood risk reduct. program. authentic learn. transform. educ. Vol. I-2015-2017* (p. 248).

Kameshwar, S., Cox, D. T., Barbosa, A. R., Farokhnia, K., Park, H., Alam, M. S., et al. (2019). Probabilistic decision-support framework for community resilience: Incorporating multi-hazards, infrastructure interdependencies, and resilience goals in a Bayesian network. *Reliability Engineering and System Safety*, *191*, 106568.

Kameshwar, S., & Padgett, J. E. (2018). Storm surge fragility assessment of above ground storage tanks. *Structural Safety*, *70*, 48–58.

Kammouh, O., Gardoni, P., & Cimellaro, G. P. (2020). Probabilistic framework to evaluate the resilience of engineering systems using Bayesian and dynamic Bayesian networks. *Reliability Engineering and System Safety*, *198*. https://doi.org/10.1016/j.ress.2020.106813, 106813.

Koliou, M., van de Lindt, J. W., McAllister, T. P., Ellingwood, B. R., Dillard, M., & Cutler, H. (2020). State of the research in community resilience: Progress and challenges. *Sustainable and Resilient Infrastructure*, *5*, 131–151.

Korswagen, P. A. (2016). *Structural damage to masonry housing due to earthquake-flood multi-hazards*. TU Delft.

Liu, C.-C., Shieh, M.-C., Ke, M.-S., & Wang, K.-H. (2018). Flood prevention and emergency response system powered by google earth engine. *Remote Sensing*, *10*, 1283.

Lo, S.-W., Wu, J.-H., Lin, F.-P., & Hsu, C.-H. (2015). Visual sensing for urban flood monitoring. *Sensors*, *15*, 20006–20029.

Mahmud, M. A., Bates, K., Wood, T., Abdelgawad, A., & Yelamarthi, K. (2018). A complete internet of things (IoT) platform for structural health monitoring (shm). In *2018 IEEE 4th world forum internet things, IEEE* (pp. 275–279).

Mobley, W., Sebastian, A., Blessing, R., Highfield, W. E., Stearns, L., & Brody, S. D. (2020). Quantification of continuous flood hazard using random forest classification and flood insurance claims at large spatial scales: A pilot study in southeast Texas. *Natural Hazards and Earth System Sciences Discuss*, 1–22.

Mosavi, A., Ozturk, P., & Chau, K. (2018). Flood prediction using machine learning models: Literature review. *Water*, *10*, 1536.

Nair, R., Avetisyan, H., & Miller-Hooks, E. (2010). Resilience framework for ports and other intermodal components. *Transportation Research Record*, *2166*, 54–65.

Padgett, J., Desroches, R., Nielson, B., Yashinsky, M., Kwon, O. S., Burdette, N., et al. (2008). Bridge damage and repair costs from Hurricane Katrina. *Journal of Bridge Engineering*, *13*, 6–14. https://doi.org/10.1061/(ASCE)1084-0702(2008)13:1(6).

Panakkal, P. (2019). *Flood impacts on emergency response accessibility and mobility in the Port of Rotterdam*. PIRE Coastal Flood Risk Reduction Program:.

Panakkal, P., Juan, A., Garcia, M., Padgett, J. E., & Bedient, P. (2019). Towards enhanced response: Integration of a flood alert system with road infrastructure performance models. In *Struct. congr. 2019 build. nat. disasters, American Society of Civil Engineers Reston, VA* (pp. 294–305).

Perumal, T., Sulaiman, M. N., & Leong, C. Y. (2015). Internet of things (IoT) enabled water monitoring system. In *2015 IEEE 4th glob. conf. consum. electron., IEEE* (pp. 86–87).

Pollard, J. A., Spencer, T., & Jude, S. (2018). Big data approaches for coastal flood risk assessment and emergency response. *Wiley Interdisciplinary Reviews: Climate Change*, *9*, e543.

Port Houston. (2019). *Statistics*. https://porthouston.com/about-us/statistics/. (Accessed 15 November 2020).

Port of Rotterdam. (2019). *The smartest port in the world*. https://www.portofrotterdam.com/sites/default/files/the-worlds-smartest-port-port-of-rotterdam-publieksfolder-2019-en.pdf. (Accessed 15 November 2020).

See, L. (2019). A review of citizen science and crowdsourcing in applications of pluvial flooding. *Front. Earth Science*, *7*, 44.

Smith, L., Liang, Q., James, P., & Lin, W. (2017). Assessing the utility of social media as a data source for flood risk management using a real-time modelling framework. *Journal of Flood Risk Management*, *10*, 370–380.

Sood, S. K., Sandhu, R., Singla, K., & Chang, V. (2018). IoT, big data and HPC based smart flood management framework. *Sustainable Computing: Informatics and Systems*, *20*, 102–117.

SSPEED Center. (2020). *FAS5-TMC rice flood alert system*. Rice Univ http://fas5.org/home.html. (Accessed 15 November 2020).

Stearns, M., & Padgett, J. E. (2012). Impact of 2008 hurricane Ike on bridge infrastructure in the Houston/Galveston region. *Journal of Performance of Constructed Facilities, 26*, 441–452.

Testa, A. C., Furtado, M. N., & Alipour, A. (2015). Resilience of coastal transportation networks faced with extreme climatic events. *Transportation Research Record, 2532*, 29–36.

Tokognon, C. A., Gao, B., Tian, G. Y., & Yan, Y. (2017). Structural health monitoring framework based on internet of things: A survey. *IEEE Internet of Things Journal, 4*, 619–635.

University LS. CERA-coastal emergency risk assessment n.d. https://cera.coastalrisk.live/ (Accessed 15 November 2020).

Van Ackere, S., Verbeurgt, J., De Sloover, L., Gautama, S., De Wulf, A., & De Maeyer, P. (2019). A review of the internet of floods: Near real-time detection of a flood event and its impact. *Water, 11*, 2275.

Wang, R.-Q., Mao, H., Wang, Y., Rae, C., & Shaw, W. (2018). Hyper-resolution monitoring of urban flooding with social media and crowdsourcing data. *Computational Geosciences, 111*, 139–147.

Zhu, C., Leung, V. C. M., Shu, L., & Ngai, E. C.-H. (2015). Green internet of things for smart world. *IEEE Access, 3*, 2151–2162.

CHAPTER 13

Understanding the impacts of the built environment on flood loss

Samuel Brody[a,b], Wesley E. Highfield[a,b], and Russell Blessing[a]
[a]Institute for a Disaster Resilient Texas, Texas A&M University, College Station, TX, United States
[b]Department of Marine and Coastal Environmental Science, Texas A&M University, Galveston Campus, Galveston, TX, United States

Increasing flood impacts are often attributed solely to changing precipitation patterns, sea level rise, and other types of environmental changes. However, a long-standing body of evidence suggests that environmental change and climate change are only partially to blame for exponential increases in losses from storm events, especially across the United States. Instead, oftentimes quicker and more powerful driver of increasing flood risk comes in the form of the human-built environment and resulting development decisions.

Several factors contribute to the role of the built environment in exacerbating or even creating flood losses. First, as population and associated development continue to expand, especially in coastal regions, more people, structures, and economic assets are being placed in flood-prone areas. Second, the expanding built environment (e.g., structures, roads, rooftops, public infrastructure, pavement, etc.) is exacerbating the problem by compromising the ability of the natural landscape to store and absorb floodwaters (Hemmati, Ellingwood, & Mahmoud, 2020). Impervious surfaces associated with new development increases the volume and velocity of stormwater runoff, causing unintended flood impacts for downstream communities. Third, engineered structures, such as sound walls, railroads, highways, and bridges, can become obstacles that back up overland flow and cause flooding when the capacity of stormwater infrastructure is overwhelmed by episodes of heavy rainfall. Lastly, aging or inadequate stormwater infrastructure and inconsistent building practices further compound the issue increasing flood risk and losses experienced across local communities.

Putting more people in harm's way

While development is, in many cases, a sign of healthy local economies, rising population in flood-prone areas, such as the upper Texas coast, is one of the largest contributors to increased flood losses. Simply put, population increase in low-lying coastal plains subject to both tidal and rainfall flood events put more people in harm's way. The dangerous confluence of people and flood hazards is nowhere more apparent than in the Texas coast.

This region has long been one of the nation's fastest growing areas. From 1980 to 2003, for example, approximately 2.5 million people moved into coastal communities, boosting the population by 52%. In the same period, Harris County ranked second among all US coastal counties for net population increase (National Oceanic and Atmospheric Administration, 2005). This trend continued into more recent decades. For example, from 2010 to 2016, the state's coastal population rose by nearly 900,000, due in large part to a strong economy and affordable housing (Texas State Data Center, 2017). More concerning is the propensity for development (mostly residential) to occur in the most flood-prone locations. Of the 300,000 acres of new development taking place in and around the Houston metroplex from 1996 to 2010, 12% were within the FEMA-defined 100-year floodplain. The fastest growing zip codes that experienced the most growth added more than 10,000 additional acres of development between 1996 and 2010. Most of these high-growth zip codes occurred around the City of Houston, creating one of the nation's largest connected expanses of impervious surface (Fig. 1).

Even in the wake of Hurricane Harvey in 2017, which devastated the Houston region in what was the largest urban flood event in United States history, new development is still taking place on the most flood-prone parcels of land. In fact, one in five new homes permitted in Houston in the year after the 2017 storm were located in a FEMA-designated floodplain. City officials also approved 260 plats in the floodplain during the same time period, setting the table for more residential development in risky areas (from https://www.houstonchronicle.com/news/houston-texas/houston/article/Even-after-Harvey-Houston-keeps-adding-new-homes-13285865.php#photo-16286110). A recent analysis estimated that nearly 750,000 additional acres are likely to be developed in the Houston-Galveston area by 2050. Moreover, the amount of development in the floodplain could nearly double from the 2001 level, the same year Tropical Storm Allison dumped on Houston what at the time was record rainfall (Texas A&M University System and Agencies, Texas A&M University, 2018).

Placing more people and structures in locations most likely to flood translates into exponentially greater economic losses. For example, another major storm striking the region, this time bringing catastrophic storm-surge-based flooding to communities surrounding Galveston Bay was Hurricane Ike in 2008. At the time, this event resulted in approximately $3 billion in residential damage. Results of a spatial analysis predicting new development shows that this same storm would inflict almost $5 billion if it occurs in 2080, a 60% increase achieved by simply adding more residents in a business-as-usual development scenario (Atoba, Brody, Highfield, & Merrell, 2018).

Spread of impervious surfaces

Heightened flood risk occurs from not just adding people and structures in exposed areas, but the impervious surface that accompanies these changes. Development, the conversion of open space to impervious surfaces (typically represented by roads, rooftops,

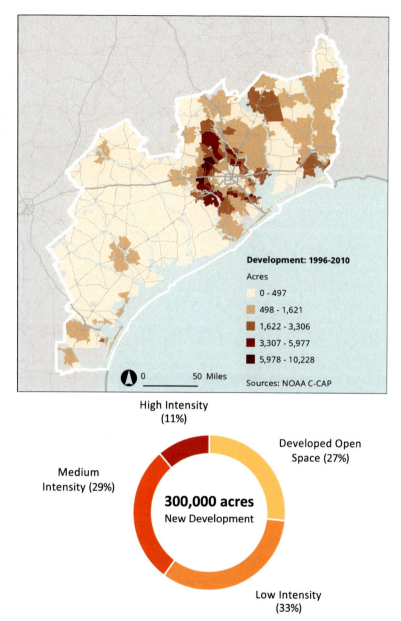

Fig. 1 Increase in development by Zip code and type of development. *(Source: Texas A&M University System and Agencies, Texas A&M University. (2018). Eye of the storm. Report of the governor's commission to rebuild Texas.)*

parking lots, and other paved surfaces), and the ability of public infrastructure to accommodate subsequent runoff is widely considered a major contributor to flood impacts. Land use and land cover (LULC) change is often linked to increased surface runoff, flooding, and associated property damage, particularly in relation to the built environment (Brody, Gunn, Highfield, & Peacock, 2011; Gori, Blessing, Juan, Brody, & Bedient, 2019). Urbanization and the proliferation of impervious surfaces across watershed units have long been considered a major contributor to adverse impacts associated with flood events (Hall, 1984).

Conversion of natural landscapes to urban or suburban developments can reduce the functionality of hydrological systems leading to reduced soil infiltration and increased surface runoff and peak discharge in nearby streams (Gill, Handley, Ennos, & Pauleit, 2007; O'Driscoll, Clinton, Jefferson, Manda, & McMillan, 2010; Paul & Meyer, 2001; Wheater & Evans, 2009). Arnold and Gibbons (1996), for example, found that stormwater runoff within a drainage basin will nearly double with only a 10%–20% increase in impervious surfaces. Because the lag time between the center of precipitation volume and runoff volume is shortened when paved surfaced become ubiquitous, floods peak more rapidly (Brezonik & Stadelman, 2002; Hey, 2002; Hirsch, Walker, Day, & Kallio, 1990; Hsu, Chen, & Chang, 2000). For example, Rose and Peters (2001) found that peak discharge increases approximately 80% in urban catchments containing greater than 50% impervious area.

Numerous studies of urbanization effects on stream runoff have also focused on rapidly expanding urban areas subject to both rainfall and tidal flooding, such Houston, TX (O'Driscoll et al., 2010). Increased stream stage and discharge variability are common responses to urbanization. Generally, urban streams behave in a flashier fashion than their rural counterparts and have a greater occurrence of extreme flow events (flows that are three or more times larger than median flows). Urbanization at a watershed scale consistently increases runoff ratios (precipitation/streamflow) and peak flows. For example, Liscum (2001) showed that in the northern part of Houston, urban development was responsible for a 40% increase in direct runoff and a 235% increase in peak yield.

Overall, there are a host of simulated studies (mostly within a single watershed unit) suggesting that increased surface runoff and resulting peak discharge from impervious surface coverage is a major driver of flood risk because it can translate into heightened frequency and severity of flooding in vulnerable areas. Until recently, little work, however, has been done to statistically relate this process to actual flood losses. Brody, Highfield, Ryu, and Spanel-Weber (2007) found that an increase in the percentage of impervious surfaces coincided with a significant increase in streamflow exceedances over a 12-year period across 85 coastal watersheds in Texas and Florida. A subsequent study of 37 coastal counties in Texas found that, on average, each square meter of impervious surface added to the landscape translated into approximately $3602 of added property damage caused by floods per year from 1997 to 2001 (Brody, Zahran, Highfield,

Grover, & Vedlitz, 2008). Houston area development, especially upstream, increase the volume of water entering downstream neighborhoods and expands the 100-year floodplain, putting more structures at risk (Gori et al., 2019; Juan, Gori, & Sebastian, 2020).

Recent observational research also indicates that flood impacts are driven not solely by the amount of impervious surface, but also by its pattern and intensity across a given landscape. That is, the specific form of the built environment is the more important trigger for flood losses over time. In general, it appears that older, denser urban development is generally located outside of areas exposed to flooding (such as the floodplain), and that high-density urban areas may be more likely to have a coordinated system of flood mitigation infrastructure to handle large amounts of runoff. In contrast, large amounts of sparsely developed areas consistent with "sprawl" actually exacerbate property damage from flooding. In this scenario, outwardly expanding, low-density development patterns can fragment hydrological systems and amplify surface runoff by spreading out impervious surfaces over a larger area (Brody, Blessing, Sebastian, & Bedient, 2014). This form of development is also more likely to infringe upon flood-prone areas that were previously left untouched. Brody et al. (2014) corroborated this finding in a parcel-level study conducted south of Houston, where a percentage increase in surrounding low-intensity development translates into, on average, approximately $1734 in additional property damage caused by floods. They found that low-density development patterns can compromise hydrological systems and amplify surface runoff by spreading out impervious surfaces across a watershed, placing more structures and residents at risk from flooding over a larger area.

Built environment as obstacles

Rapid, sprawling development that does not account for regional drainage patterns or downstream impacts can actually create flood hazards where they previously did not exist. Development-induced flood impacts are especially prominent during large-scale, widespread events that repetitively occur in the Houston-Galveston region.

During major rainfall events, features of the built environment, such as bridges, sound walls, railroad crossings, and even fences, become barriers that impede both overland and channel flow. These unintended obstacles can act as dams, backing water up into resident's homes, even if they are miles outside the known floodplain boundaries. Sound walls acting as a buffer between residential neighborhoods and highways were especially implicated during Hurricane Harvey in 2017 when many of these structures blocked the path of stormwater runoff and exacerbated flood losses in neighborhoods bordering major transportation corridors (Sypniewski, 2019).

Features of the built environment can also exacerbate the effects of storm-surge events. Seawalls, levees, and groins meant to protect certain neighborhoods can actually deflect wave-based inundation into neighboring properties. Weakly built structures (especially homes) can also become projectiles when unmoored or swept off their foundations, destroying adjacent buildings. Hurricane Ike in 2008, for example, scraped clean part of the upper Texas coast with a 15-ft storm surge, turning many beach-front homes into floating missiles targeting structures further inland.

Inadequate and aging infrastructure

In addition to alterations of the natural landscape, the infrastructure put in place to reduce flood risk and protect communities from flood impacts is often both inadequate and outdated. Since the 1950s, Houston has relied on a conveyance approach to deal with stormwater runoff. Bayous have been modified to hold and move runoff downstream. A mixture of storm drains connected to pipes and open ditch drainage are meant to carry rainfall to the Bayous. The city has also built two large reservoirs (Addicks and Barker) to the west meant to collect and temporarily store runoff during major rainfall events. In terms of storm-surge protection (a major surge event occurs approximately every 15 years in Galveston Bay), the region relies on localized seawalls or levees protecting specific neighborhoods or facilities.

In recent years, there has been a large amount of criticism of the way the Houston-Galveston region is addressing infrastructure to protect communities against both rainfall and storm-surge-based flooding. The Houston street drainage system, for example, was originally built to handle a 2-year design storm, or approximately 1.5 in of rain in 24 h, which is far below the standard of most cities in the United States. The two reservoirs, built in the 1940s to protect downtown Houston and its Ship Channel, have been designated as "high risk" by the USACE due to the potential losses that could occur if they failed. Residential development to the west of and inside of these reservoir pools is putting additional strain on the already outdated and poorly maintained structures. Finally, despite being extremely vulnerable to the adverse effects of storm surge, the Houston-Galveston region remains largely unprotected and unprepared for another major event. After Hurricane Ike in 2008 inundated much of Galveston Island and coastal communities adjacent to the bay, a Dutch-inspired coastal spine protective system was proposed, studied, and now being considered by the USACE for its effectiveness in providing region-wide protection as opposed to isolated areas (Atoba et al., 2018; Merrell, this volume Chapter 2; Jonkman, this volume Chapter 22). Regardless of the type of project being built, maintaining and updating flood infrastructure has always been problematic for the Houston region, in particular, and the United States, in general.

Looking to the Dutch for solutions

No country has addressed flooding from both storm surge and rainfall more than the Netherlands. As is consistently emphasized throughout this book, Dutch experiences and lessons learned (both positive and negative) offer important insights for increasing flood resiliency in the Houston-Galveston region. It is important to note the differences in built environment trends between the two study areas—the most obvious being growth trajectories. Houston alone is expected to increase its population by over 37% by 2040. By comparison, Amsterdam is predicted to grow only 10% during the same time period (Kim & Newman, 2019). The Houston region certainly has more physical room to grow, but also a much stronger development mindset supported by a "laissez-faire" approach favoring minimal regulation of land use and other decisions about where development can occur.

Both the Houston-Galveston region and the Netherlands, however, have the opportunity to steer development away from critical areas, protect land to support natural functions, and design the built environment to reduce the adverse effect of stormwater runoff. In this sense, projects and policies in the Netherlands provide insights for decision makers on the Upper Texas Coast.

Avoidance strategies that entail removing development or steering it away from the most vulnerable areas are considered key components of reducing flood impacts. In the United States, this technique is almost always used in response to a flood event in the form of property "buyouts" and relocation scenarios. Harris County, TX has the highest number of flood buyouts in the nation. Since 1997, the Harris County Flood Control District (HCFCD) has spent nearly $340 million, mostly in federal funds, to acquire more than 2500 properties after a major flood event has occurred (Patterson, 2018).

In contrast, the Netherlands takes a more proactive, systematic approach to removing development to support natural riverine function before a major storm event. In the now iconic project called "Room for the River," the Dutch removed human-built barriers and structures at more than 30 locations around the country to provide more space for river to expand in upstream areas, thereby reducing downstream pulses. Approximately 20% of the program's $2.6 billion budget has gone toward buying out and relocating 200 households in high-risk areas. Rather than an ad hoc approach of targeting repetitively impacted properties by the United States, the Dutch are targeting flood-vulnerable areas and removing entire neighborhoods through a consensus-building process, and then amplifying the natural capability of river corridors to accommodate increasing stormwater runoff (Zevenbergen et al., 2015). The major difference with the Dutch approach is that it does not just remove buildings from harm's way, but also uses natural features to enhance floodwater storage and reduce impacts downstream.

In addition to protecting open spaces to help natural systems function in a way that proactively reduces flood losses, the Netherlands is leading the world with their efforts to integrate flood storage into their existing urban landscapes. The Dutch uses a three-step approach to water management that involves capturing, storing, and then draining stormwater runoff (Dai, Wörner, & van Rijswick, 2018). This approach is increasingly being applied to the flood-prone built environment. For example, the city of Rotterdam has built 10 dual-purpose water plazas in recent years (with 5 more planned by 2023) that serve as recreational amenities during dry weather and flood detention facilities during rain events. The most well known of these facilities is the Benthemplein water plaza, a sunken sports court offering recreation facilities to students that can also store nearly 480,000 gallons of water channeled from a parking garage and adjacent rooftops. Multiple use projects generating a diversity of values are another mainstay of Dutch flood risk-reduction initiatives. In contrast, the vast majority of detention ponds in the United States are designed for single use only and are located in inaccessible or hidden areas within a subdivision.

Green-roof programs are another example of how the Dutch are leading the way when it comes to integrating the built environment with natural flood risk reduction strategies. Rotterdam, for example, now has more than 100 acres of rooftop rain-absorbing vegetation; its green-roof initiative has increased the city's water storage capacity by at least 1.6 million gallons. Furthermore, the Rotterdam Roofscape program aims to build 1 million m^2 of multifunctional roofs by 2030.

A final lesson learned from the Dutch is their culture of flood infrastructure maintenance, where projects are designed and funded for regular upkeep to ensure they remain operational over the long term (Jonkman, Voortman, Klerk, & van Vuren, 2018). In most cases, a percentage of the initial project funds is dedicated for long-term maintenance (Jorissen, Kraaij, & Tromp, 2016). This commitment to not just building, but maintaining flood management initiatives helps avoid the current problem in the United States as outdated infrastructure more likely to fail during storm events.

Summary and conclusions

This chapter outlines and describes the influence of the human-built environment on flood impacts along four categories: (1) placement of structures in vulnerable areas, (2) the conversion of open space to impervious surfaces, (3) development as obstacles, and (4) aging and outdated infrastructure systems. Together, these factors play a critical role in exacerbating observed flood losses in the United States, and in particular the upper Texas coast. We look to flood management approaches in Netherlands for lessons learned on how developing areas in the United States can more effectively mitigate increasing losses and pursue a more resilient approach to long-term management.

Based on the existing Dutch principles and practices, we recommend the following approaches be adopted to address increase flood losses in the Houston-Galveston region. First, more proactive and systematic implementation of avoidance strategies would better protect the natural functions of bayou systems. This policy shift would entail, for example, identifying and acquiring vacant land parcels with existing ecological value before they are built upon and/or experience flood loss (as is the current practice) (*reference to other chapters in this volume*). Second, multifunctional stormwater infrastructure projects embedded within existing developed urban areas would help reduce local ponding and at the same time add multiple values to individual projects. Green and gray infrastructure projects, such as urban plazas, green roofs, bioretention systems, etc., have been shown to be effective ways to augment existing stormwater systems that alone cannot handle increasing runoff from upstream developments. Third, making a long-term commitment to maintain and update flood infrastructure would ensure continued operational effectiveness, especially as local conditions change over time. Upfront financial allocations and planning procedures that ensure projects are monitored, maintained, and improved on a regular basis are essential for avoiding chronic and acute failures during storm events.

References

Arnold, C. L., & Gibbons, J. C. (1996). Impervious surface coverage: the emergence of a key environmental indicator. *Journal of the American Planning Association, 62*(2), 243–258.

Atoba, K. O., Brody, S. D., Highfield, W. E., & Merrell, W. J. (2018). Estimating residential property loss reduction from a proposed coastal barrier system in the Houston-Galveston region. *Natural Hazards Review, 19*(3), 05018006.

Brezonik, P. L., & Stadelman, T. H. (2002). Analysis and predictive models of stormwater runoff volumes, loads and pollutant concentrations from watersheds in the twin cities metropolitan area, Minnesota, USA. *Water Resources, 36*(7), 1743–1757.

Brody, S. D., Blessing, R., Sebastian, A., & Bedient, P. (2014). Examining the impact of land use/land cover characteristics on flood losses. *Journal of Environmental Planning and Management, 57*(9), 1252–1265.

Brody, S. D., Gunn, J., Highfield, W. E., & Peacock, W. G. (2011). Examining the influence of development patterns on flood damages along the Gulf of Mexico. *Journal of Planning Education and Research, 31*(4), 438–448.

Brody, S. D., Highfield, W. E., Ryu, H. C., & Spanel-Weber, L. (2007). Examining the relationship between wetland alteration and watershed flooding in Texas and Florida. *Natural Hazards, 40*(2), 413–428.

Brody, S. D., Zahran, S., Highfield, W. E., Grover, H., & Vedlitz, A. (2008). Identifying the impact of the built environment on flood damage in Texas. *Disasters, 32*(1), 1–18.

Dai, L., Wörner, R., & van Rijswick, H. F. (2018). Rainproof cities in the Netherlands: Approaches in Dutch water governance to climate-adaptive urban planning. *International Journal of Water Resources Development, 34*(4), 652–674.

Gill, S. E., Handley, J. F., Ennos, A. R., & Pauleit, S. (2007). Adapting cities for climate change: The role of the green infrastructure. *Built Environment, 33*(1), 115–133.

Gori, A., Blessing, R., Juan, A., Brody, S., & Bedient, P. (2019). Characterizing urbanization impacts on floodplain through integrated land use, hydrologic, and hydraulic modeling. *Journal of Hydrology, 568*, 82–95. https://doi.org/10.1016/j.jhydrol.2018.10.053.

Hall, M. J. (1984). *Urban hydrology*. London, England: Elsevier.

Hemmati, M., Ellingwood, B. R., & Mahmoud, H. N. (2020). The role of urban growth in resilience of communities under flood risk. *Earth's Future, 8*. https://doi.org/10.1029/2019EF001382. e2019EF001382.

Hey, D. L. (2002). Modern drainage design: the pros, the cons, and the future. *Hydrological Science and Technology, 18*(14), 89–99.

Hirsch, R. M., Walker, J. F., Day, J. C., & Kallio, R. (1990). The influence of main on hydrological systems. In M. G. Wolman, & H. C. Riggs (Eds.), *Vol. 1. Surface water hydrology* (pp. 329–359). Boulder, CO: Geological Society of America.

Hsu, M. H., Chen, S. H., & Chang, T. J. (2000). Inundation simulation for urban drainage basin with storm sewer system. *Journal of Hydrology, 234*(1-2), 21–37.

Jonkman, S. N., Voortman, H. G., Klerk, W. J., & van Vuren, S. (2018). Developments in the management of flood defences and hydraulic infrastructure in the Netherlands. *Structure and Infrastructure Engineering*, *14*(7), 895–910.

Jorissen, R., Kraaij, E., & Tromp, E. (2016). Dutch flood protection policy and measures based on risk assessment. In *E3S Web of Conferences (Vol. 7, p. 20016)*EDP Sciences.

Juan, A., Gori, A., & Sebastian, A. (2020). Comparing floodplain evolution in channelized and unchannelized urban watersheds in Houston, Texas. *Journal of Flood Risk Management*, *13*(2), e12604.

Kim, Y., & Newman, G. (2019). Climate change preparedness: Comparing future urban growth and flood risk in Amsterdam and Houston. *Sustainability*, *11*(4), 1048.

Liscum, F. (2001). *Effects of urban development on stormwater runoff characteristics for the Houston, Texas, metropolitan area (Vol. 1, No. 4071)*. US Department of the Interior, US Geological Survey.

National Oceanic and Atmospheric Administration. (2005). In K. M. Crosset, et al. (Eds.), *Population trends along the Coastal United States: 1980–2008*. pp. 3, 6. https://aamboceanservice.blob.core.windows.net/oceanservice-prod/programs/mb/pdfs/coastal_pop_trends_complete.pdf.

O'Driscoll, M., Clinton, S., Jefferson, A., Manda, A., & McMillan, S. (2010). Urbanization effects on watershed hydrology and in-stream processes in the southern United States. *Water*, *2*(3), 605–648.

Patterson, G. (2018). *Case studies in floodplain buyouts: Looking to best practices to drive the conversation in Harris County*. Houston, TX: February Rice University Kinder Institute for Urban Research.

Paul, M. J., & Meyer, J. L. (2001). Streams in the urban landscape. *Annual Review of Ecological Systems*, *32*, 333–365.

Rose, S., & Peters, N. (2001). Effects of urbanization on streamflow in the Atlanta area (Georgia, USA): A comparative hydrological approach. *Hydrological Proceedings*, *15*, 1441–1457.

Sypniewski, J. R. (2019). *Effects of sound walls on urban flooding*. Case Study: Hurricane Harvey (Doctoral dissertation).

Texas A&M University System and Agencies, Texas A&M University. (2018). *Eye of the storm. Report of the governor's commission to rebuild Texas*.

Texas State Data Center. (August 2017). *Texas has nation's largest annual state population growth: Births and migration push population to nearly 28 million*. https://www.census.gov/library/stories/2017/08/texas-population-trends.html.

Wheater, H., & Evans, E. (2009). Land use, water management, and future flood risk. *Land Use Policy*, *26S*, S251–S264.

Zevenbergen, et al. (2015). Room for the river: A stepping stone in adaptive delta management. *International Journal of Water Governance*, *3*(3), 1–20.

CHAPTER 14

Plan evaluation for flood-resilient communities: The plan integration for resilience scorecard

Matthew Malecha[a], Siyu Yu[a], Malini Roy[a], Nikki Brand[b], and Philip Berke[c]
[a]Department of Landscape Architecture and Urban Planning, Texas A&M University, College Station, TX, United States
[b]Department of Strategic Development, Delft University of Technology, Delft, The Netherlands
[c]Department of City and Regional Planning, University of North Carolina at Chapel Hill, Chapel Hill, NC, United States

Wise land-use planning is one of the most effective ways to prevent or reduce damage from natural hazards such as flooding. Successfully incorporating hazard mitigation across the many plans and policies that guide a city's development can be challenging, however. Communities around the world struggle with this task, to one degree or another, but those that acknowledge and plan for hazards throughout an integrated network of plans are generally more resilient than those where plans conflict and hazards are downplayed.

Researchers developed the Plan Integration for Resilience Scorecard (PIRS) method to evaluate the coordination of local plans and assess the degree to which they target areas most prone to hazards (Berke et al., 2015; Berke, Malecha, Yu, Lee, & Masterson, 2018; Berke, Yu, Malecha, & Cooper, 2019; Malecha et al., 2019). Through the *spatial evaluation* of a community's network of plan documents, a PIRS analysis helps reveal where and how plans and policies are coordinated or in conflict, and where opportunities exist to strengthen resilience. When applied in practice, this enables policymakers to address incongruities and focus more effectively on parts of the community demonstrating high vulnerability.

In a PIRS evaluation, a community is first divided into smaller districts, such as census tracts or neighborhoods (Fig. 1, map A), which can be individually assessed and compared. Zones of increased hazard risk, such as floodplains, are then defined and intersected with these districts to create a new layer of "district-hazard zones" (DHZs). Finally, documents in the community's network of plans are spatially evaluated: DHZs are assigned scores for each policy in the plans that (a) affects vulnerability, (b) influences land use, and (c) applies to a specific location(s). Policies that increase vulnerability receive a score of "−1," while those that reduce vulnerability receive a "+1" score. Scores can then be indexed for each DHZ, with higher scores indicating greater policy focus on reducing vulnerability (Fig. 1, map B). Ideally, policies are scored independently by multiple trained researchers, with intercoder agreement calculations providing feedback to ensure

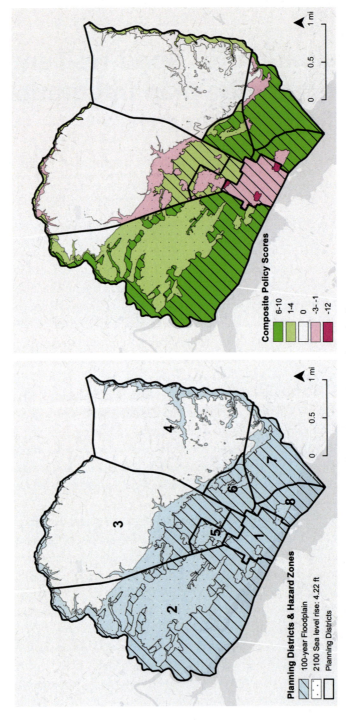

Fig. 1 Illustration of steps in the *plan integration for resilience scorecard* method.

accuracy. Cases of coder disagreement are then reconciled during a conference session, resulting in a final consensus scorecard. (For a more comprehensive description of the PIRS process, see Malecha et al., 2019.)

Transatlantic application

Though originally developed in the United States, the PIRS method was designed to be flexible, with potential for international applications. The uneven pursuit of resilience policies and conflicts among plans are by no means uniquely American phenomena. Nor is the imperative to resolve such problems: the United Nations declared in its landmark Sendai Framework for Disaster Reduction (United Nations General Assembly, 2015) that consistently integrating hazard mitigation in planning is crucial to building resilience, and that the failure of many communities to do so is a critical international concern.

The method was first applied in a sample of six cities along the US Atlantic and Gulf coastlines, including in the Houston, Texas region (Berke et al., 2018, 2019). Results revealed conflict between plan guidance in every community, but considerable variation across the study sample. Though perhaps unsurprising, given the decentralized governance structure and often limited coordination of US planning, these findings motivated an interest to apply the PIRS method in locations with better-integrated planning and hazard management. Would the method be pertinent and useful in places with less obvious plan conflict? What lessons might be learned that could help improve plan coordination for resilience the United States?

The unprecedented challenges wrought by climate change also suggest that even places with advanced planning and flood risk management systems might benefit from the perspective offered by the spatial evaluation of plans and policies. Although a global leader in water management and urban planning (Ward, Pauw, Van Buuren, & Marfai, 2013), the low-lying nation of the Netherlands is among the most flood-vulnerable countries in the world, especially in a changing climate. While comprehensiveness is a central aim of Dutch planning (Buitelaar & Sorel, 2010), flood risk management and local spatial planning practice are developed in separate silos, and have only recently started to integrate (Woltjer & Al, 2007). Land-use planning is beginning to be recognized as a method of reducing the risks and consequences of flooding (Neuvel & van den Brink, 2009), and the rigid Dutch water management strategy of attempting to prevent all flooding has begun to give way to a more flexible resilience approach, which seeks to minimize the consequences of flooding as part of a multilayer effort (Kaufmann, Mees, Liefferink, & Crabbé, 2016; Van Buuren, Ellen, & Warner, 2016).

These changing circumstances—together with an opportunity provided by a National Science Foundation Partnerships for International Research and Education (NSF-PIRE) grant—prompted researchers from Texas (Texas A&M University) and the Netherlands (UT Delft) to collaborate in applying the PIRS method in three separate studies in three Dutch cities: Rotterdam, Nijmegen, and Dordrecht. The studies were an occasion for comparisons and knowledge building, permitting the testing of the PIRS methodology in a new hazard and planning context, facilitating its continued development, and providing a novel perspective on Dutch plan integration and resilience as the country adjusts to new planning and water management challenges.

The remainder of this chapter presents summaries of the three case studies and then concludes with a brief discussion of key lessons learned from this cross-cultural research endeavor. Just as the PIRS provided a new lens to evaluate Dutch planning, the research in the Netherlands revealed insights that may help build resilience and advance flood risk management in coastal Texas and across the United States.

Feijenoord, Rotterdam

The first of the three studies (for full article, see Malecha, Brand, & Berke, 2018) was designed as a preliminary test of the generalizability of the PIRS methodology in the planning and hazard context of the Netherlands, and an exploration of how the method would suit the new situation. Feijenoord District, in central Rotterdam, was selected as the focus of the initial investigation. Successful application of the PIRS method in Feijenoord, with its different governance and hazard circumstances, provided evidence for the external validity of the PIRS method and paved the way for the subsequent studies.

Context

Feijenoord District, the second largest city in the Netherlands and the largest port in Europe, is located in central Rotterdam. Situated along a bend in the Nieuwe Maas River, this densely populated urban quarter has over 70,000 residents (Centraal Bureau voor de Statistiek [CBS], 2016) and is exposed to both storm surge and fluvial flooding (de Moel, van Vliet, & Aerts, 2014). Feijenoord's nine neighborhoods are among Rotterdam's most vulnerable (Centraal Bureau voor de Statistiek [CBS], 2016).

Like much of Rotterdam, the majority of southern Feijenoord is actually below sea level but is *embanked*—protected from river flooding from by an intricate system of dikes (City of Rotterdam, 2013). More than half of the district is located behind the bank dike (Fig. 2), where flood safety is the responsibility of the regional water authority (Correljé & Broekhans, 2015). Although a very high safety standard is maintained (Jonkman, Kok, & Vrijling, 2008), the unlikely event of a dike breach or extraordinarily high river levels would mean catastrophe for the low-lying neighborhoods (City of Rotterdam, 2013).

The remainder of Feijenoord is *unembanked*—directly exposed to the river. These parts of the district have a higher likelihood of flooding, but are elevated on higher ground. In contrast with the embanked areas, responsibilities for safety in these areas remain somewhat ambiguous (Runhaar, Mees, Wardekker, van der Sluijs, & Driessen, 2012; Ward et al., 2013). Thus, despite very high safety standards, some

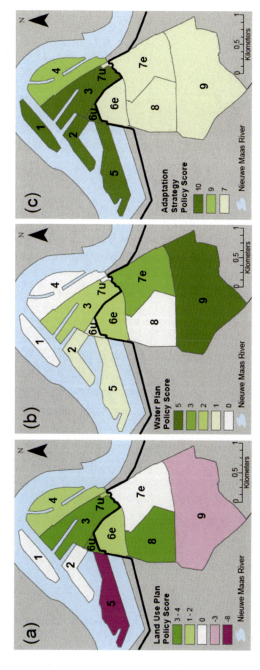

Fig. 2 Policy scores by plan type and neighborhood-hazard zone in Feijenoord district (*pink* = negative; *green* = positive): (A) land-use plans (all shown in one map), (B) submunicipal water plan, and (C) Rotterdam climate change adaptation strategy. *(Reproduced with permission from Malecha, M. L., Brand, A. D., & Berke, P. R. (2018). Spatially evaluating a network of plans and flood vulnerability using a plan integration for resilience scorecard: A case study in Feijenoord district, Rotterdam, the Netherlands. Land Use Policy, 78, 147–157. https://doi.org/10.1016/j.landusepol.2018.06.029.)*

uncertainty and vulnerability remain in Feijenoord and Rotterdam, and especially when an ever more unpredictable climate is factored in (de Moel, Bouwer, & Aerts, 2014).

Process

The local context was incorporated in all aspects of the PIRS analysis of Feijenoord District. In Rotterdam, as in the Netherlands as a whole, neighborhood-level land-use plans are important for guiding land use and planning policy, making the neighborhood the ideal subjurisdictional unit to be used in a spatial analysis. Hazard zones were delineated following the Dutch conceptualization of flood risk—which is a function of both elevation and responsibility for water management (de Moel, van Vliet, & Aerts, 2014; Jonkman et al., 2008)—and thus were defined as the embanked and unembanked areas. Within this framing, all of Feijenoord (along with much of the country, in fact) is located in at least one hazard zone, with two neighborhoods straddling both zones: 6e (embanked)/6u(unembanked) and 7e/7u (Fig. 2). With the district divided into 11 distinct neighborhood-hazard zones (NHZs), the network of plans affecting Feijenoord was spatially evaluated. The evaluation focused on local and municipal plans, including 10 neighborhood land-use plans, a Submunicipal Water Plan, and Rotterdam's Climate Change Adaptation Strategy.

Findings

When summed across all plans, scores for all NHZs were positive (overall mean = 10.4; unembanked mean = 10.4; embanked mean = 10.3), indicating that the network of plans generally emphasized vulnerability reduction across Feijenoord District. Disaggregating the scores by plan type and NHZ reveals a more nuanced picture (Fig. 2), however. Shown in aggregate, the land-use plans in Feijenoord (Fig. 2A) reflect development pressures and neighborhood goals, which vary across the district. At the time of analysis, the unembanked part of Feijenoord District was the focus of substantial development to attract affluent residents. Several neighborhoods—especially Katendrecht (#5)—were transforming from working ports to modern residential districts, and development pressures were a challenge to the prioritization of flood resilience, with some policies increasing flood vulnerability.

The submunicipal water plan (Fig. 2B) and climate change adaptation strategy (Fig. 2C) both generally reduced flood vulnerability, but affected Feijenoord in different ways. The water plan focused primarily on the embanked neighborhoods, with more resilience-building policies in Vreewijk (#9) than anywhere else. The adaptation strategy was aimed at building resilience throughout the district, but especially in unembanked neighborhood—focused on threats from the Nieuwe Maas. It appears that the water plan and adaptation strategy may have been designed to fill policy gaps—compare Fig. 2B and C to A.

Nijmegen

The second study pushed the PIRS application in the Netherlands in new directions, incorporating additional administrative scales, comparing the results to several measures of community vulnerability, and investigating the ways a national program, "Room for the River," was incorporated at the local level (for full article, see Yu, Brand, & Berke, 2020). Despite the sectoral origins of flood safety and spatial planning, the Dutch planning system was beginning to increase coordination (ESPON, 2017), and this study explored whether that trend would be apparent in a community's network of plans.

Context

Nijmegen, an inland city with a population of over 165,000 (Centraal Bureau voor de Statistiek [CBS], 2016), was selected for study in part for its location along the Waal River, which makes it naturally exposed to fluvial flooding, and also for its status as the location of the flagship project of the "Room for the River" program. The goals of the project were to (1) protect the city from future floods and (2) enhance spatial quality. Rather than raise or strengthen dikes, per the traditional approach, the ambitious project relocated part of one to create a wider floodplain and provide more room for future floodwaters, thereby reducing the threat to the city.

Process

The PIRS evaluation in Nijmegen expanded the network of plans to include national and provincial documents. To facilitate the spatial analysis, the city was divided into NHZs—there are 44 neighborhoods in Nijmegen, and hazard zones were again defined as the embanked and unembanked areas.

Several vulnerability analyses were added to the PIRS evaluation in this study. *Physical vulnerability* was determined using the mean housing value data from the Dutch Centraal Bureau voor de Statistiek (CBS). *Social vulnerability* was established by adapting the Social Vulnerability Index (Flanagan, Gregory, Hallisey, Heitgerd, & Lewis, 2011) to the Dutch context and measured using an index of indicators derived from the CBS. *Environmental vulnerability* was measured using the percentage of protected area in a NHZ as an indicator of environmental exposure (Villa & McLeod, 2002).

The PIRS analysis involved the spatial evaluation of all 14 documents in Nijmegen's network of plans, including national-, provincial-, and municipal-scale plans. Plans at all three tiers of the Dutch government can affect local decisions; the integration of some plan elements at higher administrative tiers is even mandatory in some cases (ESPON, 2017).

Findings

The network of plans in Nijmegen was shown to be generally supportive of resilience across the city. Composite policy scores from all 14 plans at the national, provincial, and local level ranged from +1 to +64. There was high variability between scores in the embanked and unembanked neighborhoods, however. A mean of 5.18 for embanked neighborhoods and 13.00 for unembanked neighborhoods suggests differences in policy emphases plans targeting different spatial areas; unembanked neighborhoods received significantly more attention for reducing risk, on average, than their embanked counterparts.

Several key findings are revealed upon closer inspection (Fig. 3). First, the national-scale *Delta Plan: Room for the River Waal* and provincial-scale *Environment Vision Plan*, created specifically for the Room for the River program, paid more attention to flood resilience in the enlarged unembanked areas. Second, local plans emphasized the building of flood resilience to accompany development in embanked areas. Third, plans at higher tiers again appeared to be filling policy gaps in the more development-focused local plans—a pattern that suggests that flood resilience may still be finding its way in the Dutch planning system.

The analysis of the vulnerability results revealed that, in general, higher physical vulnerability correlated positively with policy scores across Nijmegen NHZs, indicating a prioritization of vulnerability reduction in physically vulnerability areas than the network of plans. This suggests that Nijmegen's plans aligns with the flood safety goal of the Room

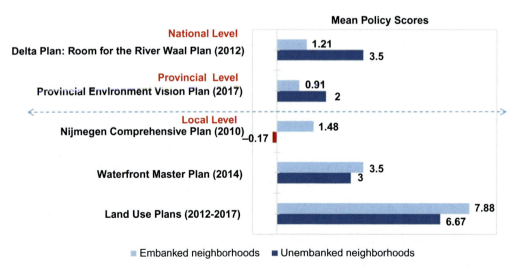

Fig. 3 Mean policy scores for plans at three administrative tiers affecting the city of Nijmegen. *(Reproduced with permission from Yu, S., Brand, A. D., Berke, P. (2020). Making room for the river: Applying a plan integration for resilience scorecard to a network of plans in Nijmegen, Netherlands. Journal of the American Planning Association, 1–14. https://doi.org/10.1080/01944363.2020.1752776.)*

for the River program to reduce physical vulnerability, with a strong focus on existing development in unembanked neighborhoods. In contrast, an inverse relationship was found between policy scores and social vulnerability. This suggests that socially vulnerable neighborhoods may not be prioritized by Nijmegen's network of plans—a potential social justice issue. Interestingly, for environmental vulnerability, correlation results are negative for embanked neighborhoods and positive for unembanked neighborhoods. Nijmegen's network of plans was successfully pursuing the preservation of nature in the enlarged unembanked neighborhoods, but missing an opportunity to encourage similar environmental-resilience-focused efforts throughout the city.

De Staart neighborhood, Dordrecht

The third study extended the application of the PIRS method in the Netherlands in yet another direction—this time focusing on uncertainty in hazard planning and evaluating a network of plans against multiple future flood risk scenarios with multiple measures of policy effectiveness (for full article, see Roy, Brand, & Berke, *forthcoming*). De Staart, a flood-vulnerable neighborhood in the city of Dordrecht, was used for this test case.

Context

The PIRS framework was used to evaluate the effectiveness of a community's network of plans against *future* flood scenarios. The analysis focused on a neighborhood in the City of Dordrecht, a highly flood-vulnerable yet critical urban center in the Rhine Delta, which is at the forefront of proactive flood risk management (Gersonius et al., 2016). Most neighborhoods in Dordrecht are protected by dikes and polders, but the neighborhood under study, De Staart, is unembanked; its primary flood defense is elevation.

The neighborhood has an average elevation of 3 m above NAP (the Amsterdam Ordnance Datum, used to indicate sea level), rendering most areas safe against a 1:2000 chance flood event. The sense of safety has catalyzed increased industrial and residential developments in recent years. However, based on the climate change scenarios (City of Dordrecht, 2009), by 2100, large areas and several critical facilities, including a fresh water supply plant and a prison, are likely to flood. Effective management of open spaces and new development complemented by the protection of the critical facilities can help avoid losses. The PIRS study was meant to provide foresight on opportunities to proactively strengthen adaptation efforts.

Process

The evaluation was focused on whether land-use planning decisions in Staart, Dordrecht were effectively anticipating future flood impact scenarios. Three future flood inundation scenarios were collected based on the middle WB21 climate scenario (Gemeente

Dordrecht, 2009)—current, 2050, and 2100. Inundation levels were aggregated from KNMI'06 river discharge report (Weiland, Hegnauer, Bouaziz, & Beersma, 2015), KNMI'06 sea-level rise and precipitation scenarios (Royal Netherlands Meteorological Institute, 2007), and flood inundation levels from a Deltares report on Staart (Asselman, 2010). The three inundation levels were spatially intersected with 10 De Staart subneighborhood districts, creating district-hazard zones (DHZs).

The PIRS methodology was then used to collect, code, and analyze the effectiveness of policies from five city and regional plans using three measures of policy effectiveness. Policies in "plan and adapt" method are theorized to anticipate future flood risk by being (1) *robust*, that is, protecting assets against multiple projected future flood risk or (2) *flexible or low-regret*, that is, providing benefits in current flood scenario and providing opportunities for modification or enhancement against future flood scenarios (Hallegatte, 2009; Stults & Larsen, 2020). Thus, robust policies reduce vulnerability to floods in all three scenarios (score = +1 in current, 2050, and 2100). Flexible policies reduce vulnerability in the current scenario, but would need to be monitored in the future (score = +1 in current, 0 in 2050 and 2100). Finally, we identified policies that induce vulnerability in current or future flood scenarios (score = −1) and called them (3) *adaptation opportunities*.

Findings

Overall, the network of plans was found to effectively reduce vulnerability in the current flood risk scenario. Out of a total of 68 policies affecting vulnerability to floods in De Staart, 25 (36.8%) were robust, that is, they reduced vulnerability to floods in current, 2050, and 2100. A smaller proportion of policies (20.6%) were low-regret policies. They reduced vulnerability in current scenarios, but would need to be monitored against changing flood impact scenarios.

Fig. 4 shows composite scores by districts for the three flood scenarios. All districts in De Staart have positive composite index scores in the current scenario, suggesting that there are more policies that reduce vulnerability than those that increase vulnerability. Several districts, such as District 10, an industrial area, remain fairly robust through 2100. Water-based industrial land uses are retained and are supported by policies that discourage building new commercial development (City of Dordrecht, 2013). On the other hand, scores in a few districts (e.g., 1, 3, 4, 5) drop to negative scores in 2050 and 2100. This is predominantly due to development regulations encouraging residential and commercial density in the proposed "urban environment" in Stadswerven region (City of Dordrecht, 2009).

This analysis provides important insights into (a) policies that need to be monitored and (b) current policy approaches across the network of plans that inadvertently eliminate opportunities for future adaptation. It serves as a critical, albeit preliminary, step to examine and manage climate impact uncertainty.

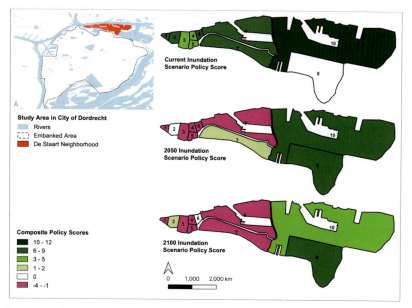

Fig. 4 Policy scores for current, 2050, and 2100 scenarios in De Staart, Dordrecht.

Conclusions

The studies described in this chapter exemplify the cross-cultural exchanges of knowledge and ideas that are at the heart of the NSF-PIRE program. A plan evaluation methodology developed in Texas to address issues of plan conflict and flood vulnerability was successfully applied and found to be generalizable in the Netherlands, a globally acknowledged leader in planning and water management. Researchers from the United States and around the world collaborated to great effect, bringing their unique expertise and understanding to bear on these important issues that exist, to one degree or another, in every country. Three case studies, in three Dutch cities, conducted over 3 years and led by three different researchers (supported by two others) resulted in findings that add to the flood risk planning and management discourse on both sides of the Atlantic.

As expected, the networks of plans in all three study locations in the Netherlands scored higher on the PIRS evaluations than did those in US communities. Long experience with the engineering and governance challenges posed to cities by flooding has led to more hazard-aware planning, on the whole. The scores were also more consistent across the Dutch cities—a likely result of stronger requirements for coordination and uniformity than currently exist in the United States. Lessons can be learned from these analyses that may advance the cause of flood resilience in both countries, however.

The PIRS method, and the concept of spatial plan evaluation, was introduced as a new tool and perspective for planners during a time of transition in the Netherlands, a global leader in urban planning and water management. Dutch planners and policy-makers were beginning to consider the usefulness of wise land-use planning and evacuation procedures as the second and third lines of defense, respectively, in a new multilayer water safety approach (Kaufmann et al., 2016; Van Buuren et al., 2016)—long neglected due to an overreliance on engineering solutions. All three case studies not only revealed strong integration of flood-resilience measures, but also noted instances where more explicit acknowledgement of the (minute, yet nonzero) plausibility of catastrophic flooding due to dam failure or overtopping would be beneficial. This very high standard, which was used in all three studies, resulted in occasional negative-scoring policies and revealed potential gaps and conflicts in plans regarding flood vulnerability. Employing this exacting lens, however, and evaluating the plans and policies spatially, may help in the reassessment as the Netherlands continues the process of updating its land use and water management approach.

The PIRS method's focus on differentiating hazard areas also brought to the fore the apparent ambiguity of responsibilities for water safety and flood mitigation in the unembanked parts of the study cities, and in the Netherlands more generally. Across all three case studies, some of the lowest plan scores were consistently found in unembanked neighborhoods, and this is especially the case for the influential neighborhood land-use plans. The lack of clarity about these areas (and the absence of a legally binding measure such as the "water test") may be leaving them significantly more vulnerable than other places, especially in a changing and increasingly volatile climate.

While unsurprising, the consistent finding of stronger plan integration toward resilience in the Netherlands provides empirical evidence for the wisdom and effectiveness of many aspects of the Dutch approach. Adopting (or at least adapting) these approaches might significantly improve plan integration and flood risk management in Texas and the United States. The first lessons to be learned relates to the value of communication in plan development—often mandated and enforced—which results in a generally complementary network of plans. This even appears to be the case with plans at different administrative scales; these preliminary spatial analyses suggest that each plan across a cities and throughout the administrative hierarchy has a specified purview and focus that acts to generally reinforce (rather than conflict with) the other plans in the network.

Secondly, the accumulation of evidence from the PIRS evaluations in the Netherlands signals the value of planners taking a leading role in preparing cities for threats from natural hazards. Planners sit at an important crossroads, and communities benefit when they are given the charge and authority to produce holistic plans that acknowledge and integrate hazard mitigation. This includes the serious consideration and candid use of scientific predictions about the likely effects of climate change to develop scenarios, plans, policies, and regulations—given equal, or even greater, weight than other drivers such as development pressures.

At the root of all of this, though, is a national mindset that appears to permeate (even dominate) government and planning around proactive, hazard-aware land use and water management—with public safety as the highest priority. This is something that is sorely needed as the climate crisis continues, and is something that American, and especially Texan, planners and decision makers would do well to emulate.

References

Asselman, N. (2010). *Flood risk in unembanked areas; part B: Flooding characteristics: Flood velocities in the downstream reaches of the river Rhine and Meuse*. Deltares. Rotterdam: Knowledge for Climate; Rotterdam Climate Initiative - Climate Proof.

Berke, P. R., Malecha, M. L., Yu, S., Lee, J., & Masterson, J. H. (2018). Plan integration for resilience scorecard: evaluating networks of plans in six US coastal cities. *Journal of Environmental Planning and Management*, *62*(5), 901–920.

Berke, P., Newman, G., Lee, J., Combs, T., Kolosna, C., & Salvesen, D. (2015). Evaluation of networks of plans and vulnerability to hazards and climate change: A resilience scorecard. *Journal of the American Planning Association*, *81*(4), 287–302. https://doi.org/10.1080/01944363.2015.1093954.

Berke, P., Yu, S., Malecha, M., & Cooper, J. (2019). Plans that disrupt development: Equity policies and social vulnerability in six coastal cities. *Journal of Planning Education and Research*. https://doi.org/10.1177/0739456X19861144.

Buitelaar, E., & Sorel, N. (2010). Between the rule of law and the quest for control: Legal certainty in the Dutch planning system. *Land Use Policy*, *27*, 983–989. https://doi.org/10.1016/j.landusepol.2010.01.002.

Centraal Bureau voor de Statistiek [CBS]. (2016). Retrieved from http://statline.cbs.nl/Statweb/.

City of Rotterdam. (2013). *Rotterdamse adaptatiestrategie [Rotterdam climate change adaptation strategy]*. Rotterdam, the Netherlands. Retrieved from http://www.deltacities.com/documents/20121210_RAS_EN_lr_versie_4.pdf.

City of Dordrecht [Gemeente Dordrecht]. (2009). *Waterplan Dordrecht 2009–2015*. Dordrecht: Gemeente Dordrecht.

City of Dordrecht [Gemeente Dordrecht]. (2013). *Structuurvisie Dordrecht 2040*. Dordrecht: Gemeente Dordrecht.

Correljé, A., & Broekhans, B. (2015). Flood risk management in the Netherlands after the 1953 flood: A competition between the public value(s) of water. *Journal of Flood Risk Management*, *8*(2), 99–115. https://doi.org/10.1111/jfr3.12087.

de Moel, H., Bouwer, L. M., & Aerts, J. C. (2014). Uncertainty and sensitivity of flood risk calculations for a dike ring in the south of the Netherlands. *Science of the Total Environment*, *473*, 224–234. https://doi.org/10.1016/j.scitotenv.2013.12.015.

de Moel, H., van Vliet, M., & Aerts, J. C. (2014). Evaluating the effect of flood damage-reducing measures: a case study of the unembanked area of Rotterdam, the Netherlands. *Regional Environmental Change*, *14*(3), 895–908. https://doi.org/10.1007/s10113-013-0420-z.

ESPON. (2017). *Comparative analysis of territorial governance and spatial planning systems in Europe*. Retrieved from https://www.espon.eu/planning-systems.

Flanagan, B., Gregory, E., Hallisey, E., Heitgerd, J., & Lewis, B. (2011). A social vulnerability index for disaster management. *Journal of Homeland Security and Emergency Management*, *8*(1), 1–22. https://doi.org/10.2202/1547-7355.1792.

Gemeente Dordrecht. (2009). *Waterplan Dordrecht 2009-2015*. Dordrecht: Gemeente Dordrecht.

Gersonius, B., Rijke, J., Ashley, R., Bloemen, P., Kelder, E., & Zevenbergen, C. (2016). Adaptive Delta Management for flood risk and resilience in Dordrecht, The Netherlands. *Natural Hazards*, *82*(2), 201–216.

Hallegatte, S. (2009). Strategies to adapt to an uncertain climate change. *Global Environmental Change*, 240–247.

Jonkman, S. N., Kok, M., & Vrijling, J. K. (2008). Flood risk assessment in the Netherlands: A case study for dike ring South Holland. *Risk Analysis*, *28*(5), 1357–1374. https://doi.org/10.1111/j.1539-6924.2008.01103.x.

Kaufmann, M., Mees, H., Liefferink, D., & Crabbé, A. (2016). A game of give and take: The introduction of multi-layer (water) safety in the Netherlands and Flanders. *Land Use Policy*, *57*, 277–286. https://doi.org/10.1016/j.landusepol.2016.05.033.

Malecha, M. L., Brand, A. D., & Berke, P. R. (2018). Spatially evaluating a network of plans and flood vulnerability using a plan integration for resilience scorecard: A case study in Feijenoord district, Rotterdam, the Netherlands. *Land Use Policy*, *78*, 147–157. https://doi.org/10.1016/j.landusepol.2018.06.029.

Malecha, M., Masterson, J. H., Yu, S., Lee, J., Thapa, J., Roy, M., et al. (2019). *Plan integration for resilience scorecard guidebook: Spatially evaluating networks of plans to reduce hazard vulnerability – Version 2.0*. College Station, TX: Institute for Sustainable Communities, College of Architecture, Texas A&M University. Retrieved from http://mitigationguide.org/wp-content/uploads/2018/03/Guidebook-2020.05-v5.pdf.

Neuvel, J. M. M., & van den Brink, A. (2009). Flood risk management in Dutch local spatial planning practices. *Journal of Environmental Planning and Management*, *52*(7), 865–880. https://doi.org/10.1080/09640560903180909.

Royal Netherlands Meteorological Institute. (2007, May 25). *KNMI'06 data per scenario for 2050 and 2100*. Retrieved from KNMI'06 Climate Scenarios: http://www.klimaatscenarios.nl/knmi06/samenvatting/#Inhoud_3.

Runhaar, H., Mees, H., Wardekker, A., van der Sluijs, J., & Driessen, P. P. (2012). Adaptation to climate change-related risks in Dutch urban areas: stimuli and barriers. *Regional Environmental Change*, *12*(4), 777–790. https://doi.org/10.1007/s10113-012-0292-7.

Stults, M., & Larsen, L. (2020). Tackling uncertainty in US local climate adaptation planning. *Journal of Planning Education and Research*, *40*(4), 416–431.

United Nations General Assembly. (2015). *United Nations world conference on disaster reduction—Sendai framework for disaster reduction 2015–2030*. Sendai, Japan. Retrieved from http://www.unisdr.org/files/43291_sendaiframeworkfordrren.pdf.

Van Buuren, A., Ellen, G. J., & Warner, J. F. (2016). Path-dependency and policy learning in the Dutch delta: Toward more resilient flood risk management in the Netherlands? *Ecology and Society*, *21*(4). Retrieved from http://www.jstor.org/stable/26270023.

Villa, F., & McLeod, H. (2002). Environmental vulnerability indicators for environmental planning and decision-making: Guidelines and applications. *Environmental Management*, *29*(3), 335–348. https://doi.org/10.1007/s00267-001-0030-2.

Ward, P. J., Pauw, W. P., Van Buuren, M. W., & Marfai, M. A. (2013). Governance of flood risk management in a time of climate change: the cases of Jakarta and Rotterdam *Environmental Politics*, *22*(3), 518–536. https://doi.org/10.1080/09644016.2012.683155.

Weiland, F. S., Hegnauer, M., Bouaziz, L., & Beersma, J. (2015). *Implications of the KNMI'14 climate scenario for the discharge of the Rhine and Meuse, comparison with earlier scenario studies*. Deltares, Ministry of Infrastructure and the Environment Delft: Royal Netherlands Meteorological Institute.

Woltjer, J., & Al, N. (2007). Integrating water management and spatial planning. *Journal of the American Planning Association*, *73*(2), 211–222. https://doi.org/10.1080/01944360708976154.

Yu, S., Brand, A. D., & Berke, P. (2020). Making room for the river: Applying a plan integration for resilience scorecard to a network of plans in Nijmegen, Netherlands. *Journal of the American Planning Association*, 1–14. https://doi.org/10.1080/01944363.2020.1752776.

Further reading

Gemeente Dordrecht *Beeldreigieplan stadswerven*. (2009). Dordrecht: Gemeente Dordrecht.

Berke, P., & Lyles, W. (2013). Public risks and the challenge to climate change adaptation: A proposed framework for planning in the age of uncertainty. *Cityscape: A Journal of Policy Development and Research*, 181–208.

Faludi, A., & van der Valk, A. (1994). *Rule and order Dutch planning doctrine in the twentieth century. Vol. 28.* Springer Science & Business Media.

Gemeente Dordrecht. (2013). *Structuurvisie Dordrecht 2040.* Dordrecht: Gemeente Dordrecht.

Haasnoot, M., Kwakkel, J. H., Walker, W. E., & ter Matt, J. (2013). Dynamic adaptive policy pathways: A method for crafting robust decisions for a deeply uncertain world. *Global Environmental Change,* 485–498.

Klijn, F., van Buuren, M., & van Rooij, S. A. (2004). Flood-risk management strategies for an uncertain future: living with Rhine river floods in the Netherlands? *AMBIO: A Journal of the Human Environment, 33*(3), 141–147. https://doi.org/10.1579/0044-7447-33.3.141.

Nadin, V., & Stead, D. (2008). European spatial planning systems, social models and learning. *disP-The Planning Review, 44*(172), 35–47. https://doi.org/10.1080/02513625.2008.10557001.

Piantadosi, S., Byar, D. P., & Green, S. B. (1988). The ecological fallacy. *American Journal of Epidemiology, 127*(5), 893–904.

Quay, R. (2010). Anticipatory governance. *Journal of the American Planning Association,* 496–511.

Roodbol-Mekkes, P. H., van der Valk, A. J., & Altes, W. K. K. (2012). The Netherlands spatial planning doctrine in disarray in the 21st century. *Environment and Planning A, 44*(2), 377–395. https://doi.org/10.1068/a44162.

Walker, W. E., Haasnoot, M., & Kwakkel, J. H. (2013). Adapt or perish: A review of planning approach for adaptation under deep uncertainty. *Sustainability, 5*(3), 955–979.

Wiering, M., & Winnubst, M. (2017). The conception of public interest in Dutch flood risk management: Untouchable or transforming? *Environmental Science & Policy, 73,* 12–19. https://doi.org/10.1016/j.envsci.2017.03.002.

CHAPTER 15

Dreaming about Houston and Rotterdam beyond oil and ship channels

Han Meyer
Department of Urbanism, Delft University of Technology, Delft, The Netherlands

Introduction

On March 9, 1872, the steamship "Richard Young" was the first sea vessel sailing through the brand new *Nieuwe Waterweg* ("New Waterway") toward the port of Rotterdam. The happening was celebrated with a festive ceremony and speeches by public dignitaries, emphasizing the importance of the new ship channel for the economic growth of the port of Rotterdam. Although it took some years to overcome some childhood diseases of the channel (initially, the channel silted up quickly. This was only under control after 20 years), it would become a main asset indeed for the growth of Rotterdam as the largest port of Europe and a main industrial center.

Something comparable took place in Houston on November 10, 1914, when the US president Woodrow Wilson opened the Houston Ship Channel by pushing an ivory button from his desk at the White House, which was wired to a cannon in Houston. The event was celebrated with great fanfare and with high expectations regarding the transshipment and processing of the new "black gold" that was found in the immediate surroundings of Houston: oil.

Both cities became iconic examples and engines of the modern industrialized world. They include the largest (Houston) and second largest (Rotterdam) petrochemical industrial complexes of the world, playing a main role in the world economy as global hubs of the production, transport, and distribution of fossil fuels. This main role of fossil fuels has influenced the spatial composition of both urban regions seriously, from the dominating role of petrochemical industries to the omnipresence of oil-related infrastructures such as highways and freeways and extensive parking lots. Both urban regions can be considered "*global petroleumscapes*" (Hein, 2018) *par excellence*.

The development of these cities and their ports as industrial hubs is strongly related to the exhaustion and increasing vulnerability of the water-dominated, swampy territories in both urban regions. Both aspects are under pressure of the need of a transition, because of climate change, rising sea levels, and more extreme weather conditions, as well as because of the need of energy transition and building a new economy, based on a *"Third Industrial Revolution"* (Rifkin, 2011).

Of course, there are also many differences between the two cities. The climate conditions are very different when comparing the moderate climate of the Netherlands with the extreme weather conditions of Houston, where hurricanes and rainstorms might occur which are unknown in Northwestern Europe. Also, the urban morphologies of both cities show more differences than similarities: Houston as the city of *"one million acres and no zoning"* (Lerup, 2011) and Rotterdam as a prototype of the Dutch postwar *"Welfare city"* (Wagenaar, 1992). As a matter of fact, these cities represent two variations of building a modern industrial society, built on two different dreams, as we will see in this chapter.

Despite these differences, there are common challenges in both regions, which are: how to create conditions for a more sustainable and adaptive territory, as well as for a necessary but still uncertain process of economic transitions. The most important challenge is the *combination* of both: how can an adaptive approach to sea level rise be combined with the transition of the economy and energy supply? How can these two developments support each other? What would it mean for the main assets which were the central key for economic growth during the last century: the ship channels?

Let us see how this common challenge might result in new solutions and strategies, based on new dreams in both cases and on a new approach to the ship channels.

Building the dream of the modern industrial urban landscape

Both industrial urban landscapes are already the result of the mutual influence and support of building a hydraulic system on the one hand and urban and regional development on the other hand. In both cases, it was also the combination of environmental disasters and new economic driving forces which was responsible for new hydraulic interventions and new economic and urban development in both regions.

The growth of Houston as a seaport, originally mainly focused on the export of cotton, was enabled by its location at Galveston Bay: accessible by water, and relatively well protected against storm surges by the stretch of barrier islands (Fig. 1). This natural condition was a problem at the same time, because of the shallow character of the bay. The first attempt to improve the accessibility of Houston for larger sea vessels started already in the mid-19th century by digging the Houston Ship Channel (Blackburn, n.d.; Bradley, 2020). However, it was not deep enough to be really competitive with Galveston, which is directly located in deep water. But around the turn of the century, the chances of both ports changed because of two main reasons: the discovery of huge amounts of oil in Texas, and the 1900 hurricane, which showed the vulnerability of Galveston. Moreover, the relatively easy connection of Houston with the continental railroad network was a great

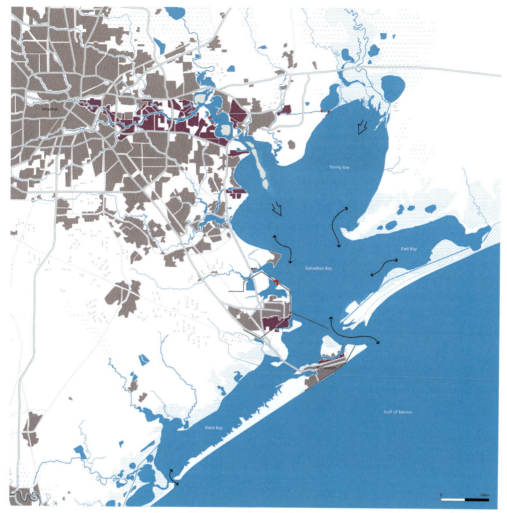

Fig. 1 Galveston Bay and Houston, 2000. *Grey*: urban, *purple*: port/industry. *(Map by MUST Stedebouw.)*

advantage. So the construction of the Houston Ship Channel was taken up again, now more seriously and radically, and supported by federal funding. The renewed Houston Ship Channel, dredged to a depth of 25 ft, was opened in 1914 with great fanfare in the city (Texas State Historical Association Website, n.d.). The channel was later hailed as "the port that built the city," making Houston the "Energy Capital of the World."

The combination of oil and the Houston Ship Channel was the key of the new, modern Houston not only because of the rise of more than five thousand energy related firms, including eight major refineries and 200 chemical plants producing a variety of synthetic products (Texas State Historical Association Website, n.d.), but also because the city became a showcase of the American dream. The infinite and extremely cheap availability of oil created the condition for the maximum independency of every American: the urban layout is a system which provides everyone an own High Chaparral, with a maximum of freedom, autonomy and accessibility with his/her own automobile.

Oil seemed to enable Houstonians to get everything under control, including the swampy natural landscape of Harris County. In order to make the American dream come true, marshlands were drained, reclaimed, and changed into a landscape suitable for building the suburban world, including thousands of homes supported by federal money like the FHA[a] homes in Oak Forest, and for the extensive network of roads and freeways.

The growth of Rotterdam as a port city and the consequences for the landscape is largely comparable with the developments in Houston. Until the mid-19th century, Rotterdam was a modest port city at the river New Meuse, at 40 km distance from the sea. The New Meuse used to be the main discharge channel of the Rhine and Meuse, but was silting up since the 17th century. Since that time, the main discharge of the rivers was moving to the southern distributaries of the Rhine-Meuse delta.

The proposal of the engineer Pieter Caland to create a new connection between Rotterdam and the sea by digging a "New Waterway" was supposed to solve both problems: it should provide an easy and quick discharge of redundant river water, as well as an open access for large sea vessels to Rotterdam. The New Waterway was opened in 1872; however, it was not earlier than 1896 that the channel was deep enough for the largest sea vessels. From that moment, the port of Rotterdam was booming and became the largest port of Europe, specialized in the transshipment of bulk like coal, ore, and cereals. The New Waterway was considered a courageous venture and presented as one of the main assets of the Netherlands at the World Exposition in Antwerp in 1930.

After the second World War, and accelerated by the Delta Works after the flood disaster of 1953, the focus of the Port of Rotterdam shifted to oil. The new Delta Works were built not only to prevent a repeat of the 1953 flood, but also to contribute to the transformation of the Netherlands into a modern industrial society. The Delta Works provided a new infrastructure of highways and inland waterways. Especially important was that the Delta Works resulted in large freshwater basins, which were considered a crucial condition for the cooling installations of the new petrochemical processing industries in the port of Rotterdam. During the postwar decades, the industrial cluster in the Rotterdam port area increased to a vast complex including five refineries and 45 chemical plants, making Rotterdam the largest port of Europe and during the period 1962–2004 also the largest port of the world (see Fig. 2).

Oil and the Delta Works together were the key of the transformation of the Netherlands into a modern industrial society and a coherent nation-state. This last point was

---

[a] FHA = Federal Housing Administration. See Wright (1981).

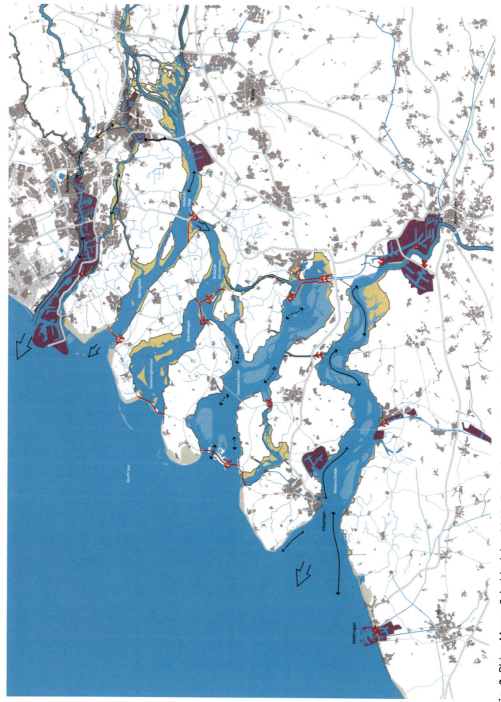

Fig. 2 Rhine-Meuse-Scheldt delta, 2000. *Grey*: urban, *purple*: port/industry. *(Map by MUST Stedebouw.)*

especially important in postwar Europe, including the Netherlands. It is true that European politics puts a strong emphasis on international cooperation, as a mean to avoid new wars between European countries. But at the same time, it was regarded necessary to pay more attention to national "team building" strategies by enhancing the national identity and pride in the different countries, as an optimistic alternative for the people, after the recent experience of the horror of war and fascist rule (Patel, 2020). In the Netherlands, an additional argument was the loss of the Dutch Indies, which gained independency in 1949. By that, the Netherlands had lost an important source of income as well as the idea of being a global empire. The Delta Works were considered an important bandage on the wound: these works provided the Dutch people a new reason for "*national pride,*" showing "*how a small country can be great.*"[b]

In the slipstream of the Delta Works, many books were published to underline the statement on the Delta Works as the new heroic achievement of the Dutch nation, with titles like "*Dredge, drain, reclaim. The art of a nation*" (Van Veen, 1948) and *"Nederland wordt groter"* ("The Netherlands is getting bigger") (Willems, 1962).

Thanks to the Delta Works, Rotterdam especially was getting bigger. Not only the reconstruction of the bombed city center and the new port areas, but also the new housing districts in the outskirts of the city became the showcases of the new postwar welfare state, built on reclaimed and intensively drained land and dominated by public housing and collective green areas.

So Houston and Rotterdam are two port cities in which two different dreams played an important role: in Houston, the American dream of maximum individual freedom; in Rotterdam, the dream of the European postwar welfare state with an emphasis on collective, national amenities. In both cases, the combination of oil and dredging, draining and reclaiming played a central role. The construct of frequent deepening of the ship channels especially was key to building two of these different showcases of modern urban society (see Table 1).

Cracks and fractures in the dream

The combination of oil and ship channels was not for everyone an obvious reason for blissful dreams. In the Rotterdam case, for example, there has always been doubts and warnings about the construction and frequent deepening of the New Waterway (Van de Ven, 2008). It was a trial-and-error intervention: experts disagreed with each other about the possible consequences of the new channel for river dynamics and influences of the sea on the hinterland. The doubts of opponents seemed justified, because the new river mouth was silting up again immediately and led to many headaches for the

---

[b] Prof. Jan Tinbergen, member of the Delta Committee, quoted in: Meyer (2017) p. 129.

Table 1 Comparison Houston-Rotterdam, 2020.

	Houston	Rotterdam
Land surface	City: 655 mi^2 (1696 km^2) Metro: 1778 mi^2 (4605 km^2)	City: 125 mi^2 (324 km^2) Metro: 315 mi^2 (817 km^2)
Population	City: 2.1 million Metro: 7.2 million	City: 650,000 Metro: 1.2 million
Gross domestic product	$472 billion (€395 billion)	$74 billion (€62 billion)
Dominating urban pattern	Low density detached houses	Medium density row houses
Average densities urban areas	10–25 persons/acre	20–80 persons/acre
Port transshipment	285 million ton	436.8 million ton
Territory	Low land (1–4 m *above* MSL) at bay area	Low land (1–4 m *below* MSL) in river delta
Climate	Warm maritime climate	Moderate maritime climate
Extreme weather conditions	With storm surges >27 ft (9 m) above MSL Extreme rainstorms	With storm surges >12 ft (4 m) above MSL Increasing draughts
Yearly precipitation	49.77 in (1264 mm)	34.65 in (880 mm)
Dominating urban pattern	Low-density detached houses	Medium-density row houses
Flood risk policy	Emphasis on emergency system, evacuation	Emphasis on flood prevention by extensive system of dikes and dams
Man-made interventions in water system	Houston Ship Channel, depth 45 ft (15 m)	Nieuwe Waterweg, depth 50 ft (16.5 m)

Based on data by public authorities Houston and Rotterdam, Port of Houston and Port of Rotterdam.

engineers of Rijkswaterstaat. It took another 24 years of intensive dredging and experimenting (including the invention and application of brand-new trailing suction hopper dredgers) before the first large-scale sea vessels could navigate through the channel and reach Rotterdam.

Also, the plan by the City of Rotterdam for building new oil terminals at the embankments of the New Waterway in the 1950s was not received with open arms by everyone. Johan van Veen, secretary of the former Delta committee and godfather of the Delta Works, was vehemently opposed to this plan. He warned about the uncontrollable consequences of making the New Waterway deeper and deeper, like salt intrusion, and increasing high water levels in the urbanized region. van Veen pleaded for new port development outside the coastline, where the coexistence of shallow sandbanks and deep channels offered excellent conditions for the construction of a new deep-sea port (Van der Ham, 2020). Moreover, the Province of South Holland was opposed to the intentions of Rotterdam, because the development of a long industrial corridor alongside the New Waterway was considered an undesirable spatial dichotomy of the provincial territory. Also, the Minister of Water Affairs was opposed to the Rotterdam plan, but relented finally for the pressure and economic arguments of the Rotterdam lobby (Lucas, 1957).

Enhancing the role of the New Waterway as the main entrance for the industrial postwar port of Rotterdam involved a step-by-step deepening of the channel, from 3 m (9 ft) in 1872 to the current depth of 16.5 m (50 ft). Next to serious pollution of the water, this process of deepening resulted in an increase of influence of the sea on the hinterland, including a growing tidal amplitude (Paalvast, 2014) and increasing salt intrusion. In order to stop salt intrusion, the discharge of the whole river system in the Netherlands is tuned to maintain maximum pressure on the New Waterway.

And finally, from the 1990s, climate change and rising sea levels accelerated the urgency to reconsider the approach to the ship channel—not only in Rotterdam, but also in Houston and many other port cities.

The year 2008 was a crucial year for Houston as well as for Rotterdam. In that year, Houston experienced the devastating consequences of Hurricane Ike, which could have been much worse if the eye of the hurricane would have passed some more miles to the west. The vulnerability of the industrial complex of refineries and chemical plants, provisionally protected with small dike rings, became visible. It made clear that the lack of a solid and collective protection against storm surge means an irresponsible risk for the life of millions of Houstonians, for the natural environment, and for the US energy supply. Although the role of the Houston Ship Channel was never questioned as a risk-increasing factor (as far as I can check), it is clear that the channel itself is also very sensitive and vulnerable to storm surges. The arduously deepened channel in the mid of the shallow Bay will be silted up seriously after a storm surge.

Hurricane Ike and also Harvey in 2017 especially addressed the vulnerability of the confluence of the Houston Ship Channel with the lower Buffalo Bayou. This is the most industrialized part of the channel and the most vulnerable at the same time, because of the compound flood events, combining the discharge of redundant rainstorm water by the Buffalo Bayou and a storm surge via the Houston Ship Channel, resulting in extremely high water levels (Couasnon, Sebastian, & Morales-Nápoles, 2018; Liu, 2017).

In the Netherlands, 2008 was the year in which a new national Delta Committee presented a report with an analysis of the possible consequences of climate change for the increase of flood risk, along with recommendations of the government. The result was the installation of the Delta Program 1 year later, with a special governmental Delta Commissioner who can dispose of an additional yearly budget of one billion euro until 2030. The adage of the Delta Committee was expressed in the title of its report:

"working together with water," referring to the need to understand and make use of the dynamics of natural water systems instead of fighting against these dynamics (Delta Committee, 2008).

One of the most important and complex questions, addressed by the Delta Committee and to be solved by the new Delta Program, was the Rhine mouth region. It is the most densely urbanized and industrialized region of the Netherlands, struggling with three water problems at the same time: sea level rise, increasing peak discharges of the rivers, and a shortage of retention capacity to be able to deal with the intensification of rainstorms.

It is true that the storm surge barrier *Maeslant*, built in the 1990s, gives protection against storm surges which lead to water levels of more than 3 m above MSL in the urban area of Rotterdam. However, there are many not embanked areas in the floodplain which will also be flooded with a water level between 2.5 and 3 m above MSL. Moreover, because of rising sea levels, the expectation is that the Maeslant barrier will have to close several times a year in the future, which will have negative consequences for navigation and the port.

Next to increasing awareness of climate change and rising sea levels, 2008 was also a pivotal year for another reason. A period of boisterous economic growth came to an end abruptly thanks to the financial and economic crisis. This crisis accelerated the discussion on the causes and effects of the environmental and climate crisis. Both the economic and the climate crises cast a dark shadow on the dreams of prosperity and welfare thanks to the industrial economy and came together in the debate on the necessity of what economist Jeremy Rifkin called the "Third Industrial Revolution" (Rifkin, 2011). Energy transition and building a circular economy should be the central goal of this revolution, with one common characteristic: a farewell to fossil fuels. For Houston as well as Rotterdam, this would have drastic consequences: how to change these two economies largely built on the trade and processing of fossil fuels?

Some other industrialized areas in the United States and Europe showed already earlier that it might be smart to develop a strategy for the time when heavy industries will change, move, or disappear. Detroit, "Motor Town" of the world, has become an example of what can happen if a city is too much dependent on one dominating industry and never thought about any strategy for economic transition. It has led not only to economic decay and mass unemployment, but also to a collapse of the city and urban life, which has become proverbial.

The German Ruhr area, once the largest industrial center of Europe, experienced the departure of the steel factories and the closing of the coal mines in the 1980s and 1990s, but transformed the industrial landscape into a large complex of public parks, museums, and theatres and attracted new industries, specialized in smart and sustainable technology. Currently, the Ruhr area is the second largest tourist destination of Germany and has developed a diversified economy with a strong presence of knowledge-based services and industries. It still is the most important contributor to the GDP of Germany (Website European Commission, n.d.).

For Houston as well as for Rotterdam, the urgent question is how to avoid a pathway like Detroit and how to prepare a pathway more similar to the Ruhr area.

Creating new perspectives: The ship channel as a leverage

The previous considerations make clear that cities like Houston and Rotterdam are facing a complex challenge, which includes as well the need of mitigation by changing the fossil-based industrial economy as the need of adaptation to the rising sea level and increasing precipitation quantities. A call for an "integrated" approach is tempting, but it will be incredibly difficult to share all different stakeholders and interests of both tasks under one umbrella. Both processes, the transition of the industrial economy and the adaptation to changing water conditions, will have their own dynamics and speeds, which cannot be organized totally parallel.

Concerning adaptation, there is a rising common sense of urgency in Rotterdam as well as in Houston. This book provides one example of this growing awareness. The responsible main actors in both cases are, for example, Rijkswaterstaat, the city of Rotterdam and the Delta program in the Netherlands; and the US Army Corps of Engineers, the city of Houston, and the Houston Resilience strategy in the Houston area, as well as a range of academic institutions. These actors are fully focused on the question of how and where adaptation measures can be applied and how they can be organized and financed.

Concerning energy transition and the change from fossil fuels toward more sustainable and recyclable energy sources, the question is not *if* it will happen but especially *when* and *how* it will happen, and how quick or slow this transition process will be. Concerning the "how" of this transition process, it is getting clear that a strategy for new economic investments and projects should fit in and contribute to a process of making the deltaic landscapes more sustainable, instead of reversed.

From this point of view, interesting proposals in this direction have been developed during the last decade.

Recently, the Dutch Delta program launched its *Kennisprogramma Zeespiegelstijging* (Knowledge program sea level rise), which addresses three fundamentally different approaches for the Netherlands to sea level rise in the future: (a) a "sea ward" approach, extending an "offensive" approach with large-scale hydraulic works; (b) a "maintaining the existing coast line" approach, with as much as possible nature-based solutions, and (c) a "withdrawal" approach, intending to move most of the urban and economic centers of the Netherlands from the lowlands in the west to the higher grounds in the east of the country (Website Deltaprogramma, n.d.). An important question is, which approach would create the best conditions for a strategy of mitigation, energy transition, and an enhancement of the natural environment and biodiversity.

The "sea ward" approach means a continuation and intensification of the approach to manipulate and control the delta as an artificial infrastructure. It will mean incredible investments in new dams, locks, pumping stations (to pump the river water to the sea) and will make the future of the country completely dependent on the maintenance and frequent enhancement of this infrastructural system.

The "withdrawal" approach is a rather fatalistic one, which will lead to unbelievably costly operations to move the whole economic, industrial, and urban infrastructure 200 km to the east.

The "maintaining the existing coastline with nature-based solutions" approach seems to be the most promising and realistic one as well as the most sustainable one. This approach tries to explore in what sense the natural dynamics of water systems can be used to create safer conditions for human settlement, building upon the adage of the Delta Committee of "working together with water."

The City of Rotterdam, Port of Rotterdam, and nature conservation organizations such as World Wildlife Fund and ARK Nature development launched a program for the restoration of tidal nature at the embankments of the river: *"The river as a tidal park"* (Website City of Rotterdam, n.d.). For the time being, this program focuses on the embankments of the river and not on the restructuring of the riverbed as a whole. But it is not so difficult to imagine that it will take just one step further to address a more radical repair of the river mouth as an estuary, including a gradual transition of land to water and a substantially more shallow riverbed.

Building on this initiative, we proposed the Delta program to investigate the possibilities and perspectives of an approach to change the New Waterway in an estuary, by allowing the natural process of sediment transport and deposits to silt up the river mouth (Meyer and ARK Natuurontwikkeling, 2020). The expectation is that it will lead to an increase in biodiversity, less extreme high-water events in the urbanized area, less salt intrusion, new possibilities for leisure and recreation areas, and new urban environments. Port and navigation activities will not disappear but should change and adapt to the new environmental conditions. It will function as an accelerator of the necessary transition of the port from an oil-based energy supplier to a supplier of zero-fossil energy (Website World Wildlife Fund, n.d.). The New Waterway will be transformed from a monofunctional industrial shipping channel into a multifunctional estuary with space for biodiversity, recreation, new urban environments, making use of natural processes of sediment transport, and creating conditions for a new type of sustainable port system (Fig. 3). This transformation of the New Waterway will mean that the main discharge of the rivers should be redirected toward the southern estuaries of the delta. The "Delta Design Studio" of the Delta program showed the possibility and desirability of this change already in 2012 (Website Rijkswaterstaat, n.d.) (Fig. 4).

Concerning the Galveston Bay area, a comparable variety of different approaches has been developed. On the one hand, there is an approach that focuses on the enhancement

Fig. 3 The Rhine mouth as an estuary, by making the riverbed of the New Waterway more shallow and wider. (*Drawing by Dirk Oomen en Peter Veldt (Bureau Stroming).*)

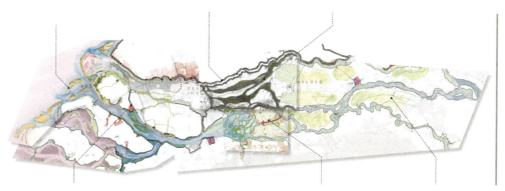

Fig. 4 Design study exploring the Haringvliet estuary as the new main discharge channel of Rhine and Meuse, reducing the role of the New Waterway in the discharge distribution. Design by H+N+S Landscape Architects, commissioned by the Dutch Delta program, 2012.

of the Texas coastline, with an emphasis on building a storm surge barrier in the sea gate between Galveston Island and Bolivar Peninsula. This approach culminated in the recently presented Coastal Texas Protection and Restoration Study by the US Army Corps of Engineers (USACE, 2020).

Another approach, focusing on the restoration of the Galveston Bay area, has been advocated by professionals and academics from Houston University, like the teams of Thomas Colbert (2014) and Peter Zweig (Zweig, Johnson, & Logan, 2020). They explored the possibilities to transform the Galveston Bay and Houston region into a multifunctional landscape, creating conditions for restoration of the natural environment, public facilities for recreation and leisure, and housing districts with public access to the beaches and wetlands of Galveston Bay. Also, student projects of TU Delft in the Houston region in 2014–15 and 2017 show interesting starts to develop nature-based approaches to create more safety against storm surges and rainstorms (Godfroy, 2017; Kothuis, Brand, Sebastian, Nillesen, & Jonkman, 2015).

The recently published *Galveston Bay Park Plan* (SSPEED Center et al., 2019) builds largely upon this idea. (Fig. 5). It shows the possibility of a new protection system by combining a new layout of the Houston Ship Channel with an extensive wetland restoration program. The renewed wetlands contribute to the safety of Houston against storm surges and also create a new publicly accessible landscape, as a condition for building a new future for Houston as a "*Post-industry, Post-oil, Post-sprawl*" urban landscape (Zweig et al., 2020).

As has been stated in the Galveston Bay Park Plan, this proposal should not be considered an alternative for or competitive with the USACE Coastal Texas study. Both plans could be regarded as complementary to each other. But for creating conditions

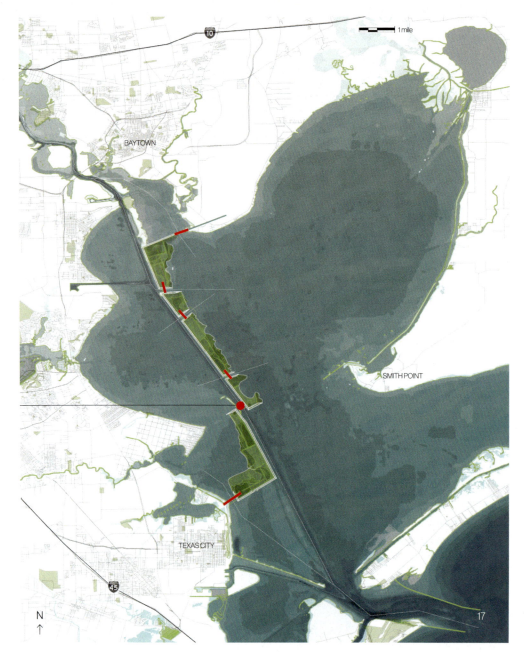

Fig. 5 Galveston Bay Park Plan, 2019. *(Map by SSPEED Center, Rogers Partners, Walter P. Moore, 2019, Galveston Bay Park plan, Houston.)*

for a radical transition of Houston and its economy in the future, the Galveston Bay Park Plan has the best cards.

Conclusions

Houston and Rotterdam both find themselves at a crossroads. Both urban regions have to wonder how they can organize two essential changes: a fundamental adaptation to climate change by nature-based solutions, and an economic transition from a primarily oil-based industrial economy to a more diversified and circular economy. In both changes, and especially in the combination of both changes, the ship channels play a key role.

The construction of both channels, more than a hundred years ago, played a key role in the realization of dreams on a new society of prosperity, built on oil and industrialization. But they also played a key role as a part of the problem of rising water levels and floods. Today, we need new dreams about a new type of urban society and economy with a new relation to the natural environment. Houston as well as Rotterdam can become guiding cities, showing how this dream can look like and how they can be realized. The *Galveston Bay Park Plan* and the *Rhine mouth as an estuary* are two first steps in making these dreams come true.

References

Blackburn, J., (n.d.) *Storm surge and the future of the Houston ship channel*, Houston: SSPEED Center, Rice University.
Bradley, B. S. (2020). *Improbable Metropolis. Houston's architectural and urban history*. Austin: University of Texas Press.
Colbert, T. (2014). Galveston Bay USA. In H. Meyer, & S. Nijhuis (Eds.), *Urbanized deltas in transition*. Amsterdam: Techne Press.
Couasnon, A., Sebastian, A., & Morales-Nápoles, O. (2018). A copula-based bayesian network for modeling compound flood hazard from riverine and coastal interactions at the catchment scale: An application to the Houston ship channel, Texas. *Water*, *10*, 1190. https://doi.org/10.3390/w10091190.
Delta Committee. (2008). *Working together with water. A living land builds for its future*. The Hague http://www.deltacommissie.com/doc/deltareport_full.pdf.
Godfroy, M. (2017). *Quantifying wave attenuation by nature-based solutions in the Galveston bay*. TU Delft (MSc thesis).
Hein, C. (2018). Oil spaces: The global petroleum scape in the Rotterdam/the Hague area. *Journal of Urban History*, *44*(5), 887–929.
Kothuis, B., Brand, N., Sebastian, A., Nillesen, A. L., & Jonkman, B. (2015). *Delft delta design. Houston Galveston Bay Region, Texas, USA*. Delft: Delft University Publishers.
Lerup, L. (2011). *One million acres and no zoning*. New York: Actar Publishers.
Liu, F. (2017). *Analyzing the influence of compound events on flooding in the downstream reach of the Houston ship channel*. Delft: TU Delft (MSc thesis).
Lucas, P. (1957). *Overzicht van de bemoeiingen van het Gemeentebestuur van Rotterdam met de totstandkoming van de havens en industrieterreinen in het gebied van de Botlek*. Rotterdam: Gemeente Archief Rotterdam.
Meyer, H. (2017). *The state of the delta. Hydraulic engineering, urban development and nation building in the Netherlands*. Nijmegen: Vantilt.

Meyer, H., & ARK Natuurontwikkeling. (2020). *De Rijnmonding als estuarium. Pleidooi voor onderzoek naar de mogelijkheid en effecten van een natuurlijke 'verondieping' van Nieuwe Waterweg en Nieuwe Maas*. Rotterdam.

Paalvast, P. (2014). *Ecological studies in a man-made estuarine environment: the port of Rotterdam*. Nijmegen: Radboud University.

Patel, K. K. (2020). *Project Europe. A history*. Cambridge: Cambridge University Press.

Rifkin, J. (2011). *The third industrial revolution. How lateral power is transforming energy, the economy, and the world*. London: Palgrave Macmillan Ltd.

SSPEED Center, Rogers Partners, & Walter P. Moore. (2019). *Galveston Bay Park plan*. Houston.

Texas State Historical Association Website. https://www.tshaonline.org/handbook/entries/houston-ship-channel.

USACE (US Army Corps of Engineers). (2020). *Coastal Texas protection and restoration feasibility study*. Washington/Houston.

Van de Ven, G. P. (2008). *De Nieuwe Waterweg en het Noordzeekanaal. Een waagstuk. Onderzoek in opdracht van de Deltacommissie*. Den Haag: Deltacommissie.

Van der Ham, W. (2020). *Johan van Veen, meester van de zee*. Amsterdam: Boom.

Van Veen, J. (1948). *Drain, dredge, reclaim. The art of a nation*. Den Haag.

Wagenaar, C. (1992). *Welvaartstad in wording. De wederopbouw van Rotterdam 1940-1952*. Rotterdam: NAi Uitgevers.

Website City of Rotterdam. https://www.rotterdam.nl/wonen-leven/getijdenpark/.

Website Deltaprogramma. https://www.deltaprogramma.nl/deltaprogramma/kennisontwikkeling/zeespiegelstijging.

Website European Commission. https://ec.europa.eu/growth/tools-databases/regional-innovation-monitor/base-profile/north-rhine-westphalia.

Website Rijkswaterstaat. https://puc.overheid.nl/rijkswaterstaat/doc/PUC_151739_31/.

Website World Wildlife Fund. https://www.wwf.nl/wat-we-doen/actueel/nieuws/meer-natuur-ondiepere-rijnmonding.

Willems, E. (1962). *Nederland wordt groter. Op zoek naar het nieuwe beeld van Nederland*. Amsterdam: De Bezige Bij.

Wright, G. (1981). *Building the dream. A social history of housing in America*. Cambridge, MA: MIT Press.

Zweig, P., Johnson, M., & Logan, J. (2020). *Houston generic city. Gulf coast urbanism post-industry, post-oil, post-sprawl*. Houston: Actar Publishers.

CHAPTER 16

A new nature-based approach for floodproofing the Metropolitan Region Amsterdam

Anne Loes Nillesen
Urban and Rural Climate Adaptation, Defacto Urbanism, Rotterdam, The Netherlands

Introduction

The Metropolitan Region Amsterdam

The Metropolitan Region Amsterdam (MRA) is one of the three metropolitan regions of the Netherlands. The MRA is a governance partnership of the Dutch provinces North-Holland and Flevoland, containing 32 municipalities including Amsterdam, the capital of the Netherlands, the historic city of Haarlem, and the 1980s "New Town" Almere. The MRA is positioned in the north-west of the Netherlands and contains ca. 2.5 million inhabitants (and 1.15 million houses) and represents 18.4% of the gross domestic value (Metropoolregio Amsterdam, 2020b).

The landscape development of the MRA is a result of the historically intertwined relation between the development of urban settlements and the region's water management and flood risk protection system. Looking at the region's height and water system map (Fig. 1), the local history of conquering and controlling the sea and rivers is still visible in the landscape structure.

Amsterdam is positioned along the IJ, a former natural river connecting the city to the North Sea and positioned in between the natural higher grounds of the dunes in the west and the Heuvelrug in the east. The Amsterdam historic city center was elevated during its construction (Hooimeijer, 2011); these century-old elevations still are an important asset in reducing the flood risk of the city. Surrounding the city, we see several polders that have been reclaimed over the centuries. The Flevopolder in the dammed former Southern Sea, which is now compartmented into Lake IJssel and the Markermeer, is the most recent reclamation.

The region is an important economic district and includes the Amsterdam harbor and the national airport Schiphol. An agglomeration of forces contributes to the popularity of the region, resulting in a major urbanization challenge. The region is aiming to develop around 230,000 additional houses until 2040 (Metropoolregio Amsterdam, 2020c), while strengthening its regional landscape and preserving historic characteristics. In

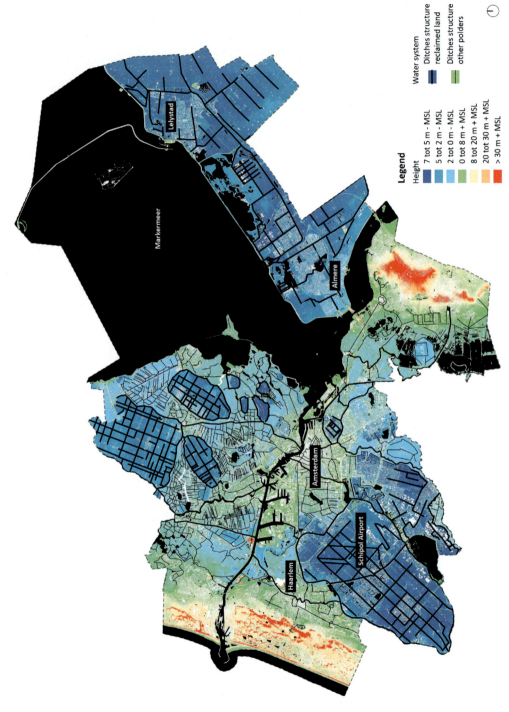

Fig. 1 Height and water system map of the Metropolitan Region Amsterdam (MRA), including the water discharge and polder system. *MSL*, mean sea level. *(Map by Defacto Urbanism, based on AHN and LIWO data.)*

addition, the energy transition and circular economy ambitions are expected to require a substantial transition (ENZH, 2020). Within this long-term context, climate change is also an important challenge, currently most tangible in the rain nuisance and land subsidence challenges.

The regional water system

A big part of the MRA is positioned below sea level, resulting in an extensive technical flood risk protection and water drainage system. When we look at the flood risk protection system, the dunes in the west combined with a sluice constructed in the North Sea Channel form the main protection against storm surges from the sea (Fig. 2). Primary levees protect the area from floods from the Markermeer, and from the more southern positioned river Lek, whose flood waters can reach the MRA in case of a levee breach. Secondary levees, the so-called "regional levees," protect the area from flooding from the many water supply and discharge channels and waterways. Freshwater for the area is supplied from the rivers Lek and IJssel, and from Lake IJssel and the Markermeer that function as a water buffer. The water supply and drainage systems often use the same channels and waterways. When draining the polders, the water is first drained by pumping into the so-called "boezem" channel system, which then drains toward the Markermeer and Nord Sea Channel. From here the water has to then be discarded into the North Sea, occasionally by gravity but mostly by pumping.

Climate change: Sea level rise and increased rainfall scenarios

The MRA faces different climate change aspects among with sea level rise, more intense rainfall, drought, heath, and land subsidence. The climate change either affects land use (e.g., buildings, agriculture, or nature) directly or indirectly through the water system by causing too much or too little water availability, increasing flood risk, or decreasing water quality.

The sea level is expected to rise even more during the next decades. The Dutch DeltaProgram, established to develop strategies that address the long-term flood risk challenges (Deltacommissie, 2008), uses scenarios that assume a 0.35–1 m sea level rise in 2100. However, recent studies show that the sea level might rise more and faster, especially after 2050: projections from the Royal Netherlands Meteorological Institute (KNMI) show a potential faster sea level rise with a bandwidth of 2–3 m sea level rise in 2100, related to a 2–4°C climate change (Haasnoot et al., 2018).

The expected climate change for the MRA is displayed in various online "climate atlases," such as the national "Climate Effect Atlas" (Klimaateffectatlas, 2020) and the "MRA Climate Atlas" (Klimaatatlas MRA, 2020). In this study, we use these two atlases to formulate the climate-related challenges for the region. While the sea level is expected to rise, the scenarios predict more extreme rainfall patterns at the same time. The rainfall

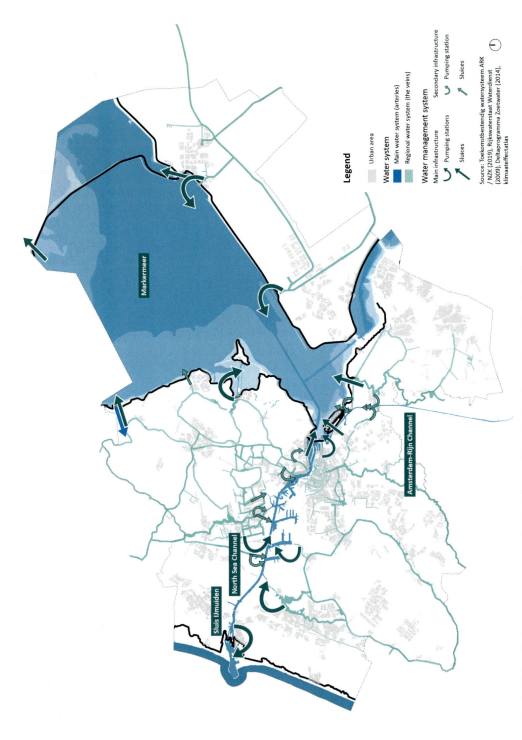

Fig. 2 Water discharge system map for the Metropolitan Region Amsterdam. *(Map by Defacto Urbanism.)*

intensity is expected to increase, resulting in more extreme rain showers that pressure the water discharge system and can, for example, result in obstruction of traffic (by flooded tunnels) and damage to crops and buildings (Must, 2018). Although the rain intensity will increase, dry periods will simultaneously become longer. Together with a rise in temperature, this will result in drought, impacting drinking water, agriculture, and nature, increasing the risk of forest fires, while reduced groundwater tables will accelerate subsidence. Within the MRA there are many areas subjected to land subsidence, reducing the suitability of these lands for agriculture and causing damage to buildings, sewerage systems, and roads, while expelling substantial amounts of CO_2 from the oxidating grounds. Increased temperatures will not only impact the urban heath effect and increase the need for cool spaces in the urban area, but also increase water temperatures, which affects water quality and thus has impacts on crops and nature.

Sea level rise, in combination with the other climate change aspects, will impact the region in different ways, especially in this region where the water system, land use, and occupation patterns are so interwoven. Based on different impact descriptions (Haasnoot et al., 2018; Mens et al., 2018; Vermeulen, Honingh, Oosterhoff, Zandvoort, & Hakvoort, 2019) and expert meetings, a first narrative on the potential impacts of climate change of the current MRA water management system is developed. The flood risk, which can be defined as "the probability × the consequence of a flood," will further increase, since both the pressure on the sea defenses as well as the flood depth in case of a flood will increase.

In time, the salt intrusion by seepage in the coastal zone is expected to further increase due to sea level rise, resulting in higher salinity levels of the polder water. Due to the temperature rise, the water will be warmer, reducing the oxygen levels and water quality (increase of phosphates). When maintaining the current land use and salinity or phosphate thresholds, this will result in the need for additional freshwater to flush the polders. However, at the same time, less water will be available. To hold more water, the water level of the Markermeer may need to be raised, which in turn will affect the required dike height.

While in dry periods the amount of available freshwater decreases, during rain events the amount of water that needs to be drained will increase. Extrapolated by the ongoing urbanization and coinciding increase of impermeable surfaces and runoff, this means that the pressure on the water discharge system will further increase; to discharge, some channels might have to be widened, and the pumping capacity will need to be increased. The North Sea Channel water level might have to be increased to be able to capture sufficient water before pumping it to the (over time even rising) sea.

Extending the system

The Netherlands' "default" response to water management problems often is the further improvement or extension of the technical water system. However, the MRA system is expected to reach some thresholds that request an extensive and cost-intensive extension of the water drainage and supply network in order to fully support the current land use and further intensification functions. In addition, this "business as usual" approach would not match the often expressed ambition to increase the robustness of the water management system, as supported by pilot projects like "Room for the River" and the so-called "Capture, Store, Drain" policy (Rijkswaterstaat, 2003). To demonstrate viable options to address the abovementioned climate challenges more robustly, the "Resilience by Design, Research by Design study" for the MRA, commissioned by the MRA and conducted by Defacto Urbanism, explores different options for nature-based climate adaptation to complement and extend the current technical water system.

MRA Resilience by Design approach

The MRA Resilience by Design project is a design-based research in which landscape designers, with the support of hydraulic and economic experts, water authorities, municipalities, and the province, explore alternative nature-based options for climate adaptation. During the project, two design teams explored demonstration projects for climate adaptive developments and investments for the MRA urban and rural areas. This chapter describes part of the research, focusing on the MRA landscape and performed by Defacto Urbanism. Within the study, different landscape zones are identified, for each of which a typical climate change challenge is further explored and addressed with a site-specific design proposal (or demonstration project). To do this, first the "business as usual" technical intervention (and associated thresholds) is described, in order to then formulate an alternative nature-based intervention (Metropoolregio Amsterdam, 2020a).

Landscape zones MRA

For the study, seven landscape types are distinguished, loosely based on the landscape categories from the "Intens" study (Vereniging Deltametropool, 2019): the Dune zone (natural coastal dunes in the west of the area), the Haarlemmermeerpolder (large polder that includes Schiphol Airport), Waterland (with its peat soils and historic concatenated polders), the Flevopolder (the most recent Netherlands polder reclaimed in the 1950s with large and productive agriculture lands), Gooi en Vechtstreek (on the edge of a landscape formed by peat excavations and the gradient of a forested hillside), and Amsterdam's green fringes (green "fingers" penetrating the Amsterdam urban area from the more rural environment).

Formulating alternative integrated strategies for climate change

Based on the available water system knowledge and climate change scenarios for each landscape type, the climate change-related challenges are identified. A representative challenge for this area is then explored further. Together with local and technical experts during multidisciplinary workshops, we defined the probable "business as usual" technical response and, where possible, identified the potential thresholds of this system. Subsequently, we explored alternative options to address the climate change challenge with integrated, often nature-based, solutions. Integrated solutions address multiple aspects (for instance, both climate change and spatial challenges) to be more effective, while nature-based solutions can be defined as strategies or measures that strategically conserve or restore nature to support gray infrastructure systems in addressing climate change risks, while providing additional benefits (World Bank and World Resources Institute, 2018). Next to the technical "business as usual" approach (Fig. 3, column 1), we formulated in this study two categories of potential alternative strategies: increasing the robustness and buffer capacity of the system (Fig. 3, column 2) and changing the land use (Fig. 3, column 3). The economist from our team would subsequently reflect on the economic feasibility of each alternative.

Demonstration projects

Propositions for alternative strategies and related demonstration projects are developed for each landscape zone. Within the MRA, we see that the flood risk challenge presents two main challenges: addressing increased flood risk from rising water levels and reducing runoff to the main water system to reduce (or not further increase) pressure on the water discharge and flood risk management system. The following sections will describe the results for four of the (in total six) demonstration projects:

1. Flevopolder: Transferring subsiding grounds into a nature connection.
2. Flevopolder: Transition to nature reserves with amphibious street furniture.
3. Heuvelrug: Restoring the natural water infiltration system to reduce runoff and drought.
4. Amsterdam forest: A new water retaining forest for the further densifying city of Amsterdam.

Demonstration project 1. Flevopolder: Transferring subsiding grounds into a nature connection

The MRA part of the Flevopolder is mainly used for agriculture and includes the nature reserve Oostvaardersplassen and the city of Almere. The polder is characterized by both deep potential inundation levels in case of a dike breach and land subsidence. Both challenges are addressed in different demonstration projects. Available scenarios show land subsidence of about 0.5 m until 2050 (Klimaateffectatlas, 2020); about 20% of this land subsidence is climate-related. After 2050, land subsidence is expected to continue at a more moderate rate until 2100, but would contain a more dominant climate component. Land subsidence is strongest in areas with already limited reclamation and thus higher relative groundwater levels. Over time, clay soils become too wet to be sufficiently efficient for agriculture.

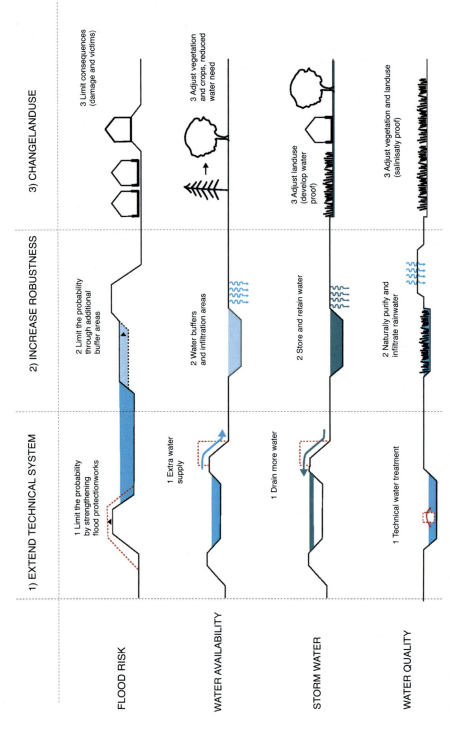

Fig. 3 Scheme showing different options for extending the water system: Expanding the technical system (Column 1), improving the robustness of the system (Column 2), or changing the land use (Column 3). *(Image by Defacto Urbanism.)*

The technical "business as usual" response would be to polder the area and drain it, that the land keeps sufficient reclamation. With this, the Flevopolder water system would have to be widened and the pumping capacity would have to be increased to be able to drain the additional water. The Province and the water authorities already expressed that this strategy would not be preferred, although no alternative strategy is yet proposed.

As a proposal for this area, a reactive approach is offered that monitors land subsidence and suitability for agriculture. Land that becomes unsuitable or less efficient can be converted into natural areas with water storage capacity (Fig. 4). An additional benefit is that rainwater of the adjacent residential area Almere Buiten can also be stored here. Depending on the speed and impact of the subsidence, areas will remain available for agriculture as long as possible, before being transferred to nature. By transferring these areas to nature, an ecological corridor can grow, eventually making the long-awaited connection of the Oostvaardersplassen nature reserve with an adjacent forest. The new nature area can also function as a recreation area for the growing urban population.

From a cost perspective, we see that nature transition can be covered to some extent by avoided costs of poldering and reclaiming the area. But when land subsidence causes most of the agriculture in this land subsidence area to become inefficient, the cost of transferring these areas to nature will exceed the cost of the "business as usual" strategy. To make it feasible, low-density housing that coincides well with nature development (Rebel Group, 2019) can be added to the program to cover additional costs.

Demonstration project 2. Flevopolder: Floodproof houses and amphibious street furniture

In case of a dike breach, the Flevopolder can inundate about 2–3 m. The polder, as most of the Netherlands, has high protection standards. With an excess probability of 1:30,000 years, the flood probability is therefore low (LIWO, 2020). However, due to the deep inundation levels, the consequences of a flood could be severe. Inundation levels may even continue to increase in the future due to a decision to store more water in adjacent lakes in response to climate change, as well as by subsiding ground levels. With the planned development of around 70,000 new houses in Almere (which is almost double the current number of residences), the flood risk would increase and either probability-reduction or consequence-reduction measures would be needed to decrease the risk. For this area, the "business as usual" strategy of reinforcing the levees is already implemented, since with the current dike reinforcements the 70,000 houses are already incorporated in the standards.

Since this is an area that is quite deeply submerged in a flood, the alternative strategy should be able to limit the consequences (damage or fatalities) of a water level of more than 2 m. On the opportunity map for flood consequence reduction, this area is marked as an opportunity area for especially reducing fatalities (Klimaateffectatlas, 2020). There are few options for vertical evacuation in this area (meaning there are few high points one

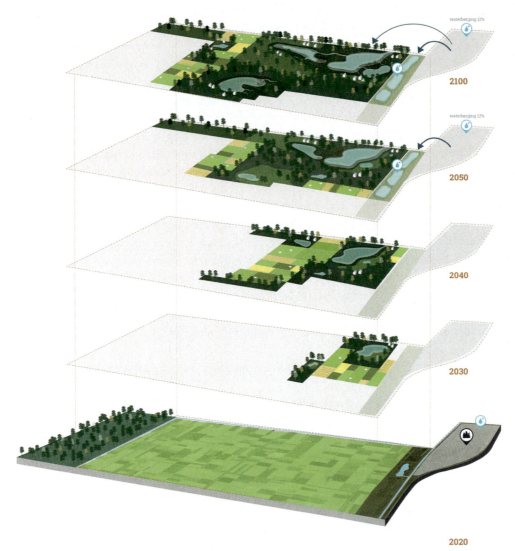

Fig. 4 Scheme illustrating the proactive strategy to address land subsidence with changing land use when necessary. This, depending on the severity and speed of climate change, results in a stepped transition from agriculture to nature-inclusive agriculture to eventually a nature area. *(Image by Defacto Urbanism.)*

could reach to save themselves in the event of a flood), so adding vertical evacuation options would reduce flood risk. For this area, amphibious housing would be an option. By building houses from lightweight materials such as polymers, they can remain afloat during a flood, preventing damage and fatalities. This concept fits well with the large proportion of the targeted single-family home type. An added benefit is that the lightweight construction without a foundation prevents subsidence-related prolapses.

Fig. 5 Collages illustrating the amphibious street furniture, which in case of flooding can serve as a lifeboat. *(Images by Defacto Urbanism.)*

However, from a financial standpoint, for larger urban areas, it is more effective to protect an area by dike rings than to adapt individual properties. Yet when no-regret options are possible to increase flood resilience without substantial additional cost, they may very well be applied to complement the dike rings. One example is amphibious street furniture that can be used on a daily basis as street furniture in rural and natural areas with poor vertical evacuation capabilities and can also function as a lifeboat during a flood (Fig. 5).

Demonstration project 3. Heuvelrug: Restoring the natural water infiltration system to reduce runoff and drought

The Heuvelrug is an area with a gradient from low-lying peat polders to a higher forested hill. The former peat extraction areas are now protected nature and recreation areas, which are important for the local economy. Several small towns can be found on the slope and ridge of the hill. They are known for their favorable settlement conditions, as open space and green surroundings are available while being close and well connected to the main cities. Due to urbanization and related paving and construction of drainage systems, rainwater runs off or is drained. The runoff causes localized stormwater nuisance in neighborhoods located in hillside pits. As less water infiltrates into the groundwater below the hillside, resulting in a reduction in available drinking water and seepage, it feeds the peat lakes at the base of the hill. In summer, drought results in reduced water quality (and damage to natural areas), increased subsidence, and a water supply limit for some functions. The expected increased temperatures will further accelerate this and also increase the urban heat island effect of the very stony settlements, as well as the risk of forest fires.

Applying the "business as usual" technical system would result in an increase of the rainwater discharge and need for pumping capacity and an additional water inlet from a discharge channel. The drained rainwater has a high phosphate content and therefore needs to be treated upon release. All in all, the system will need to be expanded and substantially invested in to meet the climate-related challenges.

The alternative proposal is an integrated land use and water system approach based on restoring infiltration of rainwater (Fig. 6). By greening the settlements and developments and collecting rainwater at the ridge and slope of the hill, water that would normally be drained can be infiltrated or reused. This reduces (or does not further increase) pressure to the drainage and water supply system and possibly prevents the need for some upcoming investments for expansion of the technical system. When more water infiltrates, this will also benefit the drinking water supply and the amount of high-quality seepage exiting into the peat pools will increase. With relatively simple, low-cost local interventions such as bioswales, permeable pavements, water retention ponds, green roofs or facades, and rain tanks, a step-by-step transition can be started. An important element of this would also be the change of vegetation; the Heuvelrug has a large planted nonnative pine population that consumes more water than native species. By slowly transitioning the vegetation to more localized deciduous trees that are diverse, vary in age, and are drought tolerant, the forest becomes more robust and therefore climate adaptive.

Demonstration project 4. Amsterdam forest: A new water retaining forest for the further densifying city of Amsterdam

The Amsterdam forest is a monumental forest in the Amsterdam urban area. The forest is popular for recreational activities, such as walking, swimming, and canoeing, but bigger festivals and theater productions also take place here. The forest was created in 1934 by famous designers Jacoba Mulder and Cornelis van Eesteren (Dupon & Van der Werf, 2019). Apart from a recreational function, the forest also has an important ecological function; for instance, the protected grass snake can be found here. With the further densification of the Amsterdam urban area and the increased temperature, the importance of the park as a cool, recreational space where people can refresh themselves will further increase. However, the water of the forest is currently already frequently covered with blue-green algae and due to the rising temperatures, the water quality will further decrease. At the moment this is being treated with de-phosphating installations. The water level of the park is kept very stable, while excess rainwater is pumped to the adjacent discharge channel. This channel discharges to the North Sea Channel. During peak rain events, the capacity of the North Sea Channel is already pressured; sea level rise, extreme rain events, or a failing pump can result in reaching the pumping limit into the North Sea Channel (Vermeulen et al., 2019).

If the technical "business as usual" system were to be applied to solve the water challenges, this would result in a combination of increasing the capacity of the drainage channels and purifying the phosphate-rich water before entering the forest. Increasing the capacity of the North Sea Channel may require a substantial investment, as not only

Ridge Strategy

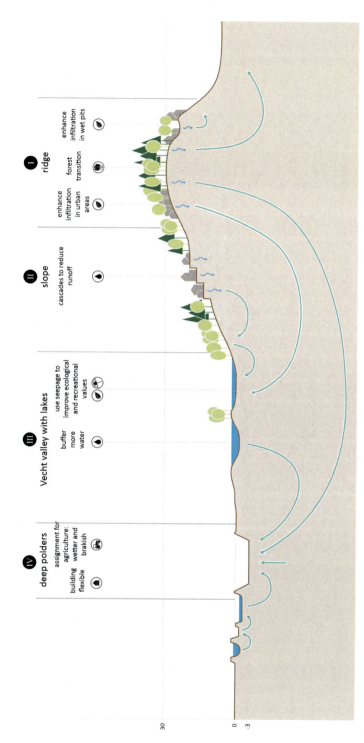

Fig. 6 Scheme showing the different zones along the gradient of the Heuvelrug. Based on an integrated land use and water system approach and strategy for each zone, different local nature-based development principles are defined; by applying these with upcoming developments and transitions, the natural water system is restored step by step. *(Image by Defacto Urbanism.)*

would the water level of these channels need to be raised, but also the pumping capacity and supply of water channels may need to be increased.

The alternative proposal is based on capturing and treating the rainwater of the Amsterdam forest within the forest. By extending the route of the water through waterways already available from the original design (rather than draining it through a shortcut) and adding green infiltration meadows, the water quality can be improved within the forest while reducing the discharge. To make sure the trees can adapt to the varying water levels, this will require a slow transition. Some areas of the forest, being the meadows and the new forest extension, can be implemented in the short term. With this change, the forest water quality will improve and the discharge capacity will be increased. Looking at the necessary investments, restoring the forest system will be a bigger investment in the short term, but in the long term much will be saved by further decreasing investments in symptom control.

However, this discharge reduction is of the sort that should be applied in many other places as well, to ultimately substantially reduce the discharge into the North Sea Channel. An additional option is to establish a new "Amsterdam Water Forest" with water storage as a specific function (Fig. 7). By creating such an additional forest, a peak water discharge area could be created, while increasing the amount of recreational and

Fig. 7 Collage illustrating how the function of water retention area and recreational and ecological space can be combined in a new Amsterdam water-forest. *(Image by Defacto Urbanism.)*

ecological space. Strengthening of the ecological structures is required anyway, as the natural part of the Amsterdam forest is expected to be pressured, due to rising temperatures and increased recreational usage.

Although creating a new forest will require large investments, especially in an area like the MRA with limited land available, there would be many beneficiaries and potential investors for this high-end project. For instance, through establishment of a forest fund, the necessary financial means could be obtained.

Conclusions

Applying an integrated land use and water system approach, which looks at the impact of economic development and climate change on the water and land use system from a long-term perspective, provides important insights into future impacts and thresholds. A long-term system strategy for expanding the current water system is essential to avoid investments that locally treat single symptoms rather than restore or improve the system as a whole (Nillesen, Koning, Stuurman, Dolman, & Gleijm, 2020).

Investments aimed at controlling symptoms in the short term can result in a lock-in situation in the long term. With a short-term cost-benefit analysis that does not yet take into account significant effects of climate change, addressing the symptom may seem preferable to a system restore that requires more investment costs in the short term, for example, in the case of the reduction of blue-green algae. However, when longer term effects of climate change are taken into account (e.g., rising temperatures 2070–2100), reduction in costs can increase exponentially over time when climate change has more impact.

In exploring the alternative options for expanding the water management system, including the potential long-term effects of climate change (system thresholds, future investments, and damages), it is essential to be able to make proper trade-offs. The integrated approach and the associated multidisciplinary workshops are essential in identifying integrated strategies in which water system expansion can be linked to spatial development projects. First, this strengthens the business case as the investments bring multiple benefits and advantages. In addition, it is essential for the feasibility of the alternative water system options (robust system and changing land use) that often require a lot of space, to explore how they can be linked to other functions and spatial developments.

For the MRA, several locations were identified where applying an alternative approach for addressing climate change (compared to the often "business as usual" technical approach) was feasible and desired. To be able to seize these opportunities, urgent action is required in some cases. By now including proactive climate change adaptation aspects in upcoming developments and investments, future damages of problem-solving investments can be reduced or prevented. In some cases, this already requires short-term investments in climate adaptation and system recovery (Amsterdam forest), while in others the investment can be stepped up following spatial development (Heuvelrug) or even be reactive (in case of the Flevopolder).

Reflection of applicability in the Houston context

In the Houston region, the water supply and flood risk management system and land use of the coast, bay, and bayous are strongly related, comparable to the functioning of these systems in the MRA. With climate change, the impacts of heat, hurricanes, and rainfall might even be more severe in Houston. However, as in the MRA, in Houston, water management and land use projects and interventions are usually not considered from a water system approach; nor do they take long-term climate change and spatial development scenarios into consideration (e.g., up to 2100). Although the regional planning policies in Texas are not as well equipped and intended to steer urban development as they are in the Netherlands, applying an integrated long-term land use and water system approach in order to explore different options for a long-term water management strategy could also provide valuable insights here.

By incorporating future urban development and climate change into the long-term scenarios for bayou water levels, the location of the floodplains along the bayous might become different. These locations can then be used to anticipate flooding by establishing development guidelines for upstream and downstream developments. For example, for new upstream developments, requirements can be set for nature-based solutions that store and/or delay stormwater runoff where necessary. Examples include rain tanks, permeable pavements, water storage ponds, greening gardens, providing water storage functions in public parks and sports fields, or expanding floodplains. Downstream flood resilient development regulations, capable of addressing future urban impacts and climate change impacts on stormwater runoff and flooding could be implemented. In this way, flood damage to urban developments (with a life span of several decades) caused by increased flooding due to climate change and urbanization can be anticipated and prevented.

References

Deltacommissie. (2008). *Working together with water, a living land builds for its future: Findings of the Deltacommissie 2008.* The Hague: Deltacommissie.

Dupon, S., & Van der Werf, J. (2019). *Het Amsterdamse Bos, een geschiedenis.* Bussum: Uitgeverij Toth (in Dutch).

ENZH. (2020). *Concept-RES NHZ. Energieregio Noord-Holland Zuid.* Available online https://energieregionhz.nl/conceptres (in Dutch).

Haasnoot, M., Bouwer, L., Diermanse, F., Kwadijk, J., van der Spek, A., Essink, G. O., et al. (2018). *Mogelijke gevolgen van versnelde zeespiegelstijging voor het Deltaprogramma. Een verkenning.* Deltares rapport 11202230-005-0002.

Hooimeijer, F. L. (2011). *The tradition of making: Polder cities.* Delft University of Technology (PhD thesis).

Klimaatatlas MRA (2020). Online viewer: https://mra.klimaatatlas.net/.

Klimaateffectatlas (2020). Online viewer: https://www.klimaateffectatlas.nl/nl/kaartverhaal-overstroming.

LIWO. (2020). *Landelijk Informatiesysteem water en Overstromingen [National information system water and floods].* Available online www.helpdeskwater.nl/onderwerpen/applicaties-modellen/applicaties-per/watermanagement/watermanagement/liwo/.

Mens, M., van der Wijk, R., Kramer, N., Hunink, J., de Jong, J., Becker, B., et al. (2018). *Hotspotanalyses voor het Deltaprogramma Zoetwater. Inhoudelijke rapportage.* Deltares rapport 11202240-004-ZWS. Delft, mei 2018.

Metropoolregio Amsterdam. (2020a). *Online program description.* https://www.metropoolregioamsterdam.nl/programma/klimaatadaptatie/resiliencebydesign/ (in Dutch).

Metropoolregio Amsterdam. (2020b). https://www.metropoolregioamsterdam.nl/over-mra/ (in Dutch).

Metropoolregio Amsterdam. (2020c). *Online description housing policy.* https://www.metropoolregioamsterdam.nl/afstemming-woningbouw-tot-2040/ (in Dutch).

Must. (2018). *Klimaatbestendige, vitale en kwetsbare functies Metropoolregio Amsterdam.* Amsterdam: Quickscan Kaartenatlas. (in Dutch). 15 November 2018 www.must.nl.

Nillesen, A. L., Koning, R., Stuurman, R., Dolman, N., & Gleijm, A. (2020). *Resilience by design Metropoolregio Amsterdam—landelijk gebied.* Available online: https://www.metropoolregioamsterdam.nl/programma/klimaatadaptatie/resiliencebydesign/ (in Dutch).

Rebel Group. (2019). *Concept eindrapport Werken met Natuur, Bouwsteen voor rivierengebied WNF* (in Dutch). Rotterdam: Rebel Group.

Rijkswaterstaat. (2003). *Het Nationaal Bestuursakkoord Water.* Available online. https://puc.overheid.nl/doc/PUC_51260_31/1 (in Dutch).

Vereniging Deltametropool. (2019). *Ruimtelijke opgaven voor de MRA gebiedsateliers.* Rotterdam: Vereniging Deltametropool.

Vermeulen, C. J., Honingh, C., Oosterhoff, C., Zandvoort, M., & Hakvoort, H. (2019). *Watersysteemanalyse ARK/NZK-gebied.* Lelystad/Delft/Rotterdam.

World Bank and World Resources Institute. (2018). *Nature-based solutions for disaster risk management.* Washington, DC: World Bank and World Resources Institute.

CHAPTER 17

Green infrastructure-based design in Texas coastal communities

Galen Newman and Dongying Li
Department of Landscape Architecture and Urban Planning, Texas A&M University, College Station, TX, United States

Introduction

Globally, more than 600 million people live in coastal regions that are less than 10 m above sea level, with 2.4 billion people living within 100 km of the coastline (United Nations, 2017). In the United States, 8% of the counties are located on the coast, with 9% (123.3 million people) of the population living in coastal counties (Wilson & Fischetti, 2010), and these numbers are projected to grow. Urban expansion due to population growth can worsen climate change conditions and enlarge hazard zones. As urban population accrues, impervious surfaces increase, resulting in amplifications in stormwater runoff and increased flood risk (Kim & Newman, 2019). The development necessary to accommodate such population growth leads to the conversion of open space into other land uses such as commercial, residential, or industrial. These land-use conversions amplify pollutant loads and the mixtures of hazardous substances during flood events (Ruckart, Borders, Villanacci, Harris, & Samples-Ruiz, 2004). Since stormwater runoff and flood event frequency are increasing due to climate change (Kim & Newman, 2020), the proper mixture of both structural (engineered) and nonstructural mechanisms (green infrastructure [GI]) in the design and planning for resilient communities has become a necessity for the appropriate mitigation of such conditions. The Netherlands, for example, has been setting a worldwide precedence in flood risk mitigation and serves as a model from which the United States can learn. The foundation for the Netherlands approach that utilizes multilayered safety for flood protection and defenses, spatial planning, and flood evacuation combines local- and regional-scale GI provisions with national and municipal-scale engineered infrastructure. As a result, coastal Texan cities have realized the importance of increasing both the quantity and quality of GI in mediating flood risk and decreasing pollutant loads. This chapter discusses the current use of GI in the design and planning of coastal Texan neighborhoods using a case study in League City, Texas, by applying landscape performance models to an urban design in a flood-prone area.

The shift toward green infrastructure for flood mitigation in coastal Texas

Coastal storm damage and related flooding threaten the Texas coastline on a regular basis. With nearly 40 significant hurricane strikes since 1900, Texas experiences a hurricane every 5 years, and a major storm every 15 years (Newman, Malecha, et al., 2020). After six major flooding events with federal disaster declarations in 5 years, the Texas coast has become a national beacon for flood risk. Flood events once projected to have less than a 2% chance of occurring have occurred annually on the Texas Coast (Inserra et al., 2018). Protection of the built environment and population is a necessary undertaking in the region as it is no longer a matter of if, but when another damaging flooding event will occur. For example, the May 2015 Memorial Day flood, the October 2015 Halloween flood, the April 2016 Tax Day flood, the August 2017 landfall of Hurricane Harvey (which produced a flood event that became one of the most damaging natural disasters in US history), and Tropical Storm Imelda in September of 2019 all serve as examples of intense flood events that are occurring with increased regularity along the Texas coast (Servidio, Shelton, & Nostikasari, 2020). The worst of these storms, Hurricane Harvey, dropped more than 50 in. of rain in some areas, resulting in the largest rainfall event in North American history. Widespread flooding inundated hundreds of thousands of houses, and more than 300,000 people were without power for days (Inserra et al., 2018). Harvey caused an estimated $125 billion in damages, tying Hurricane Katrina as the costliest disaster in US history (Servidio et al., 2020).

Decades of development in floodplains have magnified flood risk and contamination vulnerabilities in coastal Texas. The high winds, flooding, lightning, and other phenomena associated with adverse weather can cause power failures and equipment damage resulting in chemical releases or simply wash contaminated runoff across multiple neighborhoods. For example, during Tropical Storm Allison in 2001, heavy rains and flooding resulted in the release of more than 15,000,000 gal of phosphoric acid and 850,000 gal of sulfuric acid into the Houston Ship Channel (Ruckart et al., 2004). Compounding the problem, climate change has made events like Hurricane Harvey three times more likely, and, when they do occur, the storms are approximately 15% more intense (Servidio et al., 2020). NOAA recently updated the rainfall values used to define what qualifies as a 100-year or 1000-year event in Texas. This update shows that the rainfall amount previously considered as a 100-year storm event (an event with a 1% chance of happening in a given year) is now a 25-year storm event (with a 4% chance of happening within a given year) (Servidio et al., 2020).

The historic approach to flood control in coastal Texas was to simply convey water downstream as quickly as possible through channelized bayous or to construct large single-purpose detention basins to store water. More recently, Houston has begun embracing a more holistic approach that includes the practice of GI and low impact development (LID). Such developmental techniques use natural and open spaces to reduce stormwater runoff and treat flooding at its source (Hendricks, Newman, Yu, & Horney, 2018). More recently, increased investments in multifunctional applications of waterways and detention/retention basins have been made to serve flood control purposes and also as parks and trails. Such approaches have gained traction for floodplain management and buyout programs have also sprouted; this increase in GI and LID is evidenced by the passage of a $2.5 billion Harris County bond measure promoting such techniques following Hurricane Harvey (Inserra et al., 2018).

Cities and towns in the Netherlands have been pursuing similar approaches for decades and have led the way for coastal Texan cities for effective GI implementation.

For example, IJburg, an urban district in the eastern part of Amsterdam currently under development, will eventually comprise around 18,000 homes but is utilizing existing natural functions and requirements to realize the goals of the district. The extraordinarily high average density of 71 homes per hectare allows IJburg to preserve existing GI as well as create three new nature areas: Hoekelingsdam, Diemer Vijfhoek, and Zuidelijke IJmeerkust. The abundance of crustaceans and wetland species make the area a magnet for waterfowl with nearly 100 different species of birds being protected from the current development approach. New islands are also being constructed to mitigate any negative impacts from the new development. Island creation, rather than land draining, focuses the plan on natural shore conservation and solutions for coastal hydrological issues. Other smaller scale GIs are interwoven into this system to treat runoff. Precipitation on IJburg is retained for as long as possible and then allowed to infiltrate into the ground via special drains and pervious surface materials. Once infiltration has occurred, the water passes through ecological riparian zones and reed banks that further purify the water and then is discharged into the surface water. The success of such systematic networks of GI in IJburg and other similar places in the Netherlands has set the stage for LID approaches in the United States.

LID and GI as flood mitigation tools

Flooding in coastal Texas is controlled, in part, by widespread bayous and watersheds with networks of creeks and waterways that connect to the Houston Ship Channel, Galveston Bay, and the Gulf of Mexico. In Harris County (home to Houston, Texas), parks and greenways are now designed as a GI system to supplement the networks of waterways. They provide multiple functions to Houston residents, such as sustainable ecosystem services, areas for recreation, and flood control (Ahern, 2013). GI is used to help mitigate flood risk and naturally filter polluted stormwater by reducing the speed and volume of stormwater discharges primarily through trees and vegetation (Newman, Sohn, & Li, 2014). In addition to flood mitigation, GI also provides other benefits to communities, including areas for physical activities, outdoor enrichment for children, and improved air quality (US EPA, 2015).

The Houston-Galveston Metropolitan Statistical Area is projected to grow by 3.5 million people in the next 25 years (Reja, Brody, Highfield, & Newman, 2017). With that population growth will come new homes, businesses, roadways, parking lots, and sidewalks, producing billions of square feet of new impervious surface area. Simultaneously, new stormwater drainage infrastructure that will alter natural drainage patterns will also accompany this growth. Conventional stormwater management in coastal Texas includes facilities such as concrete-lined detention ponds and underground engineered infrastructure (Newman, Guo, Zhang, Bardenhagen, & Kim, 2016). The soaring costs to handle such projected growth using only engineered infrastructure have driven regional and local planners to better integrate natural and landscape features into new developments in ways that reduce infrastructure costs, improve stormwater storage capabilities, and improve water quality. As such, the Houston-Galveston Area Council (H-GAC) has developed a low impact initiative and provides permits for LID projects, which maximize the use of GI. As part of this initiative, a set of guidelines were adopted for implementing GI into new community developments.

Located in the west-central area of Harris County and maneuvering through downtown Houston is the Buffalo Bayou Watershed. With a population of nearly 450,000, the watershed occupies 102 mile^2 of drainage area and is comprised of 106 miles of open channels. Due to the recent increase in LID in Houston, a series of GI projects have begun in parks within the Buffalo Bayou Watershed, including Channel Restoration at Buffalo Bayou Park, a Demonstration Project at Memorial Park, and an Emergency Repair Project at Terry Hershey Park. Each project repairs or restores green spaces in an effort to increase their ability to both absorb floodwaters and convey stormwater to designated areas. The Channel Conveyance Restoration at Buffalo Bayou Park is a project intended to repair the banks of Buffalo Bayou as a riparian zone to help attenuate overtopping from rivers flooding during intense rains or storm surge. Another project, the Memorial Park Demonstration, is a stream restoration project, which utilizes fluvial geomorphology design principles to preserve the bayou's flood conveyance capacity, repair bank erosion, reduce sediment deposition, and improve water quality. Relatedly, the Terry Hershey Park Emergency Repair Project seeks to repair the drainage slopes of open spaces within a dozen flood-damaged areas deemed as public safety hazards in 2016.

The H-GAC promotes the use of LID techniques in new development. While there is growing interest in LID throughout the Houston-Galveston region, there are still barriers to its broad acceptance, such as lack of public awareness, misperceptions, and incompatible local development codes. Designing for Impact, a resource center for the region, provides information and strategies for educating the public about LID functions, benefits, and solutions to overcome barriers to implementation.

Planning and design promoting GI

The shift toward the use of LID and GI in coastal Texas has created more interests in design and planning strategies for regions and neighborhoods. Large-scale GIs such as parks, wetlands, and riparian areas have been proposed to break up impervious surfaces and retain stormwater while small-scale GI such as bioswales, rain gardens, and infiltration basins are being scattered and connected within new community developments. These interconnected GI systems provide multiple ecosystem services. For example, protecting open space in floodplains significantly reduces the adverse effects of flooding and helps remediate contaminated water. Such areas can also act as storm buffers to adjacent properties; a national study of localities showed average savings of approximately $200,000 per year in flood-related losses by protecting open space in the 100-year floodplain (Brody & Highfield, 2013).

GI can store, hold, and disseminate polluted floodwater, reduce peak riverine flows, suppress storm surges, and mitigate the mixtures of hazardous substances during flood events (Borsje et al., 2011). The loss of GI has been demonstrated to result in severe economic damage. In coastal Texas, Brody, Zahran, Highfield, Grover, and Vedlitz (2008) found that the loss of wetlands across 37 coastal counties from 1997 to 2001 was associated with an increased amount of property damage from floods. The loss of an acre of naturally occurring wetlands from 2001 to 2005 along the Gulf of Mexico coast increased property damage caused by flooding by an average of $7,457,549 (Brody, Peacock, & Gunn, 2012). The utilization and revitalization of riparian areas for flood retention, the allocation of additional space for water detention, and the development of architectural standards for future development in flood-prone areas are also being utilized in the region (Nillesen & Singelenberg, 2011).

Current design approaches for storm surge infrastructure in the United States primarily favor mechanical solutions to increase development potential in areas where flood risks hinder development opportunities. Few examples, however, illustrate harnessing the power of wetlands and green spaces as a means of attenuating flooding (Newman, Brody, & Smith, 2017). The appropriate use of GI in existing and new communities not only serves as a stormwater management tool but can also enhance the economic performance of real estate while improving drainage system performance. GI has been shown to help real estate projects realize higher operating income, faster lease-up or sales, higher occupancy, higher amenity values, greater lot or unit yield, and reduced drainage system costs. Simultaneously, GI can improve neighborhood resilience, reduce drainage concerns from small storms, reduce contamination in water, improve neighborhood aesthetics, and improve public health (Sohn, Kim, & Newman, 2014).

The current research aims to use League City, Texas, as a testbed to demonstrate a holistic planning and design framework that utilizes GI to achieve a myriad of environmental, social, and economic benefits. Quantifiable metrics are used to evaluate landscape performance regarding reduction in pollutants and stormwater runoff. These methods provide a replicable example of GI-based community redevelopment planning strategies that can be adapted for other flood-prone coastal communities in Texas.

Application project of GI in a community design along the Texas coast
Study area

When facing current flood risks and the eventual effects of sea-level rise, designers must think in terms of resilience and low-impact solutions. NOAA predicts that the mean sea level will rise at least 0.82 in. per year in the Gulf Coast region, reaching up to 6.29 ft by 2100 (Newman, Shi, et al., 2020). Sea-level rise has had a significant impact on coastal ecosystems and existing GI resulting in wetland loss, increased coastal erosion/inundation, and increases in the duration and frequency of flooding from storm surge. For example, in 2008, Hurricane Ike wreaked havoc on Texas, causing 113 deaths and $29.5 billion in damage; approximately 200 of the damaged homes were located in League City, Texas (Newman, Malecha, et al., 2020).

League City, connected to the Houston Ship Channel by Clear Creek, is exposed to many environmental hazards and is highly vulnerable to flood events and other issues affected by sea-level rise. With such issues in mind, a highly flood vulnerable 97-acre site on Robinson Bayou in League City, Texas (Fig. 1) poses an opportunity for a forward-thinking and resilient design using GI. League City, one of the fastest-growing cities in Texas, is forecasted to triple in population by 2040 (Horney, Dwyer, Vendrell-Velez, & Newman, 2019). The site is located in an environmentally sensitive region within Galveston County where wetlands are diminishing, saltwater is increasingly intruding, and hurricane storm surges are reaching up to 12 ft in height. As such, the primary aim of the design was to utilize GI processes to serve as flood attenuation mechanisms to protect human and nonhuman populations.

The population of League City is growing rapidly, increasing by more than 80% between 2000 and 2010, making new development all but inevitable. However, the city's comprehensive plan calls for an increase of residential land uses by nearly 80% in the next 20 years but only a 2% increase in green space; the design site is currently planned to grow as "auto dominated residential" and "urban low-density commercial

Fig. 1 Location of League City, Texas.

development." Based on these conditions, the objective of this project is to design a community that is fully protected from current and future flood events and the eventual impacts of sea-level rise, thereby reducing contamination levels in future stormwater runoff and hazardous substance mixing during future flood events.

According to GIS analysis conducted, currently, the design site will be significantly inundated by storm surge from hurricanes of Category 3 and above. Relatedly, sea-level rise will increase the vulnerability of coastal communities to hazards such as flooding and hurricanes. By 2100, nearly 50% of League City is predicted to be covered by the FEMA 100-year floodplain, should a 6 ft sea-level rise occur, with 76% of the land on the design site projected to be affected. Over 41% of League City is significantly vulnerable to flooding, and the site rests in one of the highest flood vulnerable areas in the city. With regard to the catastrophic damage these conditions have had on ecosystems, nearly 96 acres of freshwater wetlands and 154 acres of wetlands within the region have been lost since 2008 (the site has lost 43% of its wetland area in the past 20 years) (Newman, Malecha, et al., 2020).

Design strategy and layout

The concept of the master plan (see Fig. 2) was to intervene and treat stormwater issues at their source by using a series of strategies based on LID and GI principles: Treatment, Omission, Remediation, and Enhancement. *Treatment*, the application of needed GI in areas of extreme flood risk and other environmental hazards through natural detention and bioswales, improves the quality of natural resources on-site, softens the impacts of new development, and acts as a permeable ground for future flood hazards. *Omission*, the removal or lessening of flood risk in currently threatened or vulnerable areas, is

Fig. 2 Conceptual master plan for League City, Texas.

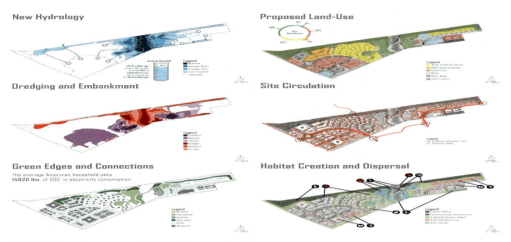

Fig. 3 Schematic representation of the master plan.

achieved by increasing the bayou's holding capacity on-site, ultimately elevating new development out of the flood plain via contour adjustments. *Remediation*, the reclamation of lost habitat, concentrates on implementing native ecologies, such as sensitive wetlands and riparian forests through habitat revitalization or protection mechanisms against human disturbance. *Enhancement*, the improvement or addition of conditions of GI within developed areas, is cultivated by the addition of outdoor spaces for living, working, commerce, events, and recreation.

The GI used on-site integrates a natural network throughout both developed and undeveloped areas combining large-scale infrastructure such as wetland reclamation and detention ponds with small-scale infrastructure such as bioswales, tree box filters, rain gardens, and filter stripes (see Fig. 3). By adopting compact development strategies, the new design not only calls for higher density residential and commercial land uses (34$ and 20%, respectively) to retain similar amounts of developed units, but also allow for the preservation of the existing and reclaiming of the lost wetland areas. The land-use plan for the design shows around 40% green space on-site, utilizing the central wetland not only as a sponge for stormwater and surge but also as a multifunctional recreational space for residents and visitors. Lower density structures are proposed closer to the floodplain, but built on stilts to allow for flooding underneath, if necessary. Mounding and increased elevation of higher density developments on the outskirts of the design site are also suggested. The master plan for the site is set up/implemented according to three strategic phases (see Fig. 4).

- *Phase 1) Excavate and establish*—focuses on placing new development in areas with higher elevation and integrating GI in newly developed areas. Dredging and other sediment removal strategies are utilized that restore the ecology of the waterway. Large-scale and linear GI projects will increase the retention capacity of the site. The

Fig. 4 Design phasing and impacts per phase.

designed medium density commercial and mixed-use spaces are connected with existing arterials integrated into the surrounding residence. These will serve as anchors to stimulate future growth. Meanwhile, the developments are strategically placed outside of flood-prone areas to mitigate risks. Public urban spaces with landscape features and permeable paving promote infiltration, retention, biological treatment, and evapotranspiration processes.

- *Phase 2) Access and Inhabit*—develops lower density residences and green infrastructure to provide protection during frequent storms. Diverse housing typologies allow for residents of broad age groups and backgrounds. Across these developments, small-scale GI projects are implemented to capture and treat rainwater on-site and reduce stormwater runoff.
- *Phase 3) Invest and Enhance*—completes and links major installation of GI and wetland preservation/reclamation to create a multifunctional system that blocks and controls heavy floods and regulates hydrologic activities during extreme hazard events. An interconnected circulation system, including pedestrian trails, boat launch points, pedestrian bridges, and bicycle paths, will improve access to the waterfront, offer recreational programs and create economic opportunities (see Fig. 5).

Landscape performance

To assess the impact of the proposed design, we applied the long-term hydrologic impact analysis (L-THIA) low impact development spreadsheet to both the city's future plan and the current land-use plan we proposed. The (L-THIA), develop from Purdue University, is a web-based urban growth analysis tool that estimates long-term runoff and nonpoint source pollution impacts of different land-use development scenarios. It provides the estimated long-term average annual runoff rather than only during extreme events based on long-term climate data for specified state and county locations in the United States. It also generates estimates of 14 types of nonpoint sources pollution loadings to waterbodies (e.g., Nitrogen, Phosphorous, Suspended particulates, etc.) based on the land-use changes. The model has been used to track land-use change in watersheds for historical land-use scenarios, identify areas sensitive to nonpoint source pollution, and evaluate land-use development for nonpoint source pollution management (Bhaduri et al., 2000).

Table 1 describes the L-THIA outputs of runoff amounts and pollutant load projections based on each plan's land-use configurations. Overall, the current proposal increases the site's capacity to capture and store rainwater, decreases future flood risk, and improves site water quality. Specifically, the current plan reduces runoff volume by 37%, runoff depth by 33%, and average overall pollutant load (when considering all 14 contaminants) by 29%, when compared to League City's comprehensive plan. With every 1% increase

Green infrastructure-based design in Texas coastal communities 237

Fig. 5 Riverine area design for League City showing capacity to adapt to sea-level rise and surge.

Table 1 L-THIA model outputs showing runoff and pollutant difference projections for each land-use plan.

Measure	Variable	League City's comprehensive plan	Current plan
Runoff	Total annual volume (acre-ft)	37.26	24.84
	Avg. annual runoff depth (in.)	4.61	3.07
Pollutants	Nitrogen (lbs)	136	90
	Phosphorous (lbs)	136	90
	Suspended solids (lbs)	5635	3757
	Lead (lbs)	1	0.88
	Copper (lbs)	1	0.981
	Zinc (lbs)	18	12
	Cadmium (lbs)	0.097	0.064
	Chromium (lbs)	1	0.676
	Nickel (lbs)	1	0.798
	BOD (lbs)	2335	1556
	COD (lbs)	11,778	7852
	Oil and grease (lbs)	913	609
	Fecal coliform (millions of coliform)	3184	2123
	Fecal strep (millions of coliform)	8308	5538

in the GI element, there is an estimated 1% decrease in runoff volume and a 0.7% decrease in pollutants. While all contaminants are predicted to decrease, fecal strip, suspended solids, chemical oxygen demand (COD), biological oxygen demand (BOD), and oil and grease show the most salient decreases in loads.

Moving forward

As climate change spurs increased flood frequency, there is an increased risk of such flooding releasing toxic pollution from contaminated sites into surrounding and downstream communities. Excessive contamination only adds to the issues faced by cities and neighborhoods experiencing flood events. For example, Hurricane Harvey's torrential downpour carried industrial chemicals, including substances that increase the risks of cancer, released from heavily contaminated industries along the San Jacinto River in Texas. Related to this circumstance, in 2019, the US Government Accountability Office issued a report that recommended that the EPA take additional actions to mitigate the risks that climate change poses to contaminated sites. The report found that 945 Superfund sites on the National Priorities List (sites posing the largest threat to public health) are threatened by climate change; of these sites, 187 are at risk of being flooded by a Category 4 or 5 hurricane (Moran, 2020).

As shown in this chapter, GI can be a significant design factor when seeking to lower flood risk and clean stormwater runoff pollutant loads. Low impact-based models of community development, which prominently feature the GI as fundamental to the development process, provide multiple ecosystem services that also help support human life, health, and well-being. Increased impervious surfaces, land-use conversion, flood frequency, and contamination levels along the Texas coast amplify the need for systematic uses of GI in planning and design. The current push toward GI and LID in Texas has slowly gained traction, but needs much more utility and impact measurement to be effective enough to appropriately reverse some of the impact of climate change.

In Houston, a large amount of the new urban area is being constructed within the 100-year floodplain, with more future urban development expected. The conversion of natural space to impervious surface increases runoff and the flood damage as water storage capabilities also diminish. To minimize flood damage, more efficient and effective structural and nonstructural flood protection mechanisms should be implemented to better manage floodplains and reduce high risk areas. Like many areas in the Netherlands, GI provisions should be considered a priority for coastal Texan communities to better protect new developments. As noted, the projected sea-level rise will impact a large amount of urban area and exacerbate the land under flood risk in Houston. Learning from the approaches taken in the Netherlands in areas such as IJburg, city and regional agencies within the Houston-Galveston Metropolitan Statistical Area must consider grander and multiscalar activities such as state-level dikes and levees. Similar to IJburg the strategic identification of future development zones, their current and future projected flood risk, and planned out mechanisms to lower this risk incorporating large-scale engineered and local-scale GI are necessary undertakings. A concerted global effort to mitigate flood damage will require each location to learn from one another, better prepare for future flood-related challenges, and decipher successful approaches to resilient urban growth.

Acknowledgments

Special thanks to Phillip Hammond, Alaina Parker, Claudia Pool, Maritza Sanchez, and Molly Morkovsky for their assistance in producing the designs and drawings for League City and their research on IJburg to assist in the site design.

References

Ahern, J. (2013). Urban landscape sustainability and resilience: The promise and challenges of integrating ecology with urban planning and design. *Landscape Ecology, 28*(6), 1203–1212.

Borsje, B. W., van Wesenbeeck, B. K., Dekker, F., Paalvast, P., Bouma, T. J., van Katwijk, M. M., et al. (2011). How ecological engineering can serve in coastal protection. *Ecological Engineering, 37*(2), 113–122.

Brody, S. D., Peacock, W. G., & Gunn, J. (2012). Ecological indicators of flood risk along the Gulf of Mexico. *Ecological Indicators, 18*, 493–500.

Bhaduri, B., et al. (2000). Assessing watershed-scale, long-term hydrologic impacts of land-use change using a GIS-NPS model. *Environmental Management*, *26*(6), 643–658.

Brody, S. D., & Highfield, W. E. (2013). Open space protection and flood mitigation: A national study. *Land Use Policy*, *32*, 89–95.

Brody, S. D., Zahran, S., Highfield, W. E., Grover, H., & Vedlitz, A. (2008). Identifying the impact of the built environment on flood damage in Texas. *Disasters*, *32*(1), 1–18.

Hendricks, M., Newman, G., Yu, S., & Horney, J. (2018). Leveling the landscape: Landscape performance as a green infrastructure evaluation tool for service-learning products. *Landscape Journal*, *37*(2), 19–39.

Horney, J., Dwyer, C., Vendrell-Velez, B., & Newman, G. (2019). Validating a comprehensive plan scoring system for healthy community design in League City, Texas. *Journal of Urban Design*.

Inserra, D., Bogie, J., Katz, D., Furth, S., Burke, M., Tubb, K., et al. (Eds.). (2018). *After the storms: Lessons from hurricane response and recovery in 2017* Heritage Foundation.

Kim, Y. J., & Newman, G. (2019). Climate change preparedness: Comparing future urban growth and flood risk in Amsterdam and Houston. *Sustainability*, *11*(4), 1048.

Kim, Y., & Newman, G. (2020). Advancing scenario planning through integrating urban growth prediction with future flood risk models. *Computers, Environment, and Urban Systems*, *82*, 101498.

Moran, G. (2020). *Toxic waste sites aren't prepared for hurricane season*. Popular Science. August 12, 2020. Retrieved from https://www.msn.com/en-us/news/technology/toxic-waste-sites-arent-prepared-for-hurricane-season/ar-BB17SEYe?fbclid=IwAR1x9zVNa62osqqlAv_OsM0Ok_jjFng-ifyHv9TXZkOYR7KyT7P4K1eN4rw.

Newman, G., Brody, S., & Smith, A. (2017). Repurposing vacant land through landscape connectivity. *Landscape Journal*, *36*, 37–57.

Newman, G., Guo, R., Zhang, Y., Bardenhagen, E., & Kim, J. H. (2016). Landscape integration for storm surge barrier infrastructure. *Landscape Architecture Frontiers*, *4*(1), 112–125.

Newman, G., Malecha, M., Yu, S., Qiao, Z., Horney, J., Lee, J., et al. (2020). Integrating a resilience scorecard and landscape performance tools into a Geodesign process. *Landscape Research*, *45*(1), 63–80.

Newman, G., Shi, T., Yao, Z., Li, D., Sansom, G., Kirsch, K., et al. (2020). Citizen science-informed community master planning: Land use and built environment changes to increase flood resilience and decrease contaminant exposure. *International Journal of Environmental Research and Public Health*, *17*(2), 486.

Newman, G., Sohn, W. M., & Li, M. H. (2014). Performance evaluation of low impact development: Groundwater infiltration in a drought prone landscape in Conroe, Texas. *Landscape Architecture Frontiers*, *2*(4), 122–133.

Nillesen, A. L., & Singelenberg, J. (2011). *Amphibious housing in the Netherlands: Architecture and urbanism on the water*. Rotterdam: BK Books.

Reja, M. Y., Brody, S., Highfield, W., & Newman, G. (2017). Hurricane recovery and ecological resilience: Measuring the impacts of wetland alteration post hurricane Ike on the upper TX coast. *Journal of Environmental Management*, *60*(6), 1116–1126.

Ruckart, P. Z., Borders, J., Villanacci, J., Harris, R., & Samples-Ruiz, M. (2004). The role of adverse weather conditions in acute releases of hazardous substances, Texas, 2000–2001. *Journal of Hazardous Materials*, *115*(1–3), 27–31.

Servidio, C., Shelton, K., & Nostikasari, D. (2020). *Community resilience initiatives: Building stronger neighborhoods in Houston*. Rice University Kinder Institute for Urban Research. https://doi.org/10.25611/7m56-j939.

Sohn, W., Kim, J. H., & Newman, G. (2014). A blueprint for stormwater infrastructure design: Implementation and efficacy of LID. *Landscape Research Record*, *2*, 50–61.

United Nations. (2017). *Factsheet: People and oceans*. https://www.un.org/sustainabledevelopment/wp-content/uploads/2017/05/Ocean-factsheet-package.pdf (12.03.2020).

US EPA. (2015). *Benefits of green infrastructure*. Overviews and Factsheets US EPA. September 30, 2015 https://www.epa.gov/green-infrastructure/benefits-green-infrastructure.

Wilson, S., & Fischetti, T. (2010). *Coastline population trends in the United States 1960 to 2008*. https://www.census.gov/prod/2010pubs/p25-1139.pdf (12.03.2020).

CHAPTER 18

Integrated urban flood design in the United States and the Netherlands

Fransje Hooimeijer, Yuka Yoshida, Andrea Bortolotti, and Luca Iuorio
Delta Urbanism Research Group, Department of Urbanism, Faculty of Architecture and the Built Environment, Delft University of Technology, Delft, The Netherlands

Introduction

Delta landscapes are characterized by a dynamic relation between land and water in which floodplains play a crucial function in sustaining an integrated balance of soil and water systems. In the Netherlands, a highly advanced water defense system has tamed, and controls the delta dynamic since 1798 (the establishment of the National Water Department *Rijkswaterstaat*) with a focus on strengthening the line between land and water through dike systems. However, while spatial planning was formalized in the Housing Act in 1902 (Anonymous, n.d.), flood defense and urban planning have traditionally been treated as two separate fields. This has been perpetuated by the idea that flood management is fundamental in offering the primary safety condition to urban development. Van der Woud (1987) calls this the *conditio* sine qua non, meaning, without dikes there is no spatial order. Technical interventions make the Dutch territory livable, and the government has a strong and coordinated responsibility for both safety and spatial planning (Hooimeijer, 2014).

In the United States, the flood risk management paradigm is somewhat inverted compared to the Netherlands; it is not focused on flood defense by a line of dikes, but rather on the reduction of consequences by evacuation and insurance (Van Hugten, Huijsman, Kok, & Rooze, 2018). In the United States, the Federal Emergency Management Agency (FEMA) is responsible for emergency preparedness and response during a disaster, and recovery is managed by the USACE (US Army Corps of Engineers). In contrast to the Netherlands, which is a welfare state with high governmental trust, Americans (especially Texans) have a large distrust in the government and are more willing to tolerate a high level of personal flood risk (Kim & Newman, 2020; Malecha, Kirsch, Karaye, Newman, & Horney, 2020; Woodruff et al., 2020). Also, the rate of risk awareness in the Netherlands is low, while it is high in the United States. This affects US planning policies in which the role of the government is limited, increasing counterproductivity in creating new public amenities. Therefore, the choice for flood safety levels in the two countries is quite different. In the Netherlands, protection ranges between 1/4000 and 1/10,000, and in the United States, there is a maximum protection of 1/100 years regardless of vulnerability level (Merrell, 2015).

The Netherlands has an increasing concern for the impacts of climate change, subsidence, and urbanization and the current flood defense paradigm is highly questioned (Van Doorn-Hoekveld & Groothuijse, 2017). This doubt is also fed by the reality that sometimes heightening a dike is spatially complicated because there are already land uses. Also, the rising awareness for ecology and socioeconomic changes related to ongoing urbanization make the Netherlands turn away from the focus on a primary flood defense line (Kaufmann, Van Doorn-Hoekveld, Gilissen, & Van Rijswick, 2015). The Multi-Layered Safety approach (RWS, 2009) and the Vulnerability approach (De Graaf, 2009) are both Netherland-based frameworks used to understand other ways of reducing risk that are including the urban area behind the dike.

The Netherlands focuses on the primary flood defense line (believed to be the most cost-effective solution), but broadening the scope of flood protection to a wider zone is now required. While the Netherlands is moving toward accepting consequence reduction as a method of risk reduction, the United States is moving toward the probability reduction paradigm. Since Hurricane Katrina (2005), the US national policy of reliance on recovery from flooding was severed through the construction of the Greater New Orleans Barrier (Merrell, 2015). The role of space and design has been part of this new paradigm, where public value is adopted as part of reducing flood risk. This was exemplified in the Dutch Dialogues (Meyer, Morris, Waggonner, Nijhuis, & Pouderoijen, 2009) and Rebuild by Design programs (Bisker, Chester, & Eisenberg, 2015).

This chapter examines the change in perspective in which spatial design integrates flood defense for the Dutch case of Vlissingen with the focus on the reduction of consequences and of the US case in Galveston as an example to expand on the current evacuation approach with spatial adaptation. In order to understand the connectivity between flood risk reduction and the spatial approach, four concepts are used: Probability Approach, Multi-Layered Safety approach, Vulnerability approach, and the Dutch Layers approach explained in the following section.

Spatial design approach

From a hydraulic engineering perspective, the main goal for flood protection is to create a safe defense line and reduce uncertainties using a dike (Van den Hoek, Brugnach, & Hoekstra, 2012). However, in spatial design, safety is but one of the aims among others to balance ecological and cultural values and interests. Ecological and cultural aims can be translated into spatial order characteristics such as site spatial and environmental quality or functionality (Dammers, Bregt, Edelenbos, Meyer, & Pel, 2014). A key to spatial design is the ability to achieve multifunctionality through the integration of the values of multiple stakeholders synthesized in a complex and interdisciplinary project (Reed, van Vianen, Barlow, & Sunderland, 2017).

The design of flood defenses has impacts on natural and cultural landscape features and people's lives both on and behind the dike (Van Loon-Steensma & Kok, 2016). Flood defense is, therefore, also "a symbol of the relationship between man and nature, an identity of people, landscapes and countries" (Palmboom, 2017). The conscious inclusion of the urban area behind the dike as a part of flood defense (e.g., Integrated urban flood design) will integrate flood defense into the urban fabric. This fabric is a complex interwoven spatial structure of systems, like infrastructure and parks, and objects, like buildings and bridges, that houses the necessary urban functions. There are multiple examples of integrated urban flood design in the Netherlands, such as the "Room for the River" project (Van den Brink et al., 2019) and the integration of parking into the dike in Katwijk (Voorendt, 2017). However, integrated urban flood design into the larger urban zone, outside of the dike itself, is still underdeveloped.

One important aspect of the spatial design approach is understanding the spatial order, in order to prioritize and organize the spatial claims and interests. An influential methodology is the Dutch Layers Approach (De Hoog, Sijmons, & Verschuuren, 1998) in which the spatial organization of the landscape is defined by three layers: the substratum, the networks and the occupation. The substratum includes water, drainage canals, soil, etc.; networks include the infrastructure, nodes, regional public transport, etc.; and the occupation layer refers to how humans build on top of both layers. These layers all have different timescales to physically transform themselves, making time a crucial planning factor. The Dutch Layers Approach was inspired by the "layer cake" approach by US landscape architect Ian McHarg (1969) in which ecology is considered the *condition* sine qua non for urban development (Whiston Spirn, 2000). Design using scales of time and a process of landscape development is operationalized through thinking in layers to negotiate between low and high dynamics in landscape development.

The Netherlands, since the Housing Act of 1901 (Anonymous, n.d.) and via a tradition of National Spatial Reports from the 1960s, has had a stronger spatial approach (Ministerie VROM, 1991). As noted, the Netherlands is both small and geographically vulnerable and the control over spatial order in which public and private interests are balanced needed to be extremely high. In contrast, the US spatial order is not controlled by a spatial approach but generally characterized by the urban grid as a means of territorial organization of land use and mobility. The grid also allowed for the rapid subdivision of large parcels of land during centuries of rapid growth (Reps, 1965). It creates a simple and straightforward division between public and private land. Spatial planning and design from the era of industrialization in the United States is very much based on a regional landscape architecture approach reflected in the current Landscape Urbanism discourse, in which natural systems are put forth as the leading operational logic (Corner, 2006; Waldheim, 2006).

The spatial design potentials of the risk approach

The primary aim of flood defense anywhere should be managing and reducing the impacts of flooding on people (including health and life), the economy, cultural heritage, and the environment (Priest et al., 2016). The question is, how the risk approach (risk = probability × consequences) can be understood spatially, not only in casualties and damage, but how probability and consequence reduction can be done by spatial design. In order to integrate flood defense (the risk approach) within a spatial approach (the Dutch Layers Approach) the concepts of vulnerability and the Multi-Layered Safety approach are needed to make clear connections.

The concept of vulnerability is a useful link to spatial components and measures. Vulnerability is the extent to which both social and physical systems at risk are able to sustain damage and to return to their initial state, or further enhance their performance (Adger & Kelly, 1999). Many authors discuss the components of vulnerability to climate change; hence, there are many opposing views on its definition and understanding. For the purpose of this work, the Vulnerability approach by De Graaf (2009) is used because it explains the components of vulnerability through the sequence of four building capacities—threshold, coping, recovery, and adaptivity.

The objective of the threshold capacity is to reduce the probability by a primary defense line. The objective of the coping and recovery capacities are related to damage reduction, coping during, and recovery after an event. These two capacities can be taken into the spatial approach. Coping can not only be done through evacuation of people but also by building housing on piles. Recovering can be supported by not only having an insurance to rebuild after the hazard but also making structures flood-proof so they will not be damaged as well. The objective of adaptive capacity is to build both an understanding and an ability to foresee development for the future in which the main goal is to establish a robust and healthy living environment, making use of all three other capacities as an ensemble (De Graaf, 2009).

Dutch flood-risk policy has broadened its scope to a Multi-Layered Safety approach (RWS, 2009). The first layer is the threshold capacity; the second layer is the reduction of vulnerability by spatial planning tools and building codes to increase coping capability, recovery, and adaptive capacity; and the third and final layer is crisis management, represented by recovery capacity. The first layer of protection has traditionally been dominant in the construction of flood defense systems. Given that the United States often prioritizes the crisis management layer, an interesting case is created to explore how the first and second layers of the Multi-Layered Safety approach could be applied in TX (Kok & Brand, 2017) (Table 1).

These concepts show how spatial design of the area behind the dike can contribute to reducing the overall risk to flooding by reducing the consequences. Simultaneously, spatial development of the urban area behind the dike increases consequences because with new housing there is increase in real estate value and number of people. This is oftentimes referred to as "the spiral of risk" (Rijcken, 2017) and needs therefore to be part of spatial planning. The measures for spatial design in integrated urban flood design can be ordered in two groups: *integration with the urban fabric and reintegration with the ecological system.*

Table 1 Relation between the risk equation, the Vulnerability Framework, Multi-Layer Safety strategy, and Dutch Layers Approach.

risk =	probability see Fig. 1A, B, C, E	x	consequences see Fig. 1F and G	
Vulnerability framework (De Graaf, (2009)	threshold	coping - recovery		adaptation
Multi-Layered Safety (RWS, 2009)	first layer	second layer		
			third layer	
Dutch Layers Approach (De Hoog et al., 1998)	infrastructure			
		occupation		
		substratum		

The spatial impacts of probability and consequence reduction are linked to Fig. 1.

Integration with the urban fabric

At first, this new risk approach considers a broader flood defense system as well as the dikes and the urban area behind the dikes that are impacted. As shown in Fig. 1, the scope of the flood defense system is extended to the urban area, dike, and foreshore, and not merely focused on the line created by a dike. By increasing safety measures in the urban fabric such as improving evacuation routes or raising the ground level of buildings, there is a reduction in the number of casualties and/or economic damage. This contributes to reducing overall consequences and could lead to a reduction of overall flood risk (Nillesen & Kok, 2015).

Reintegration with the ecological system

The "spiral of risk" can also be an opportunity to make a difference in the way flood defense is planned in dense urban areas and how it is in rural areas. The rural areas could be planned and designed as natural areas where the occasional flooding is welcome to nurture the natural system. Thus, it can impact the land use and design of landscapes on the larger scale. Reintegration with the ecological system creates a potential to use nature-based solutions for risk reduction, such as planting a willow forest in a foreshore area to dissipate energy from surge waves and unload this pressure on the dike or dune. This allows for greater possibilities to incorporate nature-based solutions as an alternative or additional strategy to the conventional gray (or nonnature based) infrastructure flood defense systems (see Fig. 1D and E).

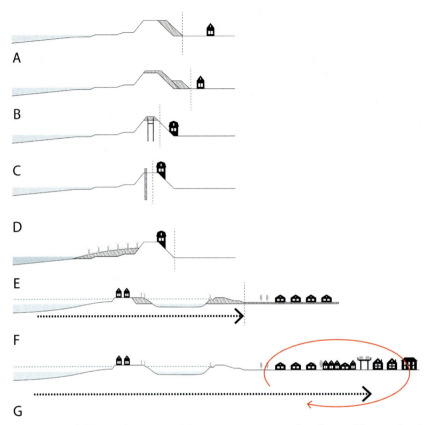

Fig. 1 (A–D) Scope of dike reinforcement; (E) wave attenuating willow forest; (F) room for the River; (G) broader scope by the new safety standard. *(Drawn by Y. Yoshida.)*

Case study: Vlissingen (Flushing)

Vlissingen is a cape town located on the coast where the North Sea and the Scheldt River intersect. The estuary acts as a funnel, which increases sea levels and flood risk. The current protection level is 1/30,000 years. This dike protecting the city is completely fixed, with buildings constructed atop it. Due to the proximity to the North Sea—Antwerp shipping route, seaward reinforcement in the foreshore is not possible. For this reason, the municipality developed the so-called "Vlissings Model" (Vlissingen & Ma.an, 2010), which includes the adaptation of the building code to allow for the possible raising of the dike.

The Vlissings Model regulates since the 1990s that new construction along the coastal boulevard must anticipate future flood risks, requiring that their foundations and ground floors be designed in such a way that are part of the future raised dike system. Some existing construction along the Boulevard (mainly hotels) has been built according to this

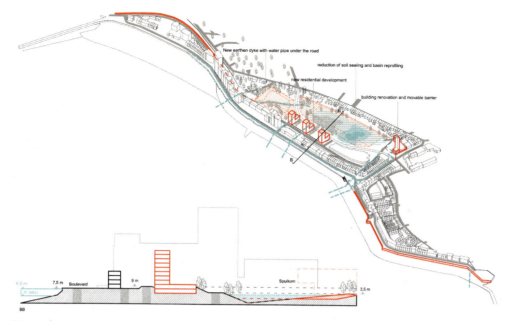

Fig. 2 The Overtopping Sump Model, making space for water (with estimated still water level at 2100 with 100,000 years RP). *(Drawn by A. Bortolotti.)*

principle, with higher ceilings on the ground floors to enable the incorporation of the ground floor into the new dike (Fig. 2).

However, the Vlissings Model may eventually lead to the replacement for the entire historical urban waterfront, an important part of the identity of the city. Therefore, a second option has been investigated to accept overtopping of seawater in the future and preserve the historical identity. The water would be conveyed to—and stored in—a former water reservoir located behind the dike, a green space called the Spuikom. This approach involves the rerouting of overtopping water from the seafront boulevard toward the Spuikom, street redesign, and basin reprofiling.

Design evaluation

Analyzed through the lens of the vulnerability concept, the Vlissings Model aims to increase the threshold capacity with spatial adaptation of the dike. The Overtopping Sump Model aims to maintain the threshold capacity and introduces the adaptive capacity of the area behind the dike. This integrated urban design is based on fine-grained and spatially explicit hydraulic models that support design decisions (e.g., the location of mobile barriers, the layout of new development, etc.). As such, it requires interdisciplinary engagement and a close collaboration between spatial designers and hydraulic engineers. Not only are hydraulic models necessary to project flood events (their extension and intensity) to suggest where to take spatial adaptation measures, but they are also an essential tool for assessing the effectiveness of such interventions within the timeframe of a flood event.

The Vlissings Model, considering the Multi-Layer Safety strategy and Dutch Layers approach, respectively, combines the first and second infrastructure and occupation layers by including private property as a part of the flood defense strategy. The Overtopping Sump Model is a new step in shifting the paradigm in which flood defense becomes part of spatial design (Fig. 3).

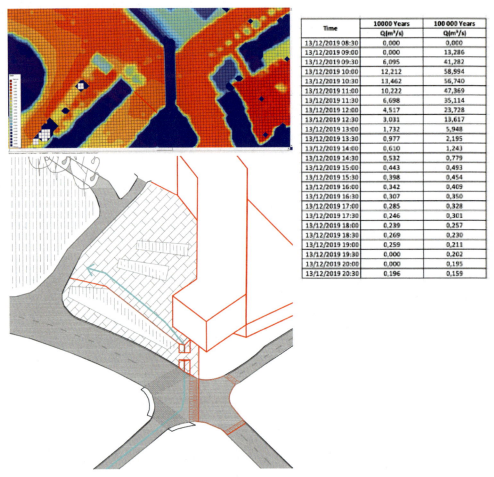

Fig. 3 Detail of the HydroNL Model for understanding volume and time sequence of the overtopping to design the street with for a probability of 1:10,000 years and 1:100,000 years in the W+ climate scenario. In the image, there is a barrier taken up in the public space design to guide the water flow to the Spuikom.

Case study: Galveston

Galveston Island, Texas, is located on a sand barrier island on the southeast coast of Texas, within the Gulf of Mexico. Surrounding the island, several bays provide important habitat, options for recreation, and access to the port of Houston. Galveston has suffered greatly during many past hurricanes (e.g., Ike, 2008; Harvey, 2017); excessive rainfall cannot always be drained, causing common local flooding complications called "nuisance flooding" (Van Hugten et al., 2018).

When a hurricane is expected, the National Weather Service (NWS) activates to recommend evacuation directions and supply essential needs (water, food, light). The USACE organizes temporary housing, and debris removal. FEMA leads the recovery afterwards for the reconstruction of properties and primary infrastructure. Due to the increasing occurrence of extreme weather events, evacuation recovery activities are becoming more complex and expensive, highlighting the need for a comprehensive strategy to mitigate the impact of urban flooding.

After Hurricane Ike in 2008, the concept of a coastal spine inspired by the Dutch Delta Works was intended to protect the Houston-Galveston area from a 10,000-year storm (Houston Chronicle, 2009; Kothuis, Brand, Sebastian, Nillesen, & Jonkman, 2015; Merrell, 2015). The so-called "Ike-Dike" raised several controversies related to its environmental impact, the economic investment (23–31 billion dollar), and the navigability of the channel (AP News, 2018). For Galveston, the strategy consists of raising the historical sea wall (built after the 1900 Great Storm) and creating new protective structures to encircle the urban development. The plan, known as the "Galveston Ring Barrier," risks could generate a "bathtub effect," which means that water would be trapped inside the barrier, creating more severe floods (Coastal Texas Study, 2020). This infrastructure is still in the early stages of planning (Powell, 2020) and open to public scrutiny, offering opportunities to think for measures, which may have a lower impact on the ecological and hydraulic balance of the bay as well as on the spatial development.

In response to the Galveston Ring Barrier, a project of TU Delft students named "Floodproof Galveston" (Van Hugten et al., 2018) was setup. Overall, 13 projects have been designed as a synergetic plan of local designs to reduce nuisance flooding and damage during hurricanes in Galveston. All designs focus on five general purposes: (1) protect evacuation infrastructure from flooding, (2) prevent inundation, when gravity discharge does not occur, (3) reduce the impact of waves, (4) facilitate recovery interventions, and (5) avoid hard infrastructure and prioritize natural solutions.

Evaluation of the design for Galveston

The plan (Fig. 4) by Van Hugten et al. (2018) for Galveston aims to mitigate nuisance flooding. Design proposals work together involving the transformation of the substratum and infrastructure layers, in order to increase the adaptation capacity of urban

Fig. 4 Integral design of mitigation measures (Van Hugten et al., 2018).

development. Coping and recovery are promoted by raising mobility hubs, installing pumps, and combining discharge and water storage measures. The probability of flood risk can be decreased by the integrated projects of partially raising the historic seawall, implementing coastal protection with dunes, sand bars, and building new wetlands.

The plan is composed from the interaction between all layers of the Multi-Layer Safety strategy and considers all components of the Dutch Layers approach, demonstrating the need for a multiscale design environment for water management. The outlined strategy examines the possibility to approach Galveston's flood issues by conceiving a more inclusive plan, where situational design proposals result from a collaboration between engineers, architects, and urban planners. The approach can be representative for a paradigm shift in the United States wherein they can move from nonspatial measures (evacuation and insurance) to the inclusion of spatial assets of coping and recovery capacities that impact and shape the urban fabric in order to improve urban sustainability.

Discussion

After introducing a spatial design strategy and defining its potential within the risk approach, two cases were studied considering the Vulnerability Concept, Multi-Layered Safety strategy, and the Dutch Layers approach. The overview shows how the two models for Vlissingen clearly differ on how to reduce vulnerability, the safety approaches, and layers of spatial dynamics. The Overtopping Sump Model is a step beyond the traditional line-based defense of Dutch flood risk management. In the Galveston case, rather than building the Ike Dike and encircling the island with sea walls (in the Dutch manner) the design demonstrates the potential for expanding the focus beyond recovery toward adaptation, which is about including all levels of the Multi-Layer Safety and Dutch Layers approaches.

In both countries, the planning systems present a challenge for integrating flood risk in spatial design. In the Netherlands, because of greater government control, integrating flood risk into spatial design means that engineers and spatial designers need to codesign, like in the Overtopping Sump Model in Vlissingen. This presents a significant shift in the design and construction of the urban fabric. The United States has the opposite issue; it is difficult to expand flood risk management due to a lack of employed local spatial design strategies. However, the urban fabric is much easier to adapt to the measures proposed.

The opportunities for both cases lie in creating spatial designs that include natural solutions and better anticipate uncertainties. In the Netherlands, this can create more awareness and adaptivity to flood risk, as new urban amenities become possible by taking out gray (engineered) infrastructure. In the United States, the recovery paradigm in which people are individually affected can take on an adaptive approach through the creation of new urban amenities, public spaces, recreational, and ecological areas (Table 2).

Table 2 Relation between the risk equation, the Vulnerability Framework, Multi-Layer Safety strategy, and Dutch Layers Approach.

risk	probability see Fig. 1A, B, C, E	consequences see Fig. 1F and G		
Vulnerability framework (De Graaf ,(2009)	threshold	coping - recovery	adaptation	adaptation
Multi-Layered Safety (RWS, 2009)	first layer		second layer	
	first	second	third layer	
Dutch Layers Approach (De Hoog et al., 1998)	infrastructure			
	occupation	occupation		
	substratum			
	infrastructure		occupation	substratum

The position of the Vlissings Model is in *light blue*, Galveston in *green*, and Overtopping Sump Model in *dark blue*.

Generally, an interdisciplinary approach in which goals, information, and measures are merged is key to integrating flood defense into spatial planning. Particularly, in the Overtopping Sump Model, the spatial design process is done in close collaboration with hydraulic engineering knowledge and tools in order to understand spatial dimensions, the volume of the water coming over the dike that needs to be stored or conveyed, and the temporal dimensions of water overtopping the dike. The result is a spatial strategy in which the consequence reduction goes hand in hand with spatial upgrading because of the re-design of public space and real estate development. Spatial design in Galveston is not necessarily about the design of public space, but it is about the design of the urban fabric as a system in which the proposed measures are integrated. This was done with the perspective of flood defense as a zone rather than just a line of defense provided by a single dike.

Conclusions

The integration of flood protection in urban development is currently undergoing a paradigm shift in which flood defense is no longer the *conditio* sine qua non, but instead urban development becomes also the *conditio* sine qua non for flood defense. This is due to the urgency caused by the environmental crises, and, in the Dutch case, related to the natural processes of subsidence, which increasingly makes the country vulnerable to floods. To keep on defending land by a dike in the Netherlands, or keep on recovering from events in the United States is too costly and missing out on the option for sustainable urban development. For the Netherlands, the paradigm shift is moving from reducing probability (the threshold capacity) to foreseeing consequences and focusing on adaptive capacity. For the United States, the shift is going from dealing with the consequences and a high recovery capacity to a spatial approach in which design is used to define a set of adaptive measures. The two shifts basically meet in the middle where integration with the urban fabric and reintegration with the ecological system are both achieved.

This study found five principles for future action in integrated urban flood design:

1. Spatial design strategies should include flood risk and consequence reduction, related to environmental conditions and vulnerabilities. This usually goes hand in hand with staying close to the *genius loci* (original natural identity) and using parameters to balance the conditions of the landscape with urban use.
2. Repurpose spatial design to prioritize ways in which to live with water. Design can create awareness and thus carry capacity; also cost and benefits of flood defense and other urban developments can be synergized.
3. Incorporate the notion of deep uncertainty like climate change into planning and design by integrating no-regret solutions, which are solutions serving multiple purposes, as a part of urban development.
4. Acknowledge and design with the natural landscape and local ecosystems as part of the sustainable development process.

5. Integrate innovative dike reinforcement with urban programs on the larger scale and supportive to the paradigm shift wherein urban development becomes also the *conditio sine qua non* for flood defense.

The integration of flood defense in urban development needs bridging. This is bridging of concepts and integrating goals, strategies, and measures. The bridging of the risk approach with the Dutch Layers approach by use of the vulnerability concept and the Multi-Layered Safety approach is very helpful to understand the integration of flood defense in urban development because it navigates through the complexity. The cases are meeting in the middle of two very different (national) approaches and thus revealing opportunities of integration and a new road to a resilient future.

References

Adger, W. N., & Kelly, P. M. (1999). Social vulnerability to climate change and the architecture of entitlements. *Mitigation and Adaptation Strategies for Global Change, 4*(3–4), 253–266.

Anonymous. (n.d.). *De woningwet 1902 – 1929: gedenkboek: samengesteld ter gelegenheid van de tentoonstelling gehouden te Amsterdam 18 – 27 oktober 1930 bij het 12 1/2-jarig bestaan van het Nederlandsch Instituut voor Volkshuisvesting en Stedebouw*. Amsterdam: Nederlandsch Instituut voor Volkshuisvesting en Stedebouw.

AP News. (2018). *Ike Dike still raises various questions*. AP News. October 30. [online] https://apnews.com/d42dd401baa34eaeb0358d8d6850f3a3 (visited 28th December 2020).

Bisker, J., Chester, A., & Eisenberg, T. (Eds.). (2015). *Rebuild by design*. http://www.rebuildbydesign.org/data/files/500.pdf.

Coastal Texas Study. (2020). *Galveston ring barrier system* (online) https://storymaps.arcgis.com/stories/f63264ce820842e2956747f502f6b97b (visited 3rd February 2021).

Corner, J. (2006). Terra Fluxus. In C. Waldheim (Ed.), *The landscape urbanism reader* (pp. 21–33). New York, NY: Princeton Architectural Press.

Dammers, E., Bregt, A. K., Edelenbos, J., Meyer, H., & Pel, B. (2014). Urbanized deltas as complex adaptive systems: Implications for planning and design. *Built Environment, 40*(2), 156–168. https://doi.org/10.2148/benv.40.2.156.

De Graaf, R. E. (2009). *Innovations in urban water management to reduce the vulnerability of cities: Feasibility, case studies and governance*.

De Hoog, M., Sijmons, D., & Verschuuren, S. (1998). *Laagland*. Amsterdam: HMD (Het Metropolitane Debat)—Herontwerp.

Hooimeijer, F. L. (2014). *The making of polder cities: A fine Dutch tradition*. Rotterdam: Jap Sam Publishers.

Houston Chronicle. (2009). *Oceanographer: 'Ike dike' could repel most storm surges*. April 7, 2009; updated July 28, 2011. Retrieved from https://www.chron.com/news/houston-texas/article/Oceanographer-Ike-Dike-could-repel-most-storm-1627411.php.

Kaufmann, M., Van Doorn-Hoekveld, W., Gilissen, H. K., & Van Rijswick, M. (2015). *Analysing and evaluating flood risk governance in the Netherlands. Drowning in safety?*. Utrecht. Retrieved from www.starflood.eu.

Kim, Y., & Newman, G. (2020). Advancing scenario planning through integrating urban growth prediction with future flood risk models. *Computers, Environment, and Urban Systems, 82*, 101498.

Kok, M., & Brand, A. D. (2017). Everything is bigger in Texas: Reflection program case 'Houston Galveston Bay, Texas'. In Kothuis, et al. (Eds.), *Integral design of multifunctional flood defenses* Delft University Publishers.

Kothuis, B. L. M., Brand, A. D., Sebastian, A. G., Nillesen, A. L., & Jonkman, S. N. (2015). *Delft delta design: The Houston Galveston Bay region, Texas, USA* (pp. 46–47). Delft University Publishers, TU Delft Library.

Malecha, M., Kirsch, K., Karaye, I., Newman, G., & Horney, J. (2020). Advancing the toxics mobility inventory: Development of a toxics mobility vulnerability index and application in Harris County, TX. *Sustainability: The Journal of Record, 13*(6), 282–291.

McHarg, I. L. (1969). *Design with nature*. Garden City, NY: Natural History Press, Doubleday (2nd ed., John Wiley & Sons, New York, 1994).

Merrell, W. J. (2015). Design of a coastal barrier for the Houston Galveston Bay region. In B. L. M. (Bee) Kothuis, A. D. (Nikki) Brand, A. G. (Antonia) Sebastian, A. L. (Anne Loes) Nillesen, S. N. (Sebastiaan) Jonkman (Eds.), *Delft Delta Design - Houston Galveston Bay Region, Texas, USA*. Delft University Publishers – TU Delft Library.

Meyer, V. J., Morris, D., Waggonner, D., Nijhuis, S., & Pouderoijen, M. T. (2009). *Virtual edition: Dutch dialogues New Orleans—Netherlands. Common challenges in urbanized deltas*. Amsterdam: SUN Architecture. ISBN 978 90 8506 7764.

Ministerie VROM. (1991). *Vierde Nota Ruimtelijke Ordening Extra*. The Hague: Ministerie VROM.

Nillesen, A. L., & Kok, M. (2015). An integrated approach to flood risk management and spatial quality for a Netherlands' river polder area. *Mitigation and Adaptation Strategies for Global Change*, *20*(6), 949–966. https://doi.org/10.1007/s11027-015-9675-7.

Palmboom, F. (2017). *IJsselmeer: A spatial perspective*. Vantilt Publishers. Retrieved from https://books.google.nl/books?id=RCFIuwEACAAJ.

Powell, N. (2020). *As twin hurricanes converge on the Gulf Coast, $31 billion 'Ike Dike' still in planning stages*. Houston Chronicles, Local, Hurricane and tropical storms. [online] https://www.houstonchronicle.com/news/houston-weather/hurricanes/article/ike-dike-proposal-hurricane-laura-marco-galveston-15510840.php. [visited 28th December 2020].

Priest, S. J., Suykens, C., Van Rijswick, H. F. M. W., Schellenberger, T., Goytia, S., & Kundzewicz, Z. W. (2016). The European Union approach to flood risk management and improving societal resilience: Lessons from the implementation of the Floods Directive in six European countries. *Ecology and Society*, *21*(4).

Reed, J., van Vianen, J., Barlow, J., & Sunderland, T. (2017). Have integrated landscape approaches reconciled societal and environmental issues in the tropics? *Land Use Policy*, *63*, 481–492. https://doi.org/10.1016/j.landusepol.2017.02.021.

Reps, J. W. (1965). *The making of urban America: A history of city planning in the United States*. Princeton University Press.

Rijcken, T. (2017). *Emergo, the Dutch flood risk system since 1986* (PhD thesis). TU Delft.

RWS. (2009). *Nationaal Waterplan 2009–2015*. The Hague: Ministry of Transport and Water Management. www.nationaalwaterplan.nl.

Van den Brink, M., Edelenbos, J., van den Brink, A., Verweij, S., van Etteger, R., & Busscher, T. (2019). To draw or to cross the line? The landscape architect as boundary spanner in Dutch river management. *Landscape and Urban Planning*, *186*(February), 13–23. https://doi.org/10.1016/j.landurbplan.2019.02.018.

Van den Hoek, R., Brugnach, M. F., & Hoekstra, A. Y. (2012). Shifting to ecological engineering in flood management: Introducing new uncertainties in the development of a Building with Nature pilot project. *Environmental Science & Policy*, *22*, 85–99. https://doi.org/10.1016/j.envsci.2012.05.003.

Van der Woud, A. (1987). *Het lege land, de ruimtelijke orde van Nederland 1798– 1848*. Amsterdam: Meulenhof.

Van Doorn-Hoekveld, W., & Groothuijse, F. (2017). Analysis of the strengths and weaknesses of Dutch water storage areas as a legal instrument for flood-risk prevention. *Journal for European Environmental & Planning Law*, *14*(1), 76–97.

Van Hugten, M., Huijsman, N., Kok, N., & Rooze, D. (2018). *Flood proof Galveston. A multidisciplinary approach project on flood risk and exploration of effective mitigation measures for the City of Galveston*. Master of Science in Civil Engineering Delft: Delft University of Technology.

Van Loon-Steensma, J. M., & Kok, M. (2016). Risk reduction by combining nature values with flood protection? In *Vol. 7. E3S Web of Conferences*. https://doi.org/10.1051/e3sconf/20160713003.

Vlissingen & Ma.an. (2010). *Structuurvisie Vlissingen Stad aan Zee—een Zee aan Ruimte*. Vlissingen.

Voorendt, M. (2017). *Design principles of multifunctional flood defences* (Doctoral thesis). TU Delft. https://doi.org/10.4233/uuid:31ec6c27-2f53-4322-ac2f-2852d58dfa05.

Waldheim, C. (Ed.). (2006). *Landscape urbanism reader*. New York: Princeton Architectural Press.

Whiston Spirn, A. (2000). Ian McHarg, landscape architecture, and environmentalism: Ideas and methods in context. In M. Conan (Ed.), *Landscape architecture*. Dumbarton Oaks: Trustees for Harvard University.

Woodruff, S., Tran, T., Lee, J., Wilkins, C., Newman, G., Ndubisi, F., et al. (2020). Green infrastructure in comprehensive plans in coastal Texas. *Environment and Planning*, *B*, 1–21.

SECTION IV

Resilient solutions for flood risk reduction—Convergence of knowledge

CHAPTER 19

Flood risk reduction for Galveston Bay: Preliminary design of a coastal barrier system

Sebastiaan N. Jonkman and Erik. C. van Berchum
Department of Hydraulic Engineering, Faculty of Civil Engineering and Geosciences, Delft University of Technology, Delft, The Netherlands

Introduction

Many coastal areas around the world are densely populated and at risk from flooding. To address coastal hazards, regions have taken different approaches, ranging from a prevention-based strategy to one that relies largely on the mitigation of impacts. The local choice of strategy is influenced by geographical factors (e.g., nature of the hazard, geography of the region, and density of population and assets) as well as financial and cultural aspects (e.g., costs and benefits of measures and role of government) and preference for collective vs individual arrangements.

The flood risk management strategy in the Netherlands is a key example of the first strategy, i.e., prevention through collective arrangements. In the 20th century, the Dutch implemented a coastal protection system based on the principle of shortening the coastline. After the flood of 1916, a 30-km closure dam (Afsluitdijk) was built to close off a large estuary. The system protects the middle of the country—including the capital of Amsterdam—from storm surge. After the 1953 storm surge disaster—which killed more than 1800 people—a series of dams and storm surge barriers were built in the southwest of the country to shorten the directly exposed coastline by hundreds of kilometers.

Whereas the Dutch approach focuses on prevention, flood management in the United States has traditionally focused on a multilayered approach The latter aims to combine prevention efforts with mitigation. The US flood management in recent decades seems to have become more prevention-based strategy. After hurricane Katrina, a large-scale flood protection system was built around New Orleans, and studies on barrier schemes are ongoing in several parts of the United States, for example, in the New York/New Jersey region (USACE, 2019).

The Galveston Bay area is also at significant risk of flooding and large-scale protection plans are being considered. Since 2012, Dutch researchers and students have participated in the exploration of a possible implementation of Dutch flood risk management concepts for this region. This effort involved over 50 scientists and students from multiple disciplines (mainly civil engineering, architecture, policy, and management) and results are summarized by Kothuis, Brand, Sebastian, Nillesen, and Jonkman (2015).

This chapter provides an overview of some of these efforts in two key areas: risk-based evaluation of strategies for the region ("Setting the scene: Risk-based evaluation of strategies" section) and a "Preliminary design of a coastal spine system" section. The chapter concludes with a "Closing discussion" section. The focus of this chapter is on coastal flooding and storm surge. Rain-induced floods are not directly considered here.

The relevance of this chapter extends beyond the study area. It addresses the challenges associated with finding an appropriate risk-reduction strategy in situations where many combinations of (interdependent) measures are possible. In addition, it highlights how a multidisciplinary "Research by Design" approach can be adopted to find a coastal strategy that matches the local situation and objectives.

Setting the scene: Risk-based evaluation of strategies

Large, complex coastal regions—such as the Galveston Bay region and the Dutch delta—often require a combination of interventions to reduce flood risk to acceptable levels. To inform planning and decision-making, multiple alternative strategies can be analyzed and evaluated based on metrics such as costs, risk reduction, and societal and environmental impacts. As part of previous work, a risk-based modeling framework for such evaluations was developed and applied to Galveston Bay area (Van Berchum et al., 2018). This rapid probabilistic model simulates and evaluates the risk reduction and costs of many flood risk reduction strategies, taking into account interdependencies between measures. The simulation includes hydraulic calculations, damage calculations, and the effects of measures for different return periods.

Many measures and strategies are possible for the Galveston Bay area, ranging from coastal defense to in-bay measures and barriers that specifically protect the Houston area—see Fig. 1 for an overview. A preliminary investigation using the risk-based model compared the coastal spine solution with a mid-bay barrier and no action. This highlighted that the coastal spine would be the most expensive alternative economically, but would provide the greatest risk reduction by maximizing the protected area. It also shortens the coastline, leading to potential cost savings compared to other strategies that would rely on perimeter protection around the bay. Therefore, in the remainder of this chapter design efforts for the coastal spine are reported. It should be noted that in-bay alternatives and features were also elaborated on as part of the collaborative studies, see Kothuis et al. (2015) and the discussion section in this chapter.

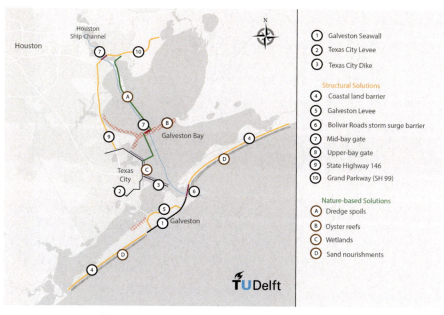

Fig. 1 Overview of potential measures for Galveston Bay (Van Berchum et al., 2018). *(Adapted after a figure developed by the SSPEED Center at Rice University. Included in a chapter by P.B. Bedient on p. 49 of Kothuis, B. L. M, Brand, A. D., Sebastian, A. G., Nillesen, A. L., & Jonkman, S. N. (2015).* Delft delta design: The Houston Galveston Bay region, Texas, USA*).*

Preliminary design of a coastal spine system

Approach and system overview

Following the devastating coastal flooding caused by hurricane Ike in 2008, a coastal spine system was proposed to protect the Galveston Bay area (Merrell, Reynolds, Cardenas, Gunn, & Hufton, 2011). It is also referred to as the Ike Dike. It would provide a barrier against storm surges into the bay and is conceptually similar to barriers built in the Netherlands to shorten the Dutch coastline. This chapter summarizes design work done on the Coastal Spine system by Dutch and American experts between 2012 and 2019. In different design stages (sketch and early conceptual design—referred to here as preliminary design), different barrier concepts were explored based on requirements for functional and engineering performance and landscape integration. The main function of the system is to prevent the inflow of the hurricane surge into the bay and to protect the areas behind it. The designs for the storm surge barriers are based on boundary conditions for navigation and environmental flow to minimize impacts on these functions. In addition, integration of the new features into the landscape and ecosystem were

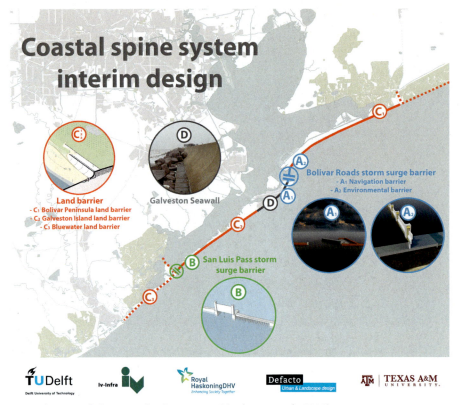

Fig. 2 Visualization of the coastal spine system (Jonkman et al., 2015).

important considerations. Further details for the concepts summarized below can be found in the reports "Coastal spine interim design" (Jonkman et al., 2015) and "Land barrier preliminary design" (Van Berchum, de Vries, & de Kort, 2016) reports.

Fig. 2 provides an overview of the system and its main elements. The barrier system includes storm surge barriers in the Bolivar Roads inlet (A1 and A2) and San Luis Pass (B). The land barriers (C1, C2, C3) are designed to protect Galveston Island and Bolivar Peninsula and stop or limit overland flow into the bay.

The total length of the coastal spine presented in Fig. 2 is 94 km (58.5 miles), consisting of 90 km (56 miles) of land barriers and two storm surge barriers with a length of 4 km (2.5 miles). Some sections of the barrier (indicated as dashed lines) require further investigation. These include the western edge of the land barrier along the Bluewater Highway and the connection of the eastern end of the land barrier near the community of High Island.

Table 1 shows the key hydraulic boundary conditions for the land and storm surge barriers. These are preliminary estimates that should be improved in further studies.

Table 1 Hydraulic boundary conditions for the design.

	Protection level	Life time (years)	Associated maximum water level in the Gulf of Mexico (h)[a] and significant wave height (H_s)
Land barrier	1/100 per year	50–100	$h = 5.7$ m and $H_s = 6.9$ m
Storm surge barriers	1/10,000 per year	100–200	$h = 7$ m and $H_s = 8.0$ s

[a]The estimate includes a somewhat conservative estimate of sea level rise of about 1 m over the 100-year lifetime. Based on Jonkman, S. N., Lendering, K. T., van Berchum, E. C., Nillesen, A., Mooyaart, L., de Vries, P., et al. (2015). *Coastal spine system—Interim design report* and Van Berchum, E. C., de Vries, P. A. L., & de Kort, R. P. J. (2016). *Galveston Bay area land barrier preliminary design*. TU Delft report.

For the storm surge barrier, a high protection level—similar to the level for Dutch coastal defenses—has been set because it serves to protect the entire metropolitan area around Galveston Bay. A lower level of protection was used for the land barrier because it is more adaptable than the storm surge barrier. Additional resilience requirements can be applied to the land barrier so that it can still limit inflow into Galveston Bay for more extreme events (e.g., 1/1000 per year). For the given 100-year protection level, it is expected that the existing Galveston seawall can still be utilized for protection—albeit with some adaptations.

Bolivar Roads storm surge barrier

The Bolivar Roads barrier consists of two sections: a navigational section with the main requirement of free passage for ships, and an environmental section for water and environmental flows.

Navigational section gates

For the navigational section, initial explorations with barrier experts considered two types of floating gates: floating sector gates (like the Dutch Maeslant barrier near Rotterdam) and a single barge gate. The barge gate was chosen as a preferred concept because a sector gate is less suitable to handle a negative head situation: water levels in Galveston Bay may be higher than those in the Gulf of Mexico due to the potential back surge during a hurricane. In such a situation, sector gates could be "pushed" out of their hinges.

A preliminary estimate of 220 m is used for the width of the navigational section. This would allow for a single, two-way shipping lane, while a Post/New Panamax tanker is assumed as a design vessel. Fig. 3 (left panel) shows the barge gate of the navigation section and the closing procedure. It would normally close during the forerunner surge to limit water levels on Galveston Bay.

The barge gate is designed as a partly floating structure that distributes the loads toward the abutments on both sides. The foundation of these concrete abutments can consist of a deep pile foundation with a mixed group batter piles (tension and pressure)

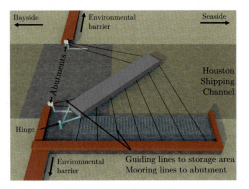 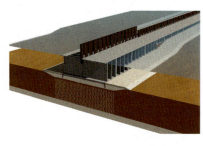

Fig. 3 Bolivar Roads storm surge barrier concepts: left: a barge gate for the navigational section (Smulders, 2014); right: a vertical liftgate for the environmental section (De Vries, 2014).

or a deep foundation of (pneumatic) caissons or (a coupled pair of) cellular cofferdams. The dynamics of the barrier in various phases of operation has been investigated in the thesis of Smulders (2014). This showed that most movements remain within tolerable limits. Critical aspects to be considered further are the landing operation and the load case under a negative head. The horizontal load transfer and dynamics of the barrier can be further optimized by considering the quantity of water allowed within the barge.

The chosen concept of floating barriers is similar to the floating barriers of the Maeslant barrier in the Netherlands in that some over- and underflow are allowed to reduce the amount of material and the cost of the structure. This concept introduces high flow velocities under the barrier gate when it is in the closed position. Under water, heavy scour protection is required to form a stable sill. It should consist of large concrete elements below the total gate width and large rock protection further away from the barrier in the perpendicular direction. During a storm closure, some overtopping of water is allowed as there is a large storage capacity in Galveston Bay. A structural design in steel was made and the gate consists of S355 steel and the weight is approximately 32,000 tons (Jonkman et al., 2015; van der Toorn et al., 2014). An alternative design in concrete was made by Karimi (2014).

Key topics for further exploration include hinges, a suitable foundation concept that is feasible in the local soft soil conditions, exact barrier dimensioning given navigation requirements and operation, and alternative barrier concepts, such as a closed barge gate system, which does not allow underflow and overflow.

Environmental section gates
The main purpose of the environmental section is to allow for tidal exchange in normal conditions and closure during storm conditions. For the environmental section, several concepts were explored at a somewhat less detailed level. These include

- Vertical lifting gates—similar to those used at the Eastern Scheldt barrier in the Netherlands. The gates can be embedded in a caisson placed on the seabed and prepared with vertical underwater drainage, placed on a shallow wide foundation.
- Vertical radial gates on a deep pile foundation.

More innovative gate concepts were also explored, such as
- A rotating gate, with a plate that is in a horizontal position in normal conditions, but rotates to a vertical position during the closure;
- Inflatable gates using rubber sheets. During closing, the sheets are filled with water and air. This concept has been applied at the Ramspol barrier in the Netherlands and has been further elaborated for Bolivar Roads by Van Breukelen (2013).

A promising concept with regard to operation, maintenance, reliability, and cost seems to be a system with vertical lift gates Fig. 3 (right). The local soil conditions are critical in further developing this solution.

In order to minimize the impacts on tidal flows, in-bay tide levels, and ecosystems, it is important that the storm surge barrier is as "open" as possible in normal conditions, i.e., the tidal flow passing through as naturally as possible. In the preliminary designs presented, the opening of the cross section at Bolivar Roads would be 70%–80% of the original cross section, leading to a limited reduction (∼5%) of the tidal range (Ruijs, 2011). Further optimization of barrier design and investigation of barrier effects on tidal flows, morphology and ecology are important topics for further studies.

A barrier solution should also be considered for the San Luis Pass. The width of this opening is about 1000 m. Without a barrier, a hurricane-induced surge will result in significant inflows into Galveston Bay and scour. A navigational gate (such as a liftgate) for smaller ships combined with a number of environmental gates could be considered as barrier solutions.

Land barrier

The land barrier plays a crucial role in protecting the Galveston Bay metropolitan area. Concepts for the land barrier have been explored based on a spatial analysis of the landscape by architects (Van Berchum et al., 2016). Fig. 4 (top panel) shows the results of this analysis. A distinction is made between (a) open landscape sections; (b) the Galveston seawall; (c) residential sections where the main island road is relatively close to the shoreline (100–150 m); and (d) residential sections where the road is further from the shoreline. Based on the landscape and available space, optimal integration of barrier solutions can be explored. Differentiation of barrier solutions and placement between locations may also be considered to allow optimal integration and alignment with the preferences of local stakeholders.

Several land barrier solutions have been developed. One option concerns a coastal dike. A preliminary design has been made assuming the hydraulic boundary conditions

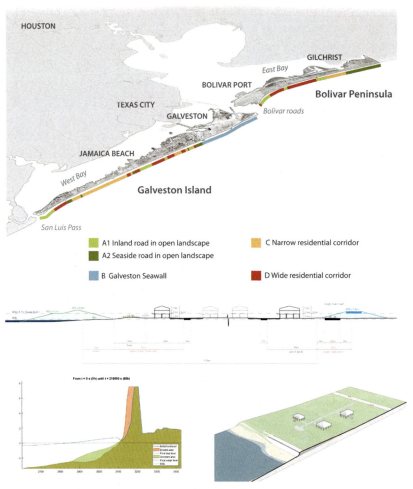

Fig. 4 Land barrier preliminary design: top: landscape analysis (Van Berchum et al., 2016); middle: cross section of the coastal dike design for the wide residential section (Van Berchum et al., 2016); bottom: alternative coastal protection concepts: dune (Rodriguez Galvez, 2019) and fortified dune (Jonkman et al., 2015).

with a return period of 100 years (Table 1). An overtopping limit of 50 L/s/m has been assumed, which in extreme cases allows for some erosion of embankments. There are two main choices for the positioning of the coastal dike, directly on the shoreline or at the location of the main road on the island (Fig. 4 middle panel). A shoreline positioning would result in a larger dike with a berm and an estimated elevation of MSL +8.8 m. It would protect all houses on the island. The dike would consist of a clay core and a hard revetment on the coastal side to withstand wave impacts. A drawback of the shoreline alternative is that it would affect the existing beach and could exacerbate erosion.

The inland option would allow for a somewhat lower elevation of the land barrier (MSL +7.5 m), but leaves houses on the southern side of the road exposed. In further exploration of land barrier alternatives, it is highly relevant to consider designing for further overtopping resilience for higher overtopping rates (>500 l/s/m), which would require armoring of the inner slope, for example, with asphalt or aggregate (e.g., Elastocoast) revetments (Van der Sar, 2016).

Alternative concepts have also been explored. In the MSc thesis of Rodriguez Galvez (2019), a **dune design** was made. During (design) storm conditions, a dune will be heavily eroded. The dune is designed so that wave overtopping is limited and the remaining profile would still be sufficient to avoid breaching and protect the area behind the dune. The numerical XBeach model was used to verify the performance of the dune in design conditions (see Fig. 4, lower left panel). This resulted in a dune profile with a crest elevation of MSL +7.5 m, width at MSL of 100 m, slopes of 1:3, and a dune crest width of 55 m. Aspects of the dune solution that require further investigation include morphological effects and long-term erosion, and associated management and maintenance practices. The availability (and costs) of large volumes of sand and the environmental impacts of sand mining and construction also requires further attention.

A very interesting land barrier solution to explore further is the **fortified dune** (or dike in dune). Such a hybrid defense would combine a structural defense with sand cover (Almarshed, Figlus, Miller, & Verhagen, 2019). This type of solution has been implemented in the Dutch coastal town of Noordwijk. It fits well into the natural coastal landscape, the footprint is smaller than for dunes, and the hard core would allow resilience to events that overtop or overflow the defense. It is recommended that the performance of fortified dune solution under storm conditions be further investigated, including potential synergies between dune erosion (leading to wave breaking) and load reduction on the hard structure. Further development of design practices and guidelines is also required to enable implementation.

Cost estimation

Due to the preliminary stage of design and a large number of factors still unknown, it is a challenge to produce a reliable and robust cost estimate. However, using assumptions and knowledge from previous projects, it is possible to provide a preliminary and indicative (but "rough") cost estimate.

Unit costs for storm surge barriers are reported in review studies for barriers around the world by Mooyaart and Jonkman (2017) and later updated in Kluijver, Dols, Jonkman, and Mooyaart (2019). Due to the complexity of the barriers, the costs are high: ranging from 1 to 4 million US$ per meter of span, with an average of 3 Million US$/m (2.5 MEuro) per meter span. This unit cost was adopted for an initial estimate of the navigational section. The environmental section is envisioned in a shallower area and allows for more repetition of elements and construction. Here a unit cost of 2 MUS$/m is

Table 2 Preliminary cost estimate for the coastal spine system features.

Element class	Location	Length (m)	Unit costs (M$/m)	Element costs (M$)	Bandwidth (%)
Storm surge barrier	Bolivar Roads (Navigational)	200	3	600	50
	Bolivar Roads (Environmental)	2800	2	5600	50
	San Luis Pass	1000	–	330	50
Land barrier	Bolivar Peninsula	40,000 (25 mile)	0.045	1800	30
	Galveston Island	30,000 (18.6 mile)	0.045	1350	30
	Bluewater Highway	20,000 (12.5 mile)	0.045	900	30
Total				**10,580**	**40**

Based on Jonkman, S. N., Lendering, K. T., van Berchum, E. C., Nillesen, A., Mooyaart, L., de Vries, P., et al. (2015). *Coastal spine system—Interim design report.*

adopted, similar to the Eastern Scheldt barrier and an additional estimate of 330 MUS$ for the San Luis Pass is included. For the land barrier, the dike alternative was considered. Unit costs from previous projects (Jonkman, Hillen, Nicholls, Kanning, & van Ledden, 2013) and a more material volume-based approach were considered. Both resulted in a cost estimate of the construction of a coastal dike of about 45 M$ per kilometer.

Table 2 presents the preliminary cost estimate for the current system at 10.6 B$ with a bandwidth between 6.3 and 14.8 B$. Again, the cost estimate will depend heavily on the designs and choices for the various system features.

The cost estimate does not yet include other potential system features, such as additional flood risk reduction measures in Galveston Bay and the Galveston ring levee. Also, many measures such as environmental restoration and mitigation and land acquisition are not yet included. Thus, it is likely and expected that the total costs of the system will be higher.

In addition, the above cost estimates do not include management and maintenance costs. These can be significant, up to 1% of construction costs on an annual basis for storm surge barriers, and in the order of US$ 100,000 per kilometer per year for dikes (Jonkman et al., 2013).

It is interesting to compare this with the costs of other large scale surge suppression systems. The hurricane protection system that was (re)built after hurricane Katrina in New Orleans had a total cost of about US$ 14 billion (Frank, 2019). The total cost of the Delta Works in the Netherlands is estimated at 5.5 billion Euros (Steenepoorte, 2014). If it is assumed that these were at the 1985 price level, the present value would be more than 11 billion Euros (US$12.5 billion).

Closing discussion

This closing section discusses a number of topics and issues related to the design of the coastal barrier along the Galveston Bay in a broader context.

The design results presented above relate to the barrier features directly on the coast. Even if a coastal barrier is in place, significant hurricane-induced surges and waves can still occur in Galveston Bay. Therefore, additional measures around the bay may be required. In particular, nature-based solutions on the western shore and at Galveston Island could contribute to surge reduction and improve the ecosystem (Godfroy, Vuik, Van Berchum, & Jonkman, 2019). In parallel, additional, or alternative structural protection measures could be considered on the west side of the bay or near Houston. One example is a local storm surge barrier closer to Houston in the Ship Channel (Schlepers, 2015).

The design studies were mainly conducted between 2012 and 2019. More recently, a conceptual design of a similar coastal barrier was published by USACE and the Texas General Land Office (USACE, 2021). A full comparison is beyond the scope of this chapter, but notable differences include the following. The USACE/GLO plan includes dunes, but much lower and smaller dunes than those proposed in this chapter, thus offering lower levels of risk reduction. Also, the navigational section employs a different storm surge barrier in the form of floating sector gates (which may be vulnerable to back surge). Moreover, the USACE/GLO plan uses higher cost estimates (including for particular features such as the storm surge barrier) and includes a ring dike around Galveston and several in-bay features for local risk reduction.

The design presented in this chapter is focused on reducing flood risks for storm surges. The area is also highly vulnerable to rainfall and runoff flooding due to hurricanes, as was illustrated during hurricane Harvey in 2017. Given the multitude of interlinked hazards, an integrated plan must be developed that addresses storm surges, wind damage, and rainfall-induced flooding.

It is expected that the presented design concepts are challenging and costly, but technically feasible. One crucial aspect that needs to be addressed is the management, maintenance, and funding of the new system. In the Netherlands, the large storm surge barriers are managed by the federal government (Rijkswaterstaat, the Dutch equivalent of USACE) and dikes are mostly managed by so-called Water Authorities, which are local water management organizations.

Planning and designing such a coastal protection strategy requires a multidisciplinary approach. This includes the collaboration of specialists within civil engineering (linking geotechnical, hydraulic and structural design), but also particularly multidisciplinary explorations between architects and civil engineers (e.g., for the land barrier) and involvement of environmental experts and other disciplines. Given the long lifetime of such coastal barrier infrastructure, its planning and design ideally involve long-term urban and development strategies.

The concepts and experiences are also applicable to other coastal regions. The risk-framework presented ("Setting the scene: Risk-based evaluation of strategies" section) can be applied to any region where multiple coastal risk reduction measures are possible. For example, it has been explored in collaboration with the World Bank to inform the development of a coastal adaptation strategy for Beira, Mozambique (Van Berchum, van Ledden, Timmermans, Kwakkel, & Jonkman, 2020). Since many urbanized areas around the world are developing coastal adaptation strategies, the plans and projects for and from Texas can also inform and inspire other regions. Several metropolitan areas such as New York and Shanghai (China) are considering the implementation of storm surge barriers. Also, given the attention to more sustainable forms of coastal protection and adaption, there is a growing interest in nature-based and hybrid solutions such as dunes and hybrid dunes.

A more overarching challenge concerns the shift from reactive to proactive: throughout history, investments in flood protection seem to be made mainly after major disasters. Examples are the construction of the Delta Works in the Netherlands after the 1953 disaster and the protection of New Orleans after hurricane Katrina in 2005. A shift to proactive planning actions and investments is required to prevent future disasters in the Netherlands, Texas, and around the world.

References

Almarshed, B., Figlus, J., Miller, J., & Verhagen, H. J. (2019). Innovative coastal risk reduction through hybrid design: Combining sand cover and structural defenses. *Journal of Coastal Research*, *36*(1), 174–188.

De Vries, P. A. L. (2014). *The Bolivar Roads Surge Barrier: A conceptual design for the environmental section* (MSc thesis). TU Delft.

Frank, T. (2019). *After a $14-billion upgrade, New Orleans' levees are sinking*. E&E News on April 11, 2019 https://www.scientificamerican.com/article/after-a-14-billion-upgrade-new-orleans-levees-are-sinking/.

Godfroy, M., Vuik, V., Van Berchum, E. C., & Jonkman, S. N. (2019). Quantifying wave attenuation by nature-based solutions in the Galveston Bay. In N. Goseberg, & T. Schlurmann (Eds.), *Coastal Structures 2019* (pp. 1008–1019). Karlsruhe: Bundesanstalt für Wasserbau. https://doi.org/10.18451/978-3-939230-64-9_101.

Jonkman, S. N., Hillen, M. M., Nicholls, R. J., Kanning, W., & van Ledden, M. (2013). Costs of adapting coastal defences to sea-level rise—New estimates and their implications. *Journal of Coastal Research*, *29*(5), 1212–1226.

Jonkman, S. N., Lendering, K. T., van Berchum, E. C., Nillesen, A., Mooyaart, L., de Vries, P., et al. (2015). *Coastal spine system—Interim design report*.

Karimi, I. (2014). *The conceptual design of the Bolivar Roads navigational surge barrier* (MSc thesis). TU Delft.

Kluijver, M., Dols, C., Jonkman, S. N., & Mooyaart, L. F. (2019). Advances in the planning and conceptual design of storm surge barriers—Application to the New York metropolitan area. In N. Goseberg, & T. Schlurmann (Eds.), *Coastal Structures 2019* (pp. 326–336). Karlsruhe: Bundesanstalt für Wasserbau. https://doi.org/10.18451/978-3-939230-64-9_033.

Kothuis, B. L. M., Brand, A. D., Sebastian, A. G., Nillesen, A. L., & Jonkman, S. N. (2015). *Delft delta design: The Houston Galveston Bay region, Texas, USA*.

Merrell, W. J., Reynolds, L., Cardenas, A., Gunn, J. R., & Hufton, A. J. (2011). The Ike dike: A coastal barrier protecting the Houston/Galveston region from hurricane storm surge. In *Macro-engineering seawater in unique environments* (pp. 691–716). Springer.

Mooyaart, L. F., & Jonkman, S. N. (2017). Overview and design considerations of storm surge. ASCE Journal of Waterway, Port, Coastal, and Ocean Engineering, 143(4). https://doi.org/10.1061/(ASCE)WW.1943-5460.0000383#sthash.H6wdGIFV.dpuf.

Rodriguez Galvez, L. (2019). *Dune based alternative to coastal spine land barrier in Galveston Bay: Conceptual design* (MSc thesis). TU Delft.

Ruijs, M. (2011). *The effects of the "Ike dike" barriers on Galveston Bay* (MSc thesis). TU Delft.

Schlepers, M. H. (2015). *A conceptual design of the Houston Ship Channel Barrier* (MSc thesis). TU Delft.

Smulders, J. (2014). *Dynamic assessment of the Bolivar Roads Navigational Barge Gate Barrier* (MSc thesis). Delft University of Technology. https://repository.tudelft.nl/islandora/object/uuid%3A05dde2aa-8c21-4fcb-a531-d541a60bc751?collection=education.

Steenepoorte, K. (2014). *De stormvloedkering in de Oosterschelde*. the Netherlands: Rijkswaterstaat Zee en Delta.

USACE. (2019). *New York—New Jersey Harbor and tributaries coastal storm risk management feasibility study—Overview of engineering work for the NYNJHATS interim report.* February 2019.

USACE. (2021). *Coastal Texas protection and restoration feasibility study.* Final Report August 2021.

Van Berchum, E. C., de Vries, P. A. L., & de Kort, R. P. J. (2016). *Galveston Bay area land barrier preliminary design.* TU Delft report.

Van Berchum, E. C., Mobley, W., Jonkman, S. N., Timmermans, J. S., Kwakkel, J. H., & Brody, S. D. (2018). Evaluation of flood risk reduction strategies through combinations of interventions. *Journal of Flood Risk Management, 12*(S2). https://doi.org/10.1111/jfr3.12506.

Van Berchum, E. C., van Ledden, M., Timmermans, J. S., Kwakkel, J. H., & Jonkman, S. N. (2020). Rapid flood risk screening model for compound flood events in Beira, Mozambique. *Natural Hazards and Earth System Sciences.* https://doi.org/10.5194/nhess-2020-56 (in review).

Van Breukelen, M. (2013). *Improvement and scale enlargement of the inflatable rubber barrier concept: A case study applicable to the Bolivar Roads barrier, Texas, USA* (MSc thesis). TU Delft.

Van der Sar, I. (2016). *Probabilistic design of the land barrier on the bolivar peninsula, Texas* (MSc thesis). TU Delft.

van der Toorn, A., Mooyaart, L., Stoeten, K. J., van der Ziel, F., Willems, A., Jonkman, S. N., et al. (2014). *Barge barrier design.* Technical report February 7, 2014.

CHAPTER 20

Design, maintain and operate movable storm surge barriers for flood risk reduction

Marc Walraven[a], Koos Vrolijk[a], and Baukje Bee Kothuis[b,c]

[a]Ministry of Infrastructure and Water Management, Rijkswaterstaat, Rotterdam, Netherlands
[b]Department of Hydraulic Engineering and Flood Risk, Faculty of Civil Engineering and Geosciences, Delft University of Technology, Delft, The Netherlands
[c]Netherlands Business Support Office, Houston, TX, United States

Introduction

After the flood disaster of 1953 in the Netherlands, it was clear that a new form of coastal defense was urgently needed. The Dutch chose to shorten the coastline, whereby part of the coastal inlet was completely closed off with dams and another part was provided with movable storm surge barriers. The so-called Delta Works were implemented, and in a period of about 45 years, six movable barriers were constructed. They were all the first of their kind prototypes of which by now various features have been applied in some form or another in several places around the globe (a.o. Daniel & Paulus, 2019). At multiple locations, possibilities for building a barrier or upgrading an existing barrier are currently being considered to protect coastal areas against storms, sea-level rise, and possible future consequences of climate change. For example, in Rotterdam, the Maeslant barrier faces a range of challenges caused by potential sea-level rise (Deltares, 2019). Similarly, in the Houston Galveston Bay region, shortening of the coastline by means of a Coastal Spine is being considered following Hurricane Ike (2008), and a movable storm surge barrier in the Houston Ship Channel has been included in the preliminary design (USACE & TGLO, 2020).

Over the years, storm surge barriers have proven to incorporate a number of very specific characteristics that have a significant impact on their management, maintenance, and operations (MMO). In the process, many lessons have been learned worldwide about the use of several types of barriers similar to those in the Delta Works, and a number of new designs have also been developed. Sharing these valuable lessons amongst barriers, worldwide, is one of the main aims of the I-STORM* network (I-STORM, 2020).

---

* I-STORM is the international knowledge-sharing network for all those working in the storm surge barrier profession. See also www.i-storm.org.

When designing a new barrier or modifying existing barriers, incorporating these lessons can be of great advantage for both practitioners and policymakers. By including the (consequences of) specific characteristics of a storm surge barrier in the design and design requirements, a number of undesirable MMO implications can be avoided. Here we first sketch the basics of movable storm surge barriers and provide a general typology. By exploring some of the specific characteristics of these barriers, the implications for MMO are highlighted, offering insights into how they might be addressed in the design along with examples.

Movable storm surge barriers

A storm surge barrier is a movable construction in an estuary or river branch that can be closed temporarily (Mooyaart & Jonkman, 2017). It is designed to protect against extreme water levels caused by storm surges. Its main function during surges is to reduce or prevent the rise of water level behind the barrier and thereby protecting the area at flood risk from inundation. A storm surge barrier is usually part of a more extensive flood protection system (Kuhn, Henao-Fernandez, Batchelor, & Morris, 2020; Walraven & Noguiera, 2018). The barrier takes the first impact of a storm, which ensures that the system behind the construction is less heavily burdened. This minimizes the required standards for flood risk protection measures in the hinterland and thus enables lowering their impact on landscape and environment. During normal daily weather conditions, movable storm surge barriers provide an open connection with the sea to enable shipping traffic to transit, for ecological reasons, and/or to allow river discharge or a gradual spillway to be uninterrupted.

In the process of transition from structural safety to intelligent safety, the variety and types of barriers have become increasingly sophisticated. Increased technological knowledge and possibilities combined with changing contextual requirements have led from human-operated constructions that mainly focused on safely controlling the surge tide, to barriers with all sorts of opening features, often operated with high-tech systems. Fig. 1 shows some exemplary storm surge barriers, with their main typology explained in Table 1 based on:

- *The movement direction*: Barriers can operate either in a horizontal or vertical direction. Horizontal movement, for example, allows a passageway for shipping that is not restricted in height. Vertical movement usually creates less stress on the bank connection; the barrier can be lowered from above the water, or rise from the waterbed.
- *The nature of the hydraulic gates*: A variety of gate types and the way they are in use, and sometimes different concepts are combined within one barrier system. Most frequently used are the vertical lift gate, vertical rising gate, segment gate, rotary segment gate, sector gate, inflatable barrier, flap gate, barge gate, and rolling gate (Dijk & Van der Ziel, 2010; Mooyaart & Jonkman, 2017).

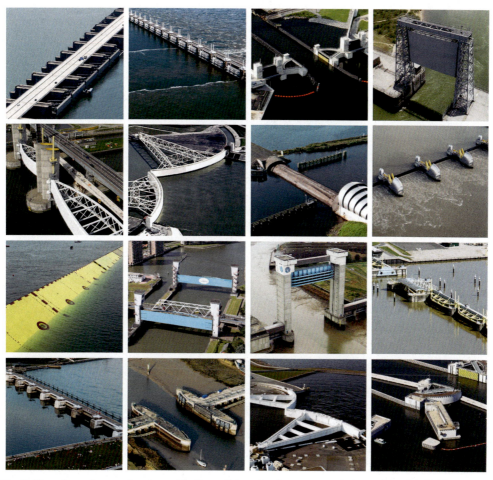

Fig. 1 Overview of exemplary movable Storm Surge Barriers, see typology explained in Table 1. From left to right, top to bottom: 1.01 Haringvliet Barrier, 1.02 Eastern Scheldt Barrier, 1.03 Seabrook Floodgate Complex, 1.04 Moses Lake Floodgate, 1.05 Hartel Barrier, 1.06 Maeslant Barrier, 1.07 Ramspol Barrier, 1.08 Thames Barrier, 1.09 MOSE Barrier, 1.10 Hollandsche IJssel Barrier, 1.11 Barking Creek Barrier, 1.12 Emssperrwerk, 1.13 Singapore Marina Barrage, 1.14 Colne Barrier, 1.15 St. Petersburg Barrier, 1.16 Lake Borgne Surge Barrier. *(Image credits: 1.01 Rijkswaterstaat/Joop van Houdt, 1.02 123RF.com/Andreas Basler, 1.03 US Army Corps of Engineers/Patrick M. Quigley, 1.04 Raven Drones/Al Barboza, 1.05 De Fotovlieger/Hans Elbers, 1.06 Rijkswaterstaat, 1.07 i4photos.nl/Paul Kaandorp, 1.08 Environment Agency, 1.09 Consorzio Venezia Nuova/Giorgio Marcoaldi, 1.10 Rijkswaterstaat/Joop van Houdt, 1.11 Environment Agency, 1.12 Bin im Garten, CC BY-SA 3.0, 1.13 123RF.com/Richard Whitcombe, 1.14 Steve Brading, 1.15 Alamy Stock Photo/Anton Vaganov, 1.16 US Army Corps of Engineers/Patrick M. Quigley.)*

Table 1 Movable storm surge barriers typology based on direction of movement and gate type.

Movable storm surge barriers typology—Based on gate type

Direction of movement	Nature of hydraulic gate	Ref. to Fig. 1	Barrier name	Location (country)	Commenced Operation	Maximum sill level (based on local ordinance datum)	Total width of barrier	Number of gates[a]
Horizontal	Segment gate	1.06	Maeslant Barrier	Rotterdam region, Netherlands	1997	−56 ft. (−17 m)	1181 ft. (360 m)	2
		1.15	St. Petersburg Barrier (S1)	St. Petersburg, Russia	2011	−52 ft. (−16 m)	656 ft. (200 m)	2
	Segment gate + Barge gate	1.16	Lake Borgne Surge Barrier	New Orleans, USA	2014	−16 ft. (−5 m)	676 ft. (206 m)	3
Vertical	Vertical lift gate	1.02	Eastern Scheldt Barrier	Province of Zeeland, Netherlands	1986	−36 ft. (−11 m)	10,171 ft. (3100 m)	62
		1.05	Hartel Barrier	Rotterdam region, Netherlands	1997	−21 ft. (−6.5 m)	591 ft. (180 m)	2
		1.04	Moses Lake Floodgate	Texas City, USA	1967	−13.4 ft. (−4.1 m)	97.4 ft. (29.7 m)	1
		1.11	Barking Creek Barrier	London, England	1983	−11.8 ft. (−3.6 m)	245 ft. (74,6 m)	4
		1.10	Hollandsche IJssel Barrier	Rotterdam region, Netherlands	1958 + 1975	−21 ft. (−6.5 m)	262 ft. (80 m)	2

	Radial lift gate	1.01	Haringvliet Barrier	Rotterdam region, Netherlands	1971	−18 ft. (−5.5 m)	3445 ft. (1050 m)	17
	Inflatable dam	1.07	Ramspol Barrrier	Lake IJssel region, Netherlands	2002	−15 ft. (−4.7 m)	1148 ft. (350 m)	3
	Rising sector gate	1.08	Thames Barrier	London, England	1982	−40 ft. (−11.3 m)	1706 ft. (520 m)	10
	Flap gate	1.09	MOSE Barrier	Venice, Italy	2020	−46 ft. (−14 m)	5.446 ft. (1660 m)	78
		1.13	Singapore Marina Barrage	Singapore	2008	−8.2 ft. (−2.5 m)	1148 ft. (350 m)	9
Combined	Vertical, moving down + Vertical, moving up	1.12	Emssperrwerk	Gandersum, Germany	2002	−29.6 ft. (−9 m)	1562 ft. (476 m)	5 + 2
	Horizontal + Vertical, moving down	1.03	Seabrook Floodgate Complex	New Orleans, USA	2013	−18 ft. (−5.5 m)	361 ft. (110 m)	2 + 2
	Horizontal + Vertical, moving down	1.14	Colne Barrier	Wivenhoe, England	1993	−7.9 ft. (−2.4 m)	24.6 ft. (max 7.5 m) 427 ft. (130 m)	2 + 13

[a] Gates in adjacent ship locks not included (often mitre gates).

Specific characteristics and their implications

One of the most important things to take into account when building or rebuilding movable storm surge barriers is their specific characteristics, as these have a major impact on the MMO of the barrier during its usually intended lifetime. Recognizing and incorporating this aspect could, for example, substantially reduce future maintenance budgets and increase safety and reliability. Nevertheless, awareness of these specific characteristics is often low, as storm surge barriers are managed by organizations that mainly manage standard structures and are set up accordingly. Unique structures, such as barriers, are not part of the standard systems and their specific characteristics then become easily overlooked (Walraven, 2020). Exploring some of the most important characteristics and highlighting their implications for MMO, can increase recognition of this important aspect.

Every storm surge barrier is a prototype

Broadly speaking, each major movable storm surge barrier can be considered a prototype, a unique structure. On the one hand, this is because every barrier system as a whole is unique when looking at the combination of the specific physical environment, specific requirements, and the specific type of barrier. On the other hand, it's because of the unique application of parts: either the use of parts which are custom made—and thus unique per definition—or/and standard parts being used in a nonstandard way—and thus unique in application.

For MMO, the uniqueness of the components and system firstly creates a need for very specific expert knowledge, a high-quality document management system, and an elaborate training program.

Secondly, incorporating unique components in the design means that in MMO many different spare parts have to be available since these parts cannot be bought 'off the shelf.' Thirdly, when 'off the shelf' parts are used differently from what they were designed for, it can cause issues. Water pumps are a common example. They are designed for continuous submerged use, but at barriers, they are mostly used infrequently in an alternating wet and mostly dry environment. This will affect their reliability and maintenance requirements as against what they were originally designed for. Fourth, political decisions on infrastructure usually do not take into account the uniqueness of storm surge barriers; they are focused on the generic situation of common infrastructure such as locks and bridges which occur in much greater numbers than storm surge barriers. Finally, the fact that barriers contain unique components has an impact on the relationship between the designer—often a market party—and the commissioner—often a government agency. These projects are often not very lucrative projects for the market, as design, construction, and MMO cannot rely on cost-saving and knowledge acquired from repetitive production.

Low occurrence of intended functioning under extremely high-reliability requirements

The combination of these two aspects is a specific characteristic for storm surge barriers, with major effects for MMO.

Extreme storm conditions are not regular occurrences. This means that many movable storm surge barriers are not often deployed regularly: usually a few times a year or in some cases, only once every few years. For example, the Ramspol Barrier closes on average 1–3 times per year, whereas the Maeslant Barrier has only operated twice in more than 20 years. The limited use has multiple effects for MMO. Firstly, the low deployment ratio limits the opportunities to gain technical and organizational experience with the barrier. In order to be sufficiently prepared, expertise must thus be obtained through simulated testing and training—also in the event of a malfunction—and by learning from experiences at other barriers. Secondly, the low deployment frequency limits possibilities to find out about unique and specific maintenance problems and the effects of maintenance on specific parts. Finally, the low frequency of intended functioning means that testing the barrier in an integrated manner is limited. When the design offers possibilities to facilitate partial tests and inspections, this can be overcome to some extent.

The reliability of a barrier must comply with legal and policy requirements throughout its entire lifecycle. The extremely high-reliability requirements have to be met in the design phase. However, MMO has to support continued compliance with these initial and/or future requirements. Firstly, a fault-tree analysis is often performed to show that the reliability requirements are met. However, it is often only used in the design and construction phase. Adjusting this approach so that it can also be used in the MMO phase is necessary because in practice you may not be able to 'measure' that the requirements are met if you have a low deployment frequency. A fault-tree analysis for MMO would make it possible to check whether requirements are being met during the entire lifecycle of the barrier. Secondly, more redundancy measures can support MMO to maintain resilient systems which provide high reliability. Design can support this, for example, by not allowing 'Single Points of Failure' or introducing some over dimensioning.

However, the reliable functioning of the barrier not only depends on its technical function reliability but also on the reliability of the operational team. This team must have appropriate knowledge and experience in order to monitor and manage (and if necessary: intervene in) the closure process in a storm call. This human factor should also be implemented in the fault-tree analysis. Usually, this kind of analysis only considers technical failure aspects; yet, humans can on the one hand correct technical failure (which adds to system reliability) but can on the other hand also make mistakes in maintenance or in operation (which makes the system less reliable).

Maintenance windows are limited

In the Netherlands, the maintenance cycle of a barrier must be adjusted accordingly to the storm season, placing high demands and constraints on maintenance planning. The availability of so-called 'safe weather windows' strongly affects MMO.

Firstly, under the impact of changing climate, the number of closures will rise and during those operations, no (major) maintenance is allowed for the Dutch barriers. Where historically the maintenance cycle was mostly related to the storm seasons, MMO is currently also looking into shorter predictable 'safe weather windows' in the storm season to be used for maintenance work. Secondly, maintenance that is not or difficult to divide into parts, requires a continuous and often long maintenance period. This might mean that including multiple smaller gates in a design could prevail over implementing one or two large gates. Thirdly, maintenance could be made more independent from the weather, for instance, by applying the 'push-through principle': replace a part, maintain extracted part, use it to replace next one, then maintain it, etc. This needs to be facilitated by design using interchangeable elements. In the Eastern Scheldt barrier, for example, this kind of maintenance is impossible because almost every lift gate has a different dimension/size; in the design of the MOSE barrier in Venice, it has been an extensive point of attention. Fourthly, enabling maintenance within a shorter time frame could be achieved by facilitating easier maintenance by design.

Static structure in a dynamic environment

Changes in the environment of a storm surge barrier strongly impact MMO. Based on empirical evidence, five categories of changes create a workable base for an inventory of these implications (Walraven, 2020) (see Table 2):
- Politics, policy, law, and regulation changes
- Organizational and process changes
- Technological changes and innovation
- Knowledge and craftsmanship changes
- Physical environment changes

Major movable storm surge barriers in the Netherlands are designed and built for a life span of about 100 years. During this period many changes take place in the organization, technique, law, society, physical environment, safety, etc. Barriers are mostly not, or only to a very limited extent, designed to adapt to these dynamics, which often involve multiple future implications for MMO. A changing organizational context, for example, more stringent environmental impact rules and regulations, means that maintenance asks for situational adjustment, and additional work is required due to the need to renew (maintenance) equipment that is still good in itself. However, by creating flexibility and adaptability in both construction and organizational design, a workable degree of agility could be obtained. In the sphere of changing political and societal demands, an

Table 2 Dynamic environment of storm surge barriers.

Type of dynamic environment	Example of dynamics
Politics, policy, law and regulation changes	• Shift from mainly focus on managing flood risk to also include attention to environmental aspects. • Awareness of the need for flood safety measures decreases the longer no flood disasters occur; this leads to paradoxical feature: high flood safety level because of high flood risk means less attention and thus (political) support for flood risk reduction measures.
Organizational and process changes	• Changing relationship between market and government. Shift from 'all technical knowledge in-house at government level' to 'obtaining technical knowledge from market parties' to 'part of the technical knowledge in-house and partly from the market'.
Technological changes and innovation	• Application of new types of materials and new design processes. For example, hydraulic laboratories used to be built and operated manually (e.g., the Mississippi River Model in Vicksburg; and Waterloopkundig Laboratory in Flevopolder) and are now built and operated mainly by computerized systems (e.g., the Lower Mississippi River Physical Model at LSU Baton Rouge, LA; and the Delta Flume at Deltares in Delft). • Shift from structural safety (on construction and material knowledge basis) to intelligent safety (including probabilistic knowledge), driven by the development of ICT technology. • New potential threats and needs for security: e.g., cyber security.
Knowledge and craftsmanship changes	• Additional requirements, for example resulting from the demand for multifunctionality, require additional and different types of knowledge. • In the Netherlands, many barriers were built after the 1953 disaster, a strong new knowledge impulse. The level of safety became so high, that for decades no new barriers have been built: knowledge and (human) knowledge carriers become obsolete or even disappear.
Physical environment changes	• Sea level rise, climate change, higher/lower levels of river discharge • Increasing amount of shipping, for example resulting in a higher chance of colliding with a storm surge barrier.

increasing amount of (potential) additional functions is becoming relevant, mostly related to creating economic, recreational, ecological, or sustainable energy benefits. Think of additional design to enable fish migration (e.g., at the Haringvliet Barrier) or to incorporate turbines to generate energy from tidal motion (e.g., at the Eastern Scheldt Barrier). For MMO, this means for the first case monitoring and maintaining an extra opening in the barrier and in the second case dependency on a third party that builds and operates the turbines. Regarding changes in knowledge and craftsmanship, it is important to secure existing knowledge. Therefore, a 'Knowledge strategy for storm surge barriers' was developed, sparking renewed focus on mastering and maintaining design knowledge and knowledge of basic principles, which is crucial for carrying out improvements or replacements.

How reasoned design could enable more efficient MMO: Three cases

The specific characteristics of storm surge barriers and their implications for MMO, could (partly) be met in reasoned design. Three cases are presented as indicative examples. We thus intend to show the relevance of design based on the specific characteristics of movable storm surge barriers and their MMO and the importance of further developing this knowledge field.

Maeslant Barrier: Replacing compression blocks

The Maeslant Barrier is the barrier with the biggest sector gate in the world, protecting the port and city of Rotterdam and its hinterland. Many components of a storm surge barrier need to be replaced sometime during their 100-year service life. These are often significant and complicated projects that (partly) could have been avoided by design. At the Maeslant barrier, the compression blocks under the gate are a good example (Figs. 2 and 3).

In the dock, the barrier rests on a structure providing some elasticity because some deformation has to be handled: rubber packages with steel plates resting on a very large concrete pole. These spring packages will need to be replaced in the coming years, yet they are under the gate and, in normal circumstances, underwater. In terms of design and construction, that was easy: build the poles, put the rubber on top, and put the gate on top. But now the rubber has to be replaced, there is a gate on top of it. And, if one concrete upstand with its spring package is removed, the gate will probably deform as a result of the excess load. However, when the barrier was designed and built, it could have been taken into account to add one extra pole. Then one pole could always be removed, upgraded, or renewed, and put back, without damaging the gate.

The maintenance complications of the compression blocks also have to do with a limited maintenance window. Maintenance of all compression blocks at once is such a major activity that it can hardly be done over one summer season. However, the design of the barrier was optimized for construction and not for MMO within a predictable safe weather window. If barrier design is considered a structure that enables splitting maintenance jobs in smaller portions of a maximum of 2–3 weeks, this kind of major maintenance could be executed, both within and outside of the storm season.

Fig. 2 Compression blocks under the gate at Maeslant Barrier. *(Image: Courtesy of Rijkswaterstaat.)*

Fig. 3 Maintenance work at Maeslant Barrier. *(Image: Courtesy of Rijkswaterstaat.)*

Ramspol Barrier: A corroded nut

The Ramspol Barrier is an inflatable flexible membrane dam or 'rubber dam' in the IJssel River system in the east of the Netherlands. During the storm season, after running a test function, one of the large water inlet valves did not open. This is part of the primary system to fill the rubber tube, without an open valve gate there is no functioning barrier. A costly, complicated procedure was followed to find out what was wrong. The 900 + kg valve had to be removed and disassembled; this involved a large, over 24-h underwater operation involving divers, expanding the valve casing, and installing extra pumps in case of leakage (Fig. 4). After disassembly, it appeared that the nut that attached the valve blade to the spindle was severely corroded. This caused a rupture, causing the blade to jam and stay shut. A new nut was not "on the shelf" because of its irregular size, and had to be custom made. The corrosion was most likely caused by the alternating wet/dry environment while the valve (a standard element from the drinking water industry) was designed for underwater use in a continuously filled pipe.

Fig. 4 Expanding the valve casing at Ramspol Barrier. *(Image: Courtesy of Rijkswaterstaat, Tycho Busnach.)*

Fig. 5 Corroded and broken nut at Ramspol Barrier. *(Image: Courtesy of Rijkswaterstaat, Tycho Busnach.)*

- This case highlights a few examples of some MMO issues in the context of design:
  - Why use a valve that shuts in case of a problem? This turned out to be the standard construction. However, for this barrier function and reliability, it would have been better to choose a valve that opens as standard.
  - The design includes two tubes with a diameter of D800. Standard is D600, therefore this valve was not easily available. Why not design three tubes of D600, so replacement pipes and valves are available off the shelf?
  - Ramspol has six of these valves, of which most showed severe stages of the corrosion process (e.g., Fig. 4). All nuts had to be replaced and an inspection strategy needed to be developed. A huge extra MMO cost in man-hours and material: valve inspections alone cost €20,000 at a time, and the extra material costs more than a quarter of the usual annual maintenance budget.
- A clear example of two characteristics: at a barrier, many components are not used in a standard fashion and the need for a reasoned choice between unique or custom-made components. None of this is a problem in the construction process, it shows only later during MMO (Fig. 5).

Bolivar Roads barrier: Preliminary design

In the Houston Galveston Region, the Coastal Texas Study, a plan to protect the Texas Coast from flooding, is currently underway (www.coastalstudy.texas.gov). One of the elements of the plan is the construction of a so-called 'coastal spine' along Galveston Bay (see also chapters by Merrell and by Jonkman; this volume). A storm surge barrier in the Bolivar Roads is an essential part of this spine, and it presents a major design challenge to meet the many design and environmental requirements. USACE Galveston held the world's first international design session under the umbrella of I-STORM to gather input for the preliminary design of this storm surge barrier in the Bolivar Roads. This exercise is the ultimate example of a modern approach: a wide spectrum of facets is taken

into account from the outset, of course, the technical and hydraulic for the construction phase, but also their possible effects on MMO. The workshop identified several aspects that (might) have an impact on future MMO, which now are taken into account by introducing them as early as this design phase:
- Redundancy in the design in order to increase reliability;
- Collision risks;
- Accessibility of various locations on and around the storm surge barrier in case of unforeseen circumstances during the closure process, but also for management and maintenance activities;
- The type of organization that needs to be prepared and trained, and to build up knowledge and experience to perform the reliable maintenance and increase the likelihood of successful closure.

Interestingly enough, the effort also yields new knowledge for other parties working in the practice of design and MMO of flood defenses. For example, a design issue about bed protection for the proposed barrier at Bolivar Roads led to a request to Rijkswaterstaat for input as they had many years of this system working in practice. An internal design session was organized locally in the Netherlands, during which Rijkswaterstaat shared their knowledge and operational experience of bed protection and MMO strategies of their own Dutch storm surge barriers. This information exchange, reviewing the proposed potential design strategies for the barrier in Galveston, greatly assisted the designers in the United States.

Preparing feedback on the preliminary design for the Bolivar Roads barrier also led to the securing of original design knowledge within RWS that had become somewhat 'out of sight' because no designs have been made for more than 20 years and many experienced designers have retired. One of the designers of the Maeslant Barrier still works at RWS, so the original designs of the Maeslant Barrier were found, and his tacit and personal knowledge about them was shared before comparing this with the draft designs of the Bolivar Roads barrier. This provided new opportunities for design and MMO not only for Galveston but also for the Netherlands: the requests for knowledge from Galveston allowed Rijkswaterstaat to dust off 'old' knowledge, actively share it, and secure it more broadly which benefited their own organization.

Conclusions

It is important to keep in mind when it comes to the design or redesign of movable storm surge barriers many design choices will have a major impact on MMO. Reasoning from the specific characteristics of these flood defenses, and based on the lessons learned from existing barriers, these implications and potential negative impact can be (partly) anticipated, reduced, and often even prevented.

However, both awareness of these characteristics and knowledge of their impact are currently limited. Most recent publications and research related to the design of storm surge barriers have either a technological character or a financial-economic approach (a.o. Dijk & Van der Ziel, 2010; Aerts, Bowman, Botzen, & de Moel, 2013a, 2013b; Mooyaart, Jonkman, De Vries, Van der Toorn, & Van Ledden, 2014; Mooyaart & Jonkman, 2017; Walraven & Noguiera, 2018; Davlasheridze et al., 2019; Daniel & Paulus, 2019; Kluijver, Dols, Jonkman, & Mooyaart, 2019). Although technological knowledge and financial-economic insights are essential for the design and adaptation of storm surge barriers, practitioners insist that knowledge of MMO aspects is as least equally as vital in this respect. Acknowledging that the most important asset in an asset management system such as a storm surge barrier is the people working on it (Kuhn et al., 2020), we recommend further research for expanding knowledge based on retrieval of empirical findings worldwide. Retrieving and translating these experiences into lessons learned to be implemented in new designs or when modifying current flood defenses, can be done in the global context of I-STORM, collaborating with various government agencies, contractors, and universities. Apart from further expanding on the specific characteristics of movable storm surge barriers, some relevant research topics to consider are

- Automated versus manual: The role of humans in the fulfillment of requirements for storm surge barriers, what is the optimum ratio 'human/automated' MMO, and how to achieve this?
- Maintenance by design: How to give maintenance and asset management of storm surge barriers the most effective and efficient place in the design and planning process?
- Innovation versus proven technology: Incorporating innovative components and processes, which are lessons learned for storm surge barrier design and MMO? How can the market be incentivized to invest beyond proven technology for one-time-only designs?
- Organizational stability in a changing environment: How to guarantee the continuity and steadiness of MMO of a storm surge barrier in changing circumstances?
- Knowledge and expertise: How to ensure that design principles and technical decisions are transferred over time as a basis for future adjustments and updates?

Acknowledgments

This research was supported by Rijkswaterstaat and I-STORM. We'd especially like to thank Andy Batchelor, Megan Fuller, Derckjan Smaling, and Tycho Busnach for their valuable contributions, comments, and suggestions, which helped us to improve the quality of the manuscript; and Alex Pitstra for the work on Fig. 1, making it aesthetically and content proof.

Rijkswaterstaat participated in NSF-PIRE research related to storm surge barriers for several PIRE students and supported research projects from a range of disciplines by providing expert knowledge and data retrieval.

References

Aerts, J., Bowman, M., Botzen, W., & de Moel, H. (2013a). Storm surge barriers for NYC and NJ. *Annals of the New York Academy of Sciences, 1294*, 49–68.

Aerts, J., Bowman, M., Botzen, W., & de Moel, H. (2013b). Cost estimates of storm surge barriers for NYC and NJ. *Annals of the New York Academy of Sciences, 1294*, 69–78.

Daniel, R., & Paulus, T. (2019). *Lock gates and other closures in hydraulic projects*. Oxford/Cambridge: Butterworth-Heinemann/Elsevier Inc.

Davlasheridze, M., Atoba, K. O., Brody, S., Highfield, W., Merrell, W., Ebersole, B., et al. (2019). Economic impacts of storm surge and the cost-benefit analysis of a coastal spine as the surge mitigation strategy in Houston-Galveston area in the USA. *Mitigation and Adaptation Atrategies for Global Change, 24*(3), 329–354. https://doi.org/10.1007/s11027-018-9814-z.

Deltares. (2019). *Strategieën voor adaptatie aan hoge en versnelde zeespiegelstijging Een verkenning*. [Strategies for adaptation to high and accelerated sea level rise. An exploration.] (In Dutch). Report commissioned by the Ministry of Infrastructure & Water—staff Delta Commissioner and Rijkswaterstaat. Retrievable from http://publications.deltares.nl/11203724_004.pdf.

Dijk, A., & Van der Ziel, F. (2010). *Multifunctionele beweegbare waterkeringen: Projectgroep Afsluitbaar open Rijnmond*. [Multifunctional movable flood defenses: Project group closable open Rhine outlet] (In Dutch) Rotterdam: Rotterdam Climate Initiative.

I-STORM. (2020). www.i-storm.org. (Accessed 15 October 2020).

Kluijver, M., Dols, C., Jonkman, S. N., & Mooyaart, L. F. (2019). Advances in the planning and conceptual Design of Storm Surge Barriers—Application to the New York metropolitan area. In N. Goseberg, & T. Schlurmann (Eds.), *Coastal structures* (pp. 326–336). Karlsruhe: Bundesanstalt für Wasserbau. https://doi.org/10.18451/978-3-939230-64-9_033.

Kuhn, M., Henao-Fernandez, H., Batchelor, A., & Morris, E. (2020). A system-level approach to managing flood defences in the River Thames estuary, UK. In *Proceedings of the Institution of Civil Engineer—Civil Engineering*. https://doi.org/10.1680/jcien.18.00056.

Mooyaart, L. F., & Jonkman, S. N. (2017). Overview and design considerations of storm surge barriers. *Journal of Waterway, Port, Coastal, and Ocean Engineering, 143*(2). https://doi.org/10.1061/(ASCE)WW.1943-5460.0000383.

Mooyaart, L., Jonkman, S. N., De Vries, P., Van der Toorn, A., & Van Ledden, M. (2014). Storm surge barriers: Overview and design considerations. In *Proceedings of the 34th International Conference on Coastal Engineering, ASCE, Reston, VA*.

USACE & TGLO. (2020). Coastal Texas study. In *2020 Draft Feasibility Report*. Available from coastal.study.gov.

Walraven, M. (2020). Storm Surge Barriers in the Netherlands & I-STORM. In *Presentation at National Strategies: Storms, Flooding & Sea Level Defense Conference 2020, Propeller Club of Northern California. November 2020*. Available at https://propellerclubnortherncalifornia.org/sfsld-webinar-videos/.

Walraven, M., & Noguiera, H. I. S. (2018). *Overview storm surge barriers*. Commissioned by Danish Coastal Authority (DCA); Deltares in collaboration with Rijkswaterstaat and I-STORM. Report # 11201883-002-ZKS-0001.

CHAPTER 21

Designing and implementing coastal dunes for flood risk reduction

Jens Figlus
Department of Ocean Engineering, Texas A&M University, College Station, TX, United States

Introduction

In the aftermath of superstorm Sandy, it became evident that in the stretches of developed shoreline along the US East Coast that were protected by a substantial dune line infrastructure and residential neighborhoods suffered less severe damage due to flooding and wave impact than those without any significant dunes (Barone, McKenna, & Farrell, 2014). The resulting increased attention by the public on dunes as a viable component of coastal flood risk reduction measures is a welcome development and part of the challenging road to more resilient coastal systems and communities worldwide. Dunes, however, have been recognized as a first line of defense against flooding from storm surge and wave attack for a long time. An increasing number of coastal risk reduction projects in the United States, the Netherlands, and other countries are incorporating sand dunes into their design as an integral component. To do this in a sustainable manner, it is critical that the morphodynamic, ecological, and social aspects of the whole coastal system are understood and that an adaptive and flexible dune management strategy is employed (Elko et al., 2016). Here we outline the basics of natural coastal dunes, explore their morphodynamic behavior during storm impact, highlight design implications for engineered coastal dunes, and offer some insights into the use of dunes in coastal risk-reduction schemes from Texas and the Netherlands.

Natural dunes

Dunes occur naturally along almost all sandy coastlines as relatively large linear sand features or entire fields of mounds separating the backshore from areas further inland (Fig. 1). Dunes are a natural habitat for a variety of species and add recreational value to a beach, attracting tourism (Bruun, 1998). The beach and dune profile at a given location is usually directly linked to the sediment that is delivered or removed from a coastal system over time (Psuty, 1988), with erosion and accretion patterns being controlled by various factors and limitations related to eolian processes (de Vries, Southgate, Kanning, &

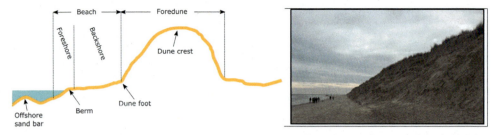

Fig. 1 Basic schematic representation of a beach and dune profile (left) and photo of a beach and dune on the island of Texel, the Netherlands. *(Photo by J. Figlus.)*

Ranasinghe, 2012), characteristics of sediment supply (Doody, 2013), availability of vegetation (Feagin et al., 2019), and duration and intensity of storms (Walker et al., 2017).

Dunes form as sediment, transported by the wind just above the beach surface, gets trapped by obstacles such as vegetation or wrack material. Especially the interaction between vegetation and wind-blown sediment allows a small embryonic dune to continue to trap and hold in place sediment to sustain its growth. The actual shape and size of dunes or dune fields are controlled by many factors that vary in space and time. The most important factors include the availability of adequate sediment, its movement via air- and water-driven sediment transport, and the ability to accumulate and stabilize said sediment with the help of vegetation. There is a direct link between the availability of dry sand on the beach in front of a dune and the dune's ability to be sustained. Dry sand is more likely to be moved toward the dune by onshore winds than wet sand. This means that a wide beach with water level changes and climate that allow for the availability of dry surface sand during times of sufficiently strong onshore wind conditions usually leads to more substantial dunes.

One of the most important aspects to keep in mind when discussing dunes as part of coastal flood-risk reduction schemes is the dynamic nature of these features. Natural dunes are not static. They are comprised of loose sediment material that is subject to forcing by wind and waves. Sediment enters and leaves the dune governed by delicate feedback processes down to individual grain scale. The overall morphodynamic behavior of the dune is the result of these processes. Over time, dunes and dune fields can grow in height, volume, and extent and also migrate as dictated by forcing conditions, available sediment, and vegetation characteristics. Growth of course is not infinite since various control mechanisms exist that limit the typical size of dunes at a specific location. These control mechanisms can be related to the type and extent of vegetation cover (Durán & Moore, 2013), the geological framework of the substrate and type of sediment available (Lentz & Hapke, 2011), and the overall coastal climate including wind characteristics, water level fluctuations, waves, and even runoff from rain events (e.g., Houser, Hapke, & Hamilton, 2008). In turn, if erosive conditions prevail and sediment availability is limited, dunes can reduce in size and become more susceptible to storm impacts. These erosive trends tend to be exacerbated by the rise in sea level.

Dunes and storm impacts

Coastal dunes, and specifically the foredune (i.e., the dune closest to the beach and open water), are located along the battlefront between ocean and land. They are an integral part of the coastal system as a sand reservoir and form a dynamic barrier of sediment acting as a buffer against storm impacts. Especially vegetated dunes protect coastal areas from flooding and wave action (Sigren, Figlus, Highfield, Feagin, & Armitage, 2018; Wootton et al., 2016) and allow shoreline oscillation during low and high energy wave conditions (French, 2001). In general, the beach and dune profile is a sand sharing system affected by variations in wave energy, sea-level rise, tidal fluctuations, currents, and mobilization of sand from one zone to another (Hanley et al., 2014; Psuty, 2004). Storms provide intermittent, short-term disturbances to the normal conditions under which the dune and beach have formed a quasi-equilibrium state with the foot of the dune situated well above the mean tide level. When water levels rise significantly above the elevation of the dune toe during a storm, sediment redistribution processes commence as the beach–dune profile is being modified toward a new equilibrium state. This new equilibrium state is never reached since the storm impact usually lasts only several hours to a few days with relatively rapid changes in water level and wave energy over that time.

Impact scale

The storm impact scale for barrier islands formulated by Sallenger Jr. (2000) includes as a central theme the relative location of hydrodynamic action to the dune crest as the highest elevation on the coast. The four regimes making up the scale are termed "swash," "collision," "overwash," and "inundation" and are used as a means to categorize the coupled water and morphodynamic processes during coastal storm impact (Table 1). The relative elevation between the toe and crest level of the dune, respectively, and the maximum extent of the wave runup over the course of a storm interacting with the dune are used to distinguish between the regimes. During the swash regime, maximum wave runup levels remain below the toe of the dune, only removing sediment from the foreshore where calm conditions after the storm are able to gradually return the sediment. During the collision regime, wave runup collides with the face of the foredune ridge, moving sediment offshore while forming a pronounced dune scarp. In this regime, no water or sediment flows over the dune crest. Recovery of the volume of sand removed from the dune is slow (on the order of months and years) since it has to rely primarily on the eolian process of dune building. The overwash regime features intermittent wave runup events that overtop the dune crest and initiate both seaward and landward water and sediment transport away from the dune. The landward transport of sediment contributes to net onshore migration of the system while eroding the dune. If the minimum elevation of swash motions exceeds the dune crest, the system is in the inundation regime. The uninhibited landward flow of water and sediment caused by the difference in water elevation in front of (ocean side) and behind (land side) the dune are the result.

Table 1 Storm impact description for dunes (based on Sallenger Jr., 2000).

REGIME	Hydrodynamic condition	Morphological response	Schematic
Swash	Maximum wave runup level remains below the dune toe	Only foreshore sediments are being reworked	
Collision	Waves attack the dune face	Sediment from the dune toe and dune face is transported offshore; a scarp forms	
Overwash	Maximum wave runup level exceeds the dune crest elevation; water overtops the crest intermittently	Sediment is transported over the dune crest to the back barrier; the dune crest erodes	
Inundation	Surge levels exceed the dune crest	Substantial washover of dune sediment to the back barrier; the dune erodes	

During this regime, the entire foredune ridge is underwater with large quantities of dune sediment being transported landward, often resulting in the complete destruction of the dune.

Once regimes 3 or 4 are reached, the dunes become susceptible to breaching. Breaches are local cuts through the dune line that form due to the removal of dune sediment by water flowing over the dune crest. Breaches tend to form at locations of local minima in dune crest elevation, at local maxima of stormwater elevation, or at locations where the dune sediment is eroded more easily than elsewhere along the dune. Once a breach has been initiated, it can quickly grow in size leading to severe flooding of the area behind the dune.

Dune evolution during a storm

In contrast to hard coastal structures such as seawalls or revetments, dunes change shape extensively during storm impact. A heavily scarped dune that has been exposed to the collision regime can actually be interpreted as a sign of success in dealing with wave-induced erosion during elevated storm surge levels. The buffer function of dunes where waves have to exert and dissipate their energy in the dune erosion process is one of the key advantages of dunes over hard structures. Sediment eroded from the dune is transported further offshore, effectively flattening the beach profile and often forming a submerged sand bar (see Fig. 1). Such a morphodynamic adjustment of the beach profile during storm impact is nothing short of an automatic "self-protecting" effect fueled in part by sediment from the dune. On a shallower beach, waves are forced to break further offshore, especially if a submerged sand bar is present. The result is that the wave energy reaching the dune is much reduced from the levels that would occur without the morphodynamic profile adjustment.

Timing is everything when assessing storm–dune interactions. Once a dune is breached or inundated, its protective capacity against flooding is exhausted. The longer a dune can withstand the attack from elevated water levels and waves, the further the flood risk is reduced. It is thus important to understand the duration of storm impact, the timing of associated water level changes, and the corresponding wave energy reaching the dune at each time step. Only armed with this knowledge can the morphodynamic evolution of the dune during the storm be estimated and an assessment be made on whether the dune will survive the storm or not. This assessment is usually conducted via process-based numerical model simulations of storm-driven dune evolution (e.g., Figlus, Kobayashi, Gralher, & Iranzo, 2011) as will be explained in more detail later. In general, a taller and wider dune (i.e., more sand volume) with mature and healthy vegetation will survive longer. In turn, a storm of a shorter duration will cause less damage to the dune than a similar storm of a longer duration.

Recovery after a storm

Immediately after a storm, the beach-dune system still displays the profile adjustments caused by the storm forcing and is thus not in equilibrium with the calmer forcing conditions that are then present again. Recovery of the beach and dune profile via net onshore sediment transport processes commences, albeit at a much slower pace than the offshore-directed net erosion that occurred during the storm. The subaqueous part of the profile can recover relatively quickly within hours or days if sand has not been moved out of the coastal system and is available for onshore transport by nearshore hydrodynamics. The subaerial part of the profile that includes the dune can take much longer to get back to prestorm conditions since eolian transport processes occur at a much slower rate. The ability of a dune to recover after a storm varies and depends on the response of the beach/dune system to sea level rise, the change in frequency and/or magnitude of storm surges (Houser et al., 2015), the supply of sediment (Bauer & Davidson-Arnott, 2003; Davidson-Arnott, 2005; Woodruff, Irish, & Camargo, 2013), the beach width, wetting and drying cycles, the growth and state of existing vegetation, as well as its capability to trap wind-blown sediment. The recovery of a dune can take years, even decades and depends on the extent of the dune erosion. Davidson-Arnott (2010) presented some approximate dune recovery durations based on the level of disturbance of the foredune during storm impact. For minor disturbances where waves reach the base of the foredune, erosion will be minor, and recovery should occur within a few months to a year. For moderate disturbance where waves erode embryonic dunes and scarp the base of the foredune, recovery occurs within 2 to 5 years. For severe disturbance where waves generate a significant vertical scarp, the recovery may range from 5 to 10 years. For catastrophic disturbance where waves breach a dune completely, the recovery may take more than 10 years. The relatively long recovery times after storm impact make dunes susceptible to repeated breaching and inundation by subsequent storms. It is thus critical to intervene and restore impacted dunes and beaches to their prestorm geometry immediately after a storm if the dunes are part of a flood risk reduction scheme. Such emergency nourishment efforts are commonly applied around the world and often complement planned recurring nourishment efforts as part of flexible dune management approaches.

Engineered dunes

The ability of dunes to protect landward assets against storm impact and flooding combined with their aesthetic and ecological qualities has led to their purposeful incorporation into coastal flood risk reduction schemes. Especially along developed coastlines, natural dunes are often reshaped or replaced by engineered ones to accommodate storm protection requirements (Nordstrom, Lampe, & Vandemark, 2000; Saye, van der Wal, Pye, & Blott, 2005). It is common in such situations that spatial constraints exist due to landward infrastructure limiting natural dune migration. Without the ability to maintain volume and shape via sufficient eolian sediment transport from a wide beach or via translating landward over time, the degradation of these engineered dune systems needs to be counteracted by adaptive nourishment strategies.

Engineered dune design parameters include the crest height, crest width, frontal dune volume, side slopes, and grain size distribution. The height and width of the crest should be sufficient to minimize overtopping and breaching during a storm. The frontal dune volume should be adequate to accommodate dune erosion due to storm impact and the grain size distribution and other characteristics of the sediment should match the existing material. Various approaches exist to determine the design dimensions of a specific dune project, most of which are based on economic optimization coupled to desired levels of flood protection. In a simple deterministic approach, surge elevation, wave setup, and wave runup extent are combined to calculate the dune's height with a factor of safety (Bruun, 1998). Although the crest width and side slopes of a dune can vary, constructability limitations and the angle of repose of the fill material affect their selection (USACE, 2008).

One of the Dutch approaches to dune design is based on the idea that for a single dune line a minimum residual dune profile should be present after design storm impact as to provide continued sufficient protection until restoration efforts can occur. The minimum residual dune profile should have at least a 1:1 seaward slope, a minimum 3-m crest width, and a 1:2 landward slope. The 1:1 seaward dune slope starts at the intersection of a 1:40 foreshore slope estimated to form under storm conditions using the equilibrium beach profile concept (Bruun, 1954) with the respective design water level. As an additional safety measure, the volume of sand eroded during the design storm is increased by 25%, effectively moving the "hypothetical" intersection point further landward. This starting point of the 1:1 dune slope is called the critical erosion point. The minimum critical height of the dune crest above the critical erosion point is then determined as

$$h_c = 0.12 \cdot T_p \sqrt{H_s} \tag{1}$$

where T_p is the peak period and H_s is the significant wave height of the design storm, respectively. The option exists, however, to compensate for reduced critical dune crest elevation by providing a wider dune, as long as the total sand volume above the design water level remains the same (Voorendt, 2017).

Guidance on engineered dune design in the US is provided by Part V of the Coastal Engineering Manual (USACE, 2008). In essence, the final engineered berm and dune parameters are determined as part of an iterative process where initial dimensions should resemble "healthy" adjacent or historic profiles under normal forcing conditions and largely depend on the purpose of the project, constrained by economics, environmental issues, or local sponsor preferences. If no data on healthy historic or adjacent berm elevations are available, the berm height can be set at the limit of wave runup under nonstorm conditions. Beach width is optimized based on storm damage reduction by computing costs and benefits of various designs and selecting the design alternative that maximizes net benefits (USACE, 1991).

While some empirical, semiempirical, and analytical dune evolution models exist, dune designs are most commonly tested via process-based numerical models. These models simulate the evolution of the beach and dune profile based on numerical formulations of the dominant physical processes. Usually, the simulations are carried out for select forcing conditions to assess the adequacy of a chosen dune design. Empirical models such as the one by Vellinga (1983) and Van Gent (2008) predict the poststorm dune profile based on storm surge level, wave conditions, and sediment grain size characteristics. Kriebel and Dean (1985) presented a numerical model capable of calculating the time-dependent erosion of beaches and dunes during storms. Larson and Kraus (1989) introduced a semiempirical model that included the ability to predict submerged bar formation during the erosion process. The latter two models both employed beach profile equilibrium concepts in their formulations. An analytical model to predict the erosion and recession of dunes utilizing a wave impact approach was presented by Larson, Erikson, and Hanson (2004). Two of the most common process-based numerical models used to test dune designs are CSHORE (Johnson, Kobayashi, & Gravens, 2012; Kobayashi, 2013) and XBeach (Roelvink et al., 2009). Based on time-varying hydrodynamic input conditions and initial profile shape, they compute the evolution of the beach and dune systems. Both models are constantly upgraded and expanded with new features such as vegetation effects, layered substrate parameterization, or the inclusion of hard structures. Harter and Figlus (2017) compared CSHORE and XBeach based on respective simulations of Hurricane Ike impact on a Texas barrier island indicating decent skill levels for both models in reproducing measured poststorm beach and dune profiles. The models are commonly calibrated using laboratory and field data of beach and dune erosion under varying conditions.

Implementation of engineered dunes in Texas and the Netherlands

Dunes in Texas and the Netherlands differ quite substantially. The two locations feature different climates, sediments, vegetation, hydrodynamic forcing conditions, and coastal policy, all of which influence the appearance of dunes. In addition, the availability of beach-quality sand as well as design philosophies for engineered dunes are different. de Vries et al. (2012) estimate mostly linear dune growth rates along the Holland coast (subsection of the Dutch coast) up to $40 \, m^3/m/year$. Although rates are site-specific, this knowledge can be used to plan and optimize engineered dune systems for flood protection over the entire design life of the system. Along the Texas coast, large differences in natural dunes exist between the southern Texas coast and the upper Texas coast (UTC). Few naturally occurring dunes exist along the UTC and the Brazos-Colorado River headland with crest heights rarely exceeding 2 m, while relatively stable dunes occur on Mustang Island and North Padre Island. Some stretches of the Texas coast with limited development and sufficient sand supply feature continuous, well-defined foredune ridges up to 6 m in height, in some rare cases even up to 12 m (Patterson, 2005). The problem for dunes along most of the Texas coastline is the limited supply of sand (especially along the UTC), relatively narrow beaches, and the fact that the difference between normal forcing conditions and extreme events (i.e., hurricanes) are very large. This means only relatively small dunes are created during normal conditions, but storms are able to easily erode them completely.

Engineered dunes are often embedded into multifunctional coastal flood defenses that include sandy beaches and even structures to form hybrid systems (Almarshed, Figlus, Miller, & Verhagen, 2020). Several examples from the Netherlands and Texas are given here.

Hondsbossche dunes

The Hondsbossche Dunes near Petten in the Netherlands are a great example of how engineered dunes can work in concert with existing hard coastal protection structures (Bodde et al., 2018). The existing basalt block sea dike fronting the North Sea was considered a weak spot in the Dutch coastal defense. A storm event with a probability of occurrence of 1:10,000 was estimated to cause appreciable overtopping and stability issues on the landward slope of the dike. The Dutch strategy is to keep coastal storm waters in the North Sea rather than allowing water to overtop dikes and dunes and flood the low-lying hinterland. Rather than raising the sea dike to the required protection standards, 35 million cubic meters of sand were placed on the ocean side of the existing dike structure along 11 km of coastline between the towns of Camperduin and Petten. By 2015, a shallow sandy foreshore and various dune habitats had been constructed to provide flood safety, spatial quality, ecosystem services, and recreational opportunities. Design constraints included a 50-year design life taking sea level rise into account, an allowable 12-m maximum dune crest elevation, and a smooth coastal shoreline contour blending the project in with adjacent stretches of coastline. The design features a wide foreshore with a vegetated dune area and a back-dune buffer area fronting the original basalt dike to minimize landward wind-driven sediment transport past the dike (Fig. 2). Several walks-over paths, a bicycle and footpath, a horse track, as well as an engineered wet dune valley freshwater habitat were incorporated into this multifunctional flood defense scheme. Planted marram grass vegetation, willow screens, and engineered dune relief features were employed to influence sediment transport patterns and help foster the dynamic nature of the dune and its continual evolution. Ongoing monitoring efforts ensure the Hondsbossche Dunes remain a viable component of the Dutch flood defense strategy over the coming decades.

Fig. 2 Photo sequence of coastal flood defenses near Petten, the Netherlands. From left to right: seaward side of the Hondsbossche Dunes (left), back-dune fronting the dike (center), and backside of the dike (right). *(Photos by J. Figlus.)*

Katwijk dunes

Another example of engineered dunes used in multifunctional coastal risk reduction schemes (Voorendt, 2017) is the dunes fronting the seaside boulevard of the town Katwijk aan Zee in the Netherlands. Approximately 1 km of seaside boulevard did not meet the Dutch safety level for elevation. In this case, a hybrid approach consisting of a relatively low but wide dune was chosen. It covers an embedded basalt block dike including a sand core, geotextile, and filter layer. The dune extends landward past the dike all the way to the boulevard and covers an underground parking garage (Fig. 3) with the intent to optimize the use of space near the coastal boulevard (ARCADIS, 2012; Voorendt, 2015). One of the design constraints was posed by the viewshed from the boulevard. To avoid blocking the view of the ocean with a high dune, the dune crest height was limited to +8.00 m (NAP—Normaal Amsterdams Peil). To maintain sufficient flood protection and avoid overtopping of the boulevard during storm conditions, the sand volume available for erosion in the beach and dune system was considered. The reduced dune height was "balanced" by an increased dune width, while the overall volume available for erosion was preserved. The width of the dune from the boulevard to the dune toe is 120 m, an increase in dune width of 90 m compared to the situation prior to the construction of the project. The embedded dike provides further risk reduction against wave overtopping contributing to the viability of the reduced dune crest height. Over the 11-m-long crest of the dike, the vegetated dune sand layer only has a thickness of 0.5 m. As with all

Fig. 3 Photo showing the Katwijk aan Zee seaside boulevard and the entrance to the parking garage under the dunes. *(Photo by J. Figlus.)*

dune systems, the idea is that during elevated water levels and wave attack a portion of the frontal dune volume is moved offshore. This volume of sand forms a submerged bar, dissipates incoming wave energy, reduces overtopping over the dike, and protects the remainder of the dune and the landward assets as confirmed via time-dependent process-based numerical modeling simulations. Even if the dike is exposed, the storm cross-shore profile of the beach and dune helps limit the wave overtopping rate to a safe level of less than 1 L/s/m. Future adaptations to changed forcing conditions (i.e., increased sea level rise rates) can be accomplished by adding sediment to the dune component of the system or by adding flood wall elements at the outer edge of the seaside boulevard. The flexibility and adaptability of the dune to changing conditions highlights the utility of dunes as a viable option for flood protection schemes.

Galveston "Seabale" dunes

The UTC is particularly vulnerable to flooding from storm surges and waves since limited sand supply has contributed to continual erosion, narrow beaches, and small or nonexistent dunes. Flood risk is particularly high in the highly developed Houston–Galveston metropolitan area along with the southern and western parts of Galveston Bay including Galveston Island and adjacent barrier islands. Innovative strategies to grow and strengthen dunes along the Texas coastline are needed to support ongoing nourishment efforts. This pilot project (Figlus, Sigren, Webster, & Linton, 2015) made use of floating seaweed wrack material (i.e., Sargassum) that had washed up onto Galveston beaches in large quantities in the summer of 2014 (up to 1-m-high deposits). While low-volume Sargassum landings are a common occurrence all along the Gulf of Mexico during the summer months, this particular event caused extremely hazardous conditions on the beach for visitors and residents and Sargassum wrack material had to be removed mechanically in many places. A viable beach management option was created by collecting the seaweed from the beach, compacting it to manageable dense bales, and using these "seabales" to build and strengthen local dunes. Test dune segments totaling 250 m in length were constructed in the extension of the existing dune line on Galveston Island's east end using a local beach sediment source. The dunes mimicked existing dune geometry in the area and were built to 2 m in height and 7 m in base width including a pronounced dune berm (Fig. 4). A row of seabales was placed inside the dune berm for half of the test dune

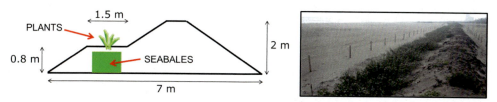

Fig. 4 Schematic representation of vegetated seabale dune (left) and photo (right) 9 weeks after vegetation planting. *(Photo by J. Figlus.)*

length and vegetation was planted on top of the berm for certain test segments to evaluate the respective effect of seabales and vegetation on dune evolution. In particular, the dune segments with seabales and vegetation faired remarkably well and grew to almost twice their initial height over 5 years, developing into diversely vegetated ecosystems. While seabales decomposed within the first 2 months, they provided nutrients and moisture to the plants, allowing them to take root quickly, grow tall, and trap wind-blown sediment effectively to help grow the dunes. Part of the success of this project can be attributed to the fact that it was built on one of the few beaches of Galveston Island that are accreting, but the concept could be just as successful for eroding beaches if made part of smart nourishment schemes.

Houston-Galveston coastal spine

A coastal spine of approximately 100-km length is currently planned as the primary line of defense against flood damage from storm surge and wave attacks along a stretch of the UTC including the Bolivar Peninsula and Galveston Island as discussed in various sections of this book. Up to 80 km of the coastal spine are conceptualized to include an engineered dune system as part of an extensive beach and dune nourishment scheme. While designs are far from being finalized, a system with two closely spaced dune ridges fronted by a nourished beach has been proposed to complement the already existing Galveston seawall and anticipated surge and navigation gates at the entrance to Galveston Bay. This development shows that in the US engineered dunes are valued more and more as a viable alternative to pure structural protection measures due to their flood-risk reduction capacity, flexibility, natural aesthetics, and ecosystem functionality. The presented examples from the Netherlands show that this approach can be very successful, especially if dunes are used in concert with structural measures in a true hybrid fashion to combine the benefits of soft and hard coastal engineering concepts into one system.

Conclusions

Coastal dunes have long been known for their ability to reduce flood risk from storm surge and wave attacks. They are a natural barrier providing elevation and sediment volume above the normal water line and are activated during storm conditions. Their ability to erode under wave attack and provide sediment to the subaqueous beach in the process fulfills a critical buffer function. In addition, the presence of dunes along the coast enhances aesthetics and ecosystem functionality. The deliberate inclusion of dunes into coastal flood risk reduction schemes is accompanied by some challenges that need to be understood to allow for adequate design. It is also important to understand the inherent vulnerabilities to erosion that come with dune systems to avoid creating a false sense of security against flood damage and an incentive to increase development behind dunes. The dynamic nature of dunes under both normal and storm conditions is one of the most critical aspects that needs to be accounted for and assessed in a site-specific manner. The natural tendency of dunes to migrate and evolve according to local coastal system conditions is often drastically constrained when they are part of coastal protection systems where space is limited. In such situations, careful planning and adaptive and flexible dune management approaches are crucial.

References

Almarshed, B., Figlus, J., Miller, J., & Verhagen, H. J. (2020). Innovative coastal risk reduction through hybrid design: Combining sand cover and structural defenses. *Journal of Coastal Research*, *36*(1), 174–188. https://doi.org/10.2112/JCOASTRES-D-18-00078.1.

ARCADIS. (2012). *Ontwerp-projectplan kustversterking Katwijk* (p. 89). ARCADIS (in Dutch).

Barone, D. A., McKenna, K. K., & Farrell, S. C. (2014). Hurricane Sandy: Beach-dune performance at New Jersey Beach profile network sites. *Shore & Beach*, *82*(4), 13–23.

Bauer, B. O., & Davidson-Arnott, R. G. D. (2003). A general framework for modeling sediment supply to coastal dunes including wind angle, beach geometry, and fetch effects. *Geomorphology*, *49*(1), 89–108. https://doi.org/10.1016/S0169-555X(02)00165-4.

Bodde, W., Jansen, M., Smit, M., Scholl, M., Lagendijk, G., Kuiters, L., et al. (2018). Sand nourishment—Hondsbossche dunes (Monitoringsrapportage 2017 no. DDT169-13/18-000.148; HPZ Innovatieproject, p. 84). In *Witteveen + Bos and Wageningen marine research* (in Dutch).

Bruun, P. (1954). *Coast erosion and the development of beach profiles*. U.S. Beach Erosion Board.

Bruun, P. (1998). Dunes—Their function and design. *Journal of Coastal Research*, *SI*(26), 26–31.

Davidson-Arnott, R. G. D. (2005). Conceptual model of the effects of sea level rise on sandy coasts. *Journal of Coastal Research*, *216*, 1166–1172. https://doi.org/10.2112/03-0051.1.

Davidson-Arnott, R. G. D. (2010). *Introduction to coastal processes and geomorphology*. Cambridge University Press.

de Vries, S., Southgate, H. N., Kanning, W., & Ranasinghe, R. (2012). Dune behavior and aeolian transport on decadal timescales. *Coastal Engineering*, *67*, 41–53. https://doi.org/10.1016/j.coastaleng.2012.04.002.

Doody, J. P. (2013). In J. P. Doody (Ed.), *Sand dune conservation, management and restoration*. Netherlands: Springer. https://doi.org/10.1007/978-94-007-4731-9_4.

Durán, O., & Moore, L. J. (2013). Vegetation controls on the maximum size of coastal dunes. *Proceedings of the National Academy of Sciences*, *110*(43), 17217–17222.

Elko, N., Brodie, K., Stockdon, H. F., Nordstrom, K. F., Houser, C., McKenna, K., et al. (2016). Dune management challenges on developed coasts. *Shore & Beach*, *84*(1), 1–14.

Feagin, R. A., Furman, M., Salgado, K., Martinez, M. L., Innocenti, R. A., Eubanks, K., et al. (2019). The role of beach and sand dune vegetation in mediating wave run up erosion. *Estuarine, Coastal and Shelf Science*, *219*, 97–106. https://doi.org/10.1016/j.ecss.2019.01.018.

Figlus, J., Kobayashi, N., Gralher, C., & Iranzo, V. (2011). Wave overtopping and overwash of dunes. *Journal of Waterway, Port, Coastal, and Ocean Engineering*, *137*(1), 26–33.

Figlus, J., Sigren, J. M., Webster, R., & Linton, T. (2015). *Innovative technology seaweed prototype dunes demonstration project*. (p. 47) [Final Technical Report for CEPRA Cycle 8 Project 1581 to the Texas General Land Office] Texas A&M University at Galveston.

French, P. (2001). *Coastal defences: Processes, problems and solutions*. Routledge.

Hanley, M. E., Hoggart, S. P. G., Simmonds, D. J., Bichot, A., Colangelo, M. A., Bozzeda, F., et al. (2014). Shifting sands? Coastal protection by sand banks, beaches and dunes. *Coastal Engineering*, *87*, 136–146.

Harter, C., & Figlus, J. (2017). Numerical modeling of the morphodynamic response of a low-lying barrier island beach and foredune system inundated during Hurricane Ike using XBeach and CSHORE. *Coastal Engineering, 120*, 64–74. https://doi.org/10.1016/j.coastaleng.2016.11.005.

Houser, C., Hapke, C., & Hamilton, S. (2008). Controls on coastal dune morphology, shoreline erosion and barrier island response to extreme storms. *Geomorphology, 100*(3–4), 223–240. https://doi.org/10.1016/j.geomorph.2007.12.007.

Houser, C., Wernette, P., Rentschlar, E., Jones, H., Hammond, B., & Trimble, S. (2015). Post-storm beach and dune recovery: Implications for barrier island resilience. *Geomorphology, 234*, 54–63.

Johnson, B. D., Kobayashi, N., & Gravens, M. B. (2012). Cross-shore numerical model CSHORE for waves, currents. In *Sediment transport and beach profile evolution* Engineer Research and Development Center. Technical Report ERDC/CHL-TR-12-22.

Kobayashi, N. (2013). *Cross-shore numerical model CSHORE 2013 for sand beaches and coastal structures* (Research Report CACR-13-01). University of Delaware.

Kriebel, D. L., & Dean, R. G. (1985). Numerical simulation of time-dependent beach and dune erosion. *Coastal Engineering, 9*(3), 221–245.

Larson, M., Erikson, L., & Hanson, H. (2004). An analytical model to predict dune erosion due to wave impact. *Coastal Engineering, 51*(8), 675–696.

Larson, M., & Kraus, N. C. (1989). *SBEACH: Numerical model for simulating storm-induced beach change. Report 1. Empirical foundation and model development*. DTIC Document.

Lentz, E. E., & Hapke, C. J. (2011). Geologic framework influences on the geomorphology of an anthropogenically modified barrier island: Assessment of dune/beach changes at Fire Island, New York. *Geomorphology, 126*(1), 82–96. https://doi.org/10.1016/j.geomorph.2010.10.032.

Nordstrom, K. F., Lampe, R., & Vandemark, L. M. (2000). Reestablishing naturally functioning dunes on developed coasts. *Environmental Management, 25*(1), 37–51. https://doi.org/10.1007/s002679910004.

Patterson, J. (2005). *Coastal dunes—Dune protection and improvement manual for the Texas coast* (5th ed.). Texas General Land Office (Manual).

Psuty, N. P. (1988). Sediment budget and dune/beach interaction. *Journal of Coastal Research, 3*, 1–4.

Psuty, N. P. (2004). The coastal foredune: A morphological basis for regional coastal dune development. In M. L. Martínez, & N. P. Psuty (Eds.), *Coastal dunes: Ecology and conservation* (pp. 11–27). Springer. https://doi.org/10.1007/978-3-540-74002-5_2.

Roelvink, D., Reniers, A., van Dongeren, A., van Thiel de Vries, J., McCall, R., & Lescinski, J. (2009). Modelling storm impacts on beaches, dunes and barrier islands. *Coastal Engineering, 56*(11 – 12), 1133–1152.

Sallenger, A. H., Jr. (2000). Storm impact scale for barrier islands. *Journal of Coastal Research, 16*(3), 890–895.

Saye, S. E., van der Wal, D., Pye, K., & Blott, S. J. (2005). Beach–dune morphological relationships and erosion/accretion: An investigation at five sites in England and Wales using LIDAR data. *Geomorphology, 72*(1), 128–155. https://doi.org/10.1016/j.geomorph.2005.05.007.

Sigren, J. M., Figlus, J., Highfield, W., Feagin, R. A., & Armitage, A. R. (2018). The effects of coastal dune volume and vegetation on storm-induced property damage: A hurricane Ike case study. *Journal of Coastal Research, 34*(1), 164–173.

USACE. (1991). *National economic development procedures manual—Coastal storm damage and erosion* (Institute of Water Resources Report No. 91-R-6; p. 274). U.S. Army Corps of Engineers.

USACE. (2008). *Beach fill design* (Engineer Manual EM 1110-2-1100 Part V Chapter 4; p. 113). U.S. Army Corps of Engineers.

Van Gent, M. (2008). Large-scale tests to analyse the influence of collapsed dune revetments on dune erosion. *Coastal Engineering Proceedings, 3*, 2583–2595.

Vellinga, P. (1983). Predictive computational model for beach and dune erosion during storm surges. *Specialty Conference on Coastal Structures Proceedings*, 806–819.

Voorendt, M. Z. (2015). *Examples of multifunctional flood defences*. (p. 72) [Working Report] Delft University of Technology, Department of Hydraulic Engineering.

Voorendt, M. Z. (2017). *Design principles of multifunctional flood defences*. [Ph.D. Dissertation]. Delft University of Technology.

Walker, I. J., Davidson-Arnott, R. G. D., Bauer, B. O., Hesp, P. A., Delgado-Fernandez, I., Ollerhead, J., et al. (2017). Scale-dependent perspectives on the geomorphology and evolution of beach-dune systems. *Earth-Science Reviews*, *171*, 220–253. https://doi.org/10.1016/j.earscirev.2017.04.011.

Woodruff, J. D., Irish, J. L., & Camargo, S. J. (2013). Coastal flooding by tropical cyclones and sea-level rise. *Nature*, *504*(7478), 44–52. https://doi.org/10.1038/nature12855.

Wootton, L., Miller, J., Miller, C., Peek, M., Williams, A., & Rowe, P. (2016). *New Jersey Sea Grant Consortium Dune Manual*. (p. 77).

CHAPTER 22

A proactive approach for the acquisition of flood-prone properties in advance of flood events

Kayode O. Atoba
Institute for a Disaster Resilient Texas, Texas A&M University, College Station, TX, United States

The acquisition or buyout of flooded properties plays a vital role in flood resiliency. In the United States, however, buyout programs are driven by a cost-benefit approach and primarily reactive, occurring only after a flood event. The poor structure of a buyout policy also makes it contentious for some homeowners who refuse to participate. By proposing a more proactive approach, this chapter presents a framework that identifies flood-prone properties and vacant land that are economically viable for acquisition well in advance of a storm event. Coupling proximity-based geospatial analysis with hazard modeling, land transformation modeling, and a Geodesign framework, this chapter evaluates the potential economic benefits of acquiring built and vacant properties in the Houston region. Results indicate that cumulative avoided flood losses exceed the cost of property acquisition. This study emphasizes the benefits of a proactive approach that avoids the social and institutional problems associated with traditional buyout programs. This chapter also highlights the contextual differences between buyouts in the United States and the Netherlands' Room for the River program.

Property acquisition for flood resiliency in the United States

In the United States, flooding remains the natural hazard that causes the most impact on lives and property. The rise in flood events and their associated impacts has led to calls for adopting more effective risk reduction strategies (Brody et al., 2020). One strategy gaining increasing acceptance is called *avoidance*, which involves removing properties from high-risk locations and returning them to vacant open spaces. Buyouts are increasingly becoming an essential mitigation and adaptation strategy in the United States (Siders, 2019a). For example, over 41,000 residential properties have been purchased nationwide, primarily through the Federal Emergency Management Agency's (FEMA) programmatic funding (Mach et al., 2019). Once purchased, these properties are demolished and the land remains open space in perpetuity, thereby preventing repetitive flood losses and protection of environmental assets (Conrad, McNitt, & Stout, 1998; Zavar, 2015). A recent national study shows that buying out 1 million homes from flood-prone areas in the United States could cost about $1 trillion but generate $1.16 trillion in savings over a 100-year period (NIBS, 2019).

Although removing flood-prone properties through acquisition can result in reduced flood losses, existing buyout programs in the United States face serious challenges. The execution of buyouts as a flood risk reduction strategy in the United States is reactionary at best. Flood-related buyouts are mainly driven by a cost-benefit calculus that leads to an ad hoc selection process conducted after a flood event has already occurred (Atoba, Brody, Highfield, Shepard, & Verdone, 2020; Siders, 2019b), leading to an uncoordinated checkerboard pattern of open spaces. Because the acquisition occurs postflood, buyout participants are burdened by socioeconomic and psychological distress experienced by households that need to relocate after a buyout occurs (Binder, Baker, & Barile, 2015; Binder, Greer, & Zavar, 2020). Additionally, there is limited transparency in the selection process, potentially contributing to social equity issues and reduced participation, especially for socially vulnerable households (Greer & Brokopp Binder, 2017; Siders, 2019b). Other problems include the loss of existing infrastructure and utility services and the cost of maintaining small, isolated buyout parcels (Atoba et al., 2020; Blanco et al., 2009; Freudenberg, Calvin, Tolkoff, & Brawley, 2016; Maly & Ishikawa, 2013; Zavar & Hagelman III, 2016). Moreover, in the United States, accepting a buyout is voluntary, and many property owners are emotionally attached to their homes and reluctant to leave, even under dire circumstances (Maldonado, Shearer, Bronen, Peterson, & Lazrus, 2013; Marino, 2018). These problems necessitate a more proactive approach to identifying existing properties for buyouts, as well as targeting open space clusters in flood-prone areas as having the potential for vacant property buyouts.

Vacant open spaces also provide an opportunity for proactive flood risk reduction if they are considered for acquisition in advance of development interest, thereby preventing any form of future flood risk. "Vacant land" accounts for almost 17% of land in each large US city (Newman, Bowman, Jung Lee, & Kim, 2016). A recent nationwide study found that by 2070, cumulative avoided future flood damages may exceed the costs of land acquisition by up to a factor of five to one if properties are left undeveloped across different flood zones (Johnson et al., 2019). Other studies at smaller scales have also evaluated the benefits of conserving vacant lands within the floodplain for reduced flood impact (Atoba et al., 2021; Kousky & Walls, 2014; Newman et al., 2019). When vacant lands in flood-prone areas remain vacant, they automatically record no direct damages from flooding and can even store floodwaters, thereby reducing flood losses for surrounding development (Brody, Blessing, Sebastian, & Bedient, 2014; Brody, Highfield, Blessing, Makino, & Shepard, 2017; Highfield, Brody, & Shepard, 2018). When left undeveloped, vacant properties also provide other opportunities such as restoring native vegetation, wetlands, and floodplains (Conrad et al., 1998; Harter, 2007). They also have the potential of increased green space and opportunities for recreation

(Crompton, 2005; Geoghegan, 2002), biodiversity conservation (Hausmann, Slotow, Burns, & Di Minin, 2016), all of which improve the quality of life of nearby residents.

While there is an existing framework for acquiring residential buildings through FEMA programs, there is currently no existing comprehensive framework for acquiring vacant land as buyouts. Although these vacant properties represent an enormous opportunity for acquisition for flood risk reduction, the reactive approach of US flood risk reduction greatly impacts them as seriously being considered in advance as candidates for buyouts. Moreover, development patterns in the United States are driven by a series of market forces and responses to policies that further increase the likelihood of development occurring in vulnerable locations.

This chapter addresses these challenges by proposing a proactive framework that identifies candidate parcels for acquisition along a spectrum of ecological and proximity variables that should be considered in advance of storm events. This approach will identify both existing properties and vacant open spaces that can be acquired as a more proactive method of reducing the impact of flooding in coastal communities. It will further make a case for property buyouts by comparing the cost of acquisition to the potential benefits.

Contextual differences between buyouts in the United States and the Netherlands

There is a basic contextual difference between flood risk reduction in the Netherlands compared to the United States In the Netherlands, authorities deal with flood safety not merely as a regulation, but by integrating water management policies into spatial plans across different spatial and municipal scales (Woltjer & Al, 2007; ESPON, 2017), indicative of a more proactive approach. Additionally, the Dutch decision-making model is distinguished by elements of consultation, consensus, and compromise (Kolen, Hommes, Trainor, Yuiten, & Shaw, 2012), with flood risk being a major priority in emergency management. There is also a command-and-control approach from a central government in their approach to flood risk reduction. In the United States, on the other hand, the federal government influences flood risk reduction by providing advice, placing restrictions, or providing incentives in funding aimed at facilitating the adoption of specific types of programs (Kolen et al., 2012).

The contextual differences between flood risk reduction in the Netherlands and the United States are also seen in each country's approach to property acquisition. Unlike the United States, Dutch buyouts are proactive, occurring before flooding to make room for floodwaters. It is part of an integrated water management approach rather than a reactive approach of buying out properties 'after big flood disasters' (Wendland, 2020). Flood-related property acquisition in the Netherlands is also administered as part of a holistic approach that complements existing flood defenses and reduces the risk of future flooding (Jan Goossen, 2018). In the Netherlands, buyouts hardly play a significant role in flood risk reduction, primarily because of safe development policies, which limits the construction of residences in flood-prone areas without adequate structural or policy protections. On the other hand, US buyouts are carried out in a reactionary ad hoc approach across different communities in response to extreme flood events and decades of poor development and land management choices.

In the Netherlands, one of the most popular avenues for property acquisition was through the Room for the River program (RftR). The RftR program was set up by the Dutch Water management authority, spending almost € 2.3 billion to convert residential and agricultural developments in 34 flood-prone areas back to open space along the IJssel, Waal, Nederrijn, and Lek rivers (Rijkswaterstaat, 2019). About 95% of the RftR funding was directed toward capital works improvement and flood infrastructure, while the remaining funds were spent on relocating residents and providing new social amenities and infrastructure (Jan Goossen, 2018). One of the most popular RftR programs was the Nijmegen project that integrated a holistic spatial planning approach to flood risk reduction by acquiring about 50 properties and converting them to vacant open space floodplains (ESPON, 2017; Yu, Brand, & Berke, 2020).

In general, the RftR program was largely successful as the dozens of residents who lived in the floodplain supported the government acquisition efforts and volunteered their homes for acquisition after receiving compensation to move to a safer area in the city (Bently, 2016). This is not surprising since Dutch citizens traditionally support their government's effort toward flood risk reduction (Jan Goossen, 2018). Of course, similar to US buyouts, the quest for relocating residents is met with resistance by some Dutch residents (see examples in Jan Goossen, 2018; Wendland, 2020; van Alphen, 2020). However, despite these challenges, Dutch officials promote a high level of engagement with local municipalities and a great deal of flexibility for actualizing buyouts in the RftR program (Jan Goossen, 2018; van Alphen, 2020). This flexibility and citizen engagement paid off; in some cases, residents even chair the efforts for buyouts and help convince neighbors to participate in the program (see examples in Wendland, 2020). This is partly possible because unlike the red tape and bureaucratic setback for buyout participation and funding that exist in the United States, the RftR program treated buyout eligibility differently. Their approach is more nuanced with lots of local buy-ins and support at the community level (Jan Goossen, 2018). The program created incentives for cooperation between residents, local officials, and the central government toward a more resilient ecological system (De Bruijn, De Bruijne, & Ten Heuvelhof, 2015). The bottom-up approach of the RftR program in general engendered local leadership and active lobbying to enable government authorities to cooperate with local stakeholders (Edelenbos, Van Buuren, Roth, & Winnubst, 2017).

Another difference between the United States and Dutch approaches to buyouts is the scale of buyout implementation and their motivation for property acquisition. While US buyout programs focus on avoided flood losses usually for each property being bought out (Atoba et al., 2020), the Dutch RftR buyout program embeds buyout costs and benefits with integrated flood risk management, thereby assessing benefits and cost at a larger ecological scale (van Alphen, 2020). Additionally, US buyouts are administered mainly to reduce future spending on repeatedly flooded properties in floodplains and only acquire these properties if a benefit–cost analysis (BCA) shows them to be economically viable candidates for acquisition. The Dutch RftR buyouts, on the other hand, focus on improving overall spatial quality. The RftR program defined spatial quality as the protection of land and assets from flooding, a robust ecological system that requires little maintenance, and enhanced the aesthetics of the existing landscape (Klijn, de Bruin, de Hoog, Jansen, & Sijmons, 2013).

The selection criterion for the RftR buyout program is starkly different from the general US buyout selection criteria. US buyouts rely on strict federal and local requirements, with most of these requirements lacking transparency at a national scale (Atoba et al., 2020; Siders, 2019a). The Dutch approach under the RftR program, on the other hand, applied a less bureaucratic, bottom-up approach toward determining buyout eligibility. In many cases, the Dutch water management agency "would sit down at the 'kitchen table' with residents to look for individual solutions. In addition, families were offered the opportunity to sell their homes at market value (before RftR); and farmers were helped in their search for new farmland" (van Alphen, 2020 p. 318).

A model for prioritizing ecological gains for property acquisition in the United States

Buyout selection and eligibility in the United States is mainly driven by the potential economic benefits of avoided flood losses. To ensure the efficient disbursement of federal funds, local communities can purchase properties with limited federal funds as well as a local match to acquire a handful of properties that have flooded in the past. In Harris County, TX, to qualify for buyouts through federal grants, the property must: (i) be located deep within the 10- or 100-year floodplain and mainly affected by inland flooding, (ii) have been repetitively flooded in the past, thereby being cost-effective for acquisition, and (iii) have community support based on the voluntary nature of US property acquisition (HCFCD, 2017). Most programs are also full of bureaucratic red tape that slows down the process, which means that it can take up to 2 years after a homeowner indicates interest in a buyout before the buyout actually occurs. Qualifying for buyouts is usually one of the most complicated parts of the process.

There is presently no existing framework for identifying vacant properties for buyouts. The US Army Corps of Engineers is primarily responsible for granting permits for wetland alteration or construction of large flood reservoirs. However, development policies for land in the floodplain are usually the responsibilities of local jurisdictions. Private lands can be sold for conversion to residential neighborhoods or as stipulated in zoning ordinances, while public lands like parks may also be leveraged for use as flood control measures and serve a dual purpose as retention or detention ponds. Prioritizing private vacant lands in flood-prone locations is an essential proactive approach to preventing development in vulnerable locations. Applying an ecologically driven approach to identifying vacant land for acquisition provides additional opportunities for conserving vacant land.

This chapter discusses a proactive framework for acquiring both vacant and existing flood-prone properties for acquisition in advance of flood events. As shown in Fig. 1, rather than beginning with a cost–benefit calculus for property buyouts, eligibility is

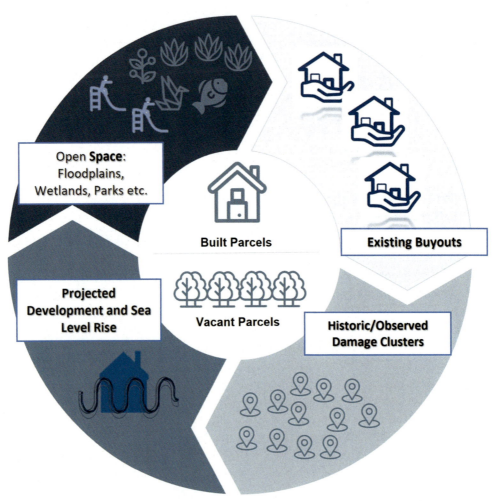

Fig. 1 Selecting built and vacant buyout candidates based on proximity to ecological features and historic flood risk.

initially determined by the proximity of either vacant or built-upon properties to existing open spaces, flood damage, previously bought out areas, and parcels projected for development that are also prone to sea-level rise. This buyout selection approach offers the potential to capture additional benefits that would have otherwise been lost if the selection criteria were based on a strictly cost-benefit calculus.

In selecting built-upon properties as candidate buyouts, it is important to look beyond the regulatory floodplain and consider proximity to other ecological features. This approach becomes even more critical as studies suggest that there is a growing disconnect between the 100-year floodplain and the distribution of observed flood losses in the United States (Atoba et al., 2020; Blessing, Sebastian, & Brody, 2017). Wetlands are one of the most important considerations for identifying buyouts. A recent study of a Houston watershed found that residential properties within 152.4 m of wetlands recorded less flood damage than those that were unsheltered and/or at distances beyond this threshold (Highfield et al., 2018). Other studies also indicate that close proximity to existing open spaces, such as other buyouts, encourages open space clustering, helps maintain viable habitats, reduces ecosystem fragmentation, and amplifies flood attenuation (Harter, 2007). Purchasing flood-prone homes adjacent to existing parks and wetlands could over the long term allow for the parcels to be annexed into those existing parks or revegetated to restore and expand wetland areas.

The selection of vacant properties for buyouts in this chapter is informed by the Geodesign framework (Atoba et al., 2021; Newman et al., 2019; Steinitz, 2012). This model systematically integrates concepts from geographic, spatial, and statistical-based disciplines into site selection. This framework begins with *representation*, which involves identifying the spatial scope of vacant properties in an area. Second, *risk and exposure* of the selected vacant properties are analyzed through several proximity variables such as distance to floodplains, wetlands, and historic flood damages. The *change* model focuses on identifying areas prone to future development, while the *impact* model tests a case scenario of a vacant test site in a selected change area. The final part of the model focuses on the *decision* to purchase a vacant site for buyouts now or develop it in the future.

Case study analysis and results

The proactive approach to built and vacant buyouts proposed in this chapter (Fig. 1) was tested in the Clear Creek Watershed as a case study. The watershed has tidally influenced water bodies that flow west–east toward the Galveston Bay with approximately 154 miles of stream (Atoba, 2018). It is historically the most flood-impacted watershed in the Houston-Galveston region (Atoba et al., 2020). As shown in Fig. 2, the application of the proactive framework led to the selection of both vacant and built-upon properties as candidate buyouts in the study region. The selected properties are in close proximity to existing buyouts, future floodplains, and areas of future development (see Fig. 2B).

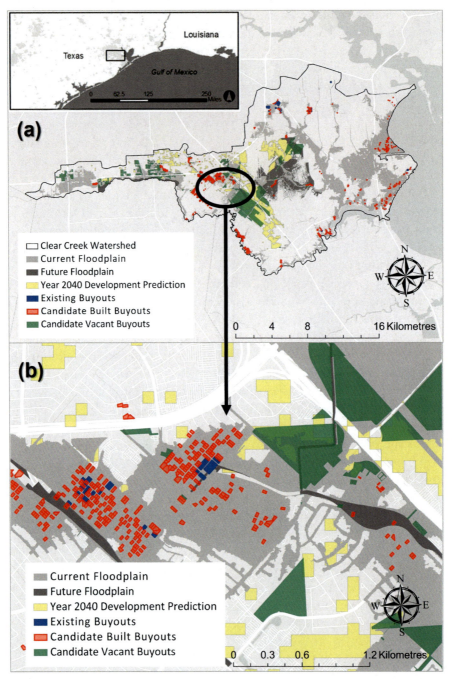

Fig. 2 Locations of vacant and built buyouts in (A) the Clear Creek Watershed and (B) a sample area within the watershed.

Table 1 Ecological summary of buyout scenarios.

Scenario		No of properties	Average BCR	Max BCR	Min BCR	Buyout distance (ft)	Wetland distance (ft)	Park distance (ft)
1	Floodplain only	400	1.74	19.15	0.75	5188	363	1699
2	Floodplains and wetlands	442	1.14	4.03	0.75	512	395	1885
3	All features	498	0.52	2.60	0.11	8899	443	519

When proximity to open spaces is factored into the selection of buyouts, the acquired properties can be better used for additional flood control infrastructure such as detention or retention ponds to store floodwaters and reduce the risk of flooding for adjacent developments. In addition to identifying candidate buyouts, a BCA further helps decision-makers to better understand the financial implications (Mobley, Atoba, & Highfield, 2020). This analysis generates a benefit–cost ratio (BCR) which is determined by comparing acquisition cost with avoided future flood damages. A ratio higher than 1 often indicates that the property would have a net financial benefit for acquisition.

For built-upon properties, the selection criteria prioritize residential properties within 1000 ft. of wetlands, floodplains, parks, and existing buyouts. The results of the BCR analysis are shown in Table 1. The first scenario follows existing requirements for buyouts by restricting buyouts to properties within the floodplain only. Unsurprisingly, this scenario generates the highest return on investments for built buyouts in the Clear Creek watershed with an average of $1.74 for every $1 spent on acquiring a residential property. This BCR can reach upward of $19 for every dollar spent on acquisition of selected properties. Surprisingly, extending the buyout requirements to include properties outside the floodplain and in proximity to existing wetlands also yields a significant return on investments. The second scenario resulted in 442 candidate buyouts with an average BCR of 1.14.

The BCA results of vacant properties in the watershed suggest that vacant parcels represent the most proactive approach to flood risk reduction. Following the proactive property acquisition framework, Table 2 summarizes the results of prioritizing vacant properties for acquisition in advance of development. This test site consists of 11 vacant parcels with the potential of about 1103 residential buildings being located in this test site. The results suggest that buyouts are generally more favorable for vacant properties than for built-upon properties. The analysis shows that there are savings of $10.37 for every $1 spent on vacant buyouts at the test site. On the other hand, if the buyouts occur in 30 years after the properties are already developed, only $1.64 is saved for every $1 spent. Granted that these benefits may be limited for other areas in flood zones, they still represent the potential benefits that proposed development may have if adequate policies are not in place to limit or prevent development in vulnerable locations.

Table 2 BCR of vacant buyouts and proposed development on 11 vacant test sites in the Clear Creek Watershed.

Buyout vacant property now		Buyout built parcels in 30 years	
Cost ($ million)		**Cost ($ million)**	
Land acquisition cost	11.32	Property acquisition cost	240.07
Land demolition cost	0	Building demolition cost	18.06
30-yr maintenance cost	4.12	30-yr maintenance cost	4.12
Total cost	15.44	Total cost	262.25
Benefit ($)		**Benefit ($)**	
Discounted 30-yr loss	160.18	Discounted 30-yr loss	160.18
Total benefit	160.18	Total benefit	160.18
Vacant BCR	**10.372**	**Built BCR**	**1.64**

Discussions and conclusions

The major take-away from this chapter regarding a proactive approach to property buyouts for existing development is that property acquisition can prioritize open space conservation and still be economically justifiable. This outcome complements past research on the positive economic value of floodplain and wetland conservation (Watson, Ricketts, Galford, Polasky, & O'Niel-Dunne, 2016). Of course, the selection criteria are up to decision-makers who must cater to local contextual factors. This chapter nevertheless puts forward a systematic, data-driven approach to buyout selection for both built and vacant properties. The methods proposed can build on the framework of an existing government program, but call on local governments to adopt a more proactive and intuitive approach to maximize both natural features and economic benefits. This strategy will enable communities to stretch buyout dollars and make acquisition programs as effective as possible.

This chapter highlights that a proactive approach of purchasing vacant properties in risky zones outweighs a reactionary, postdevelopment approach when considering future flood risk and exposure. For the selected site, these benefits are about 10 times greater than the cost of buying out these properties postdevelopment and provide a large return on investment. These benefits could be even higher if officials also consider the cost of infrastructure to support residential building, as well as the monetary loss of abandoning or even demolishing those infrastructures after the area is abandoned postbuyout.

In addition to a data-driven framework, local officials need to identify the best approach for estimating future flood conditions in their jurisdiction. Additionally, they must recognize that incentives are also needed for landowners to volunteer their lands for acquisition to NGOs or environmental groups, rather than sell to the highest bidder or succumb to market forces. Landowners are more likely to bow to the pressure of selling their land to developers due to rising tax rates (Lee & Newman, 2019; Lee, Newman, &

Park, 2018), irrespective of interest in retaining these lands as open spaces for flood risk reduction. Furthermore, local officials need to reconcile the loss of their tax base with the benefits accrued from avoided flood losses from keeping their lands as vacant open spaces. These costs can be offset by the additional value of open space. For example, previous research found that when lands are repurposed as open space, they can increase the property value of surrounding residential development (Crompton, 2005; Kousky & Walls, 2014), subsequently addressing issues relating to the loss of tax revenue due to vacant lots.

The contextual differences between the US and Dutch approaches to buyouts and, by extension, flood risk reduction reveal several areas where the United States needs to improve. Integrating buyouts with a large-scale flood risk management strategy rather than a checkerboard of buyouts will result in better flood risk reduction outcomes. Federal funding agencies such as FEMA will also need to reduce, if not eliminate, the red tape and bureaucratic roadblocks that prevent homeowners from participating in buyouts at the community level. The US buyout programs can also apply the bottom-up approach of the RftR program by fostering cooperation among federal agencies, local officials, and community members. This will help officials better communicate flood risk and management strategies and enable homeowners to make better decisions regarding buyouts. Some buyout programs in the United States have already applied similar strategies. For example, the New York Rising and New Jersey Blue Acres buyout programs recognized the disadvantages of checkerboards and incentivized large housing blocks to volunteer their property for acquisition after Hurricane Sandy. Policy learning from the success/failures of previous buyout programs can also improve future buyout efforts in the United States (Greer & Brokopp Binder, 2017).

Planners and decision-makers can also apply smart growth and development strategies without compromising on tax benefits. A proactive incentive-based approach to open space management can encourage landowners to participate in a vacant buyout program. Examples include transfer of development rights, density bonuses, development clustering, and so on (see Brody et al., 2020). The framework and analysis in this chapter enables planners and decision-makers to create a decision model to determine if existing properties should be bought out given their potential ecological benefits and proximity to natural systems. It also informs important decisions on vacant properties as to whether development should occur, property should remain vacant, or development should be moved to other suitable locations. By evaluating the economic implications of proactive open space acquisition, this chapter promotes an open space conservation culture which coastal communities dealing with rising seas and increasing development can apply to best support the mitigation and adaptation of flood risk for better future benefits.

References

Atoba, K. (2018). *Fill and floods: Analysis of the impact of parcel fill on residential flood damages* (Doctoral dissertation). College Station: Texas A&M University.

Atoba, K., Brody, S. D., Highfield, W. E., Shepard, C. C., & Verdone, L. (2020). Strategic property buyouts to enhance flood resilience: A multi-criteria spatial approach for incorporating ecological values into the selection process. *Environmental Hazards*, 1–19.

Atoba, K., Newman, G., Brody, S., Highfield, W., Kim, Y., & Juan, A. (2021). Buy them out before they are built: Evaluating the proactive acquisition of vacant land in flood-prone areas. *Environmental Conservation*, 48(2), 118–126. https://doi.org/10.1017/S0376892921000059.

Bently S. (2016) Holland is relocating homes to make more room for high water. The World-Environment. Retrieved from https://www.pri.org/stories/2016-06-22/holland-relocating-homes-make-more-room-high-water on Nov 10, 2020.

Binder, S. B., Baker, C. K., & Barile, J. P. (2015). Rebuild or relocate? Resilience and postdisaster decision-making after Hurricane Sandy. *American Journal of Community Psychology*, 56(1–2), 180–196.

Binder, S. B., Greer, A., & Zavar, E. (2020). Home buyouts: A tool for mitigation or recovery? *Disaster Prevention and Management: An International Journal.* https://doi.org/10.1108/DPM-09-2019-0298.

Blanco, et al. (2009). Shaken, shrinking, hot, impoverished and informal: Emerging research agendas in planning. *Progress in Planning*, 72(4), 195–250.

Blessing, R., A. Sebastian, and S.D. Brody, 2017. Flood risk delineation in the United States: How much loss are we capturing? Natural Hazards Review, 04017002.

Brody, S., Atoba, K., Sebastian, A., Highfield, W., Blessing, R., Mobley, W., Stearns, L. (2020 in print). A comprehensive framework for coastal flood risk reduction: Charting a course towards resiliency. In Kousky, C., Fleming, B., and Berger, A. (Eds.) A Blueprint for Coastal Adaptation. Island Press.

Brody, S., Blessing, R., Sebastian, A., & Bedient, P. (2014). Examining the impact of land use/land cover characteristics on flood losses. *Journal of Environmental Planning and Management*, 57(8), 1252–1265.

Brody, S. D., Highfield, W. E., Blessing, R., Makino, T., & Shepard, C. C. (2017). Evaluating the effects of open space configurations in reducing flood damage along the Gulf of Mexico coast. *Landscape and Urban Planning*, 167, 225–231.

Conrad, D. R., McNitt, B., & Stout, M. (1998). *Higher ground: A report on voluntary property buyouts in the Nation's floodplains: A common ground solution serving people at risk, taxpayers and the environment.* National Wildlife Federation.

Crompton, J. L. (2005). The impact of parks on property values: Empirical evidence from the past two decades in the United States. *Managing Leisure*, 10(4), 203–218.

De Bruijn, J.A., De Bruijne, M.L.C. and Ten Heuvelhof, E.F., 2015. The politics of resilience in the Dutch 'Room for the River'-project. In Procedia Computer Science, 44, 2015; Conference on Systems Engineering Research. Elsevier.

Edelenbos, J., Van Buuren, A., Roth, D., & Winnubst, M. (2017). Stakeholder initiatives in flood risk management: Exploring the role and impact of bottom-up initiatives in three 'room for the River' projects in the Netherlands. *Journal of Environmental Planning and Management*, 60(1), 47–66.

European Spatial Planning Observation Network (ESPON). (2017). *Comparative analysis of territorial governance and spatial planning systems in Europe.* https://www.espon.eu/planning-systems.

Freudenberg, R., Calvin, E., Tolkoff, L., & Brawley, D. (2016). *Buy-in for buyouts: The case for managed retreat from flood zones.* Lincoln Institute of Land Policy.

Geoghegan, J. (2002). The value of open spaces in residential land use. *Land Use Policy*, 19, 91–98.

Greer, A., & Brokopp Binder, S. (2017). A historical assessment of home buyout policy: Are we learning or just failing? *Housing Policy Debate*, 27(3), 372–392.

Harter, J. L. (2007). *Riparian restoration: An option for voluntary buyout lands in New Braunfels, TX.* Texas State University-San Marcos https://digital.library.txstate.edu/bitstream/handle/10877/3371/fulltext.pdf (Assessed 07/28/2020).

Hausmann, A., Slotow, R. O., Burns, J. K., & Di Minin, E. (2016). The ecosystem service of sense of place: Benefits for human well-being and biodiversity conservation. *Environmental Conservation*, 43(2), 117–127.

HCFCD. (2017). *Harris County flood control district. Harris County flood control district buyout interest areas.* https://www.hcfcd.org/Portals/62/Home-Buyout-Program/HomeBuyouts_BuyoutProgram420.pdf?ver=2020-04-09-142059-087 (Accessed 05/14/2020).

Highfield, W. E., Brody, S. D., & Shepard, C. (2018). The effects of estuarine wetlands on flood losses associated with storm surge. *Ocean and Coastal Management, 157*, 50–55.

Jan Goossen W. (2018). Interview—The Dutch make room for the river. Retrieved from https://www.eea.europa.eu/signals/signals-2018-content-list/articles/interview-2014-the-dutch-make. Accessed on 01/08/2020.

Johnson, K., et al. (2019). A benefit–cost analysis of floodplain land acquisition for US flood damage reduction. *Nature Sustainability, 3*(1), 56–62.

Klijn, F., de Bruin, D., de Hoog, M., Jansen, S., & Sijmons, D. (2013). Design quality of room-for-the-river measures in the Netherlands: Role and assessment of the quality team (Q-team). *International Journal of River Basin Management, 11*, 287–299.

Kolen, B., Hommes, S., Trainor, J., Yuiten, K., & Shaw, G. (2012). Flood response, an introduction in flood preparedness in the Netherlands: A US perspective. In B. Kolen, S. Hommes, & E. Huijskes (Eds.), *Netherlands US water crisis research network* (pp. 7–31). NUWCReN–HKV–DELTARES.

Kousky, C., & Walls, M. (2014). Floodplain conservation as a flood mitigation strategy: Examining costs and benefits. *Ecological Economics, 104*, 119–128. https://doi.org/10.1016/j.ecolecon.2014.05.001.

Lee, R. J., & Newman, G. (2019). A classification scheme for vacant urban lands: Integrating duration, land characteristics, and survival rates. *Journal of Land Use Science, 14*(4–6), 306–319.

Lee, J., Newman, G., & Park, Y. (2018). A comparison of vacancy dynamics between growing and shrinking cities using the land transformation model. *Sustainability, 10*(5), 1513–1530.

Mach, K. J., Kraan, C. M., Hino, M., Siders, A. R., Johnston, E. M., & Field, C. B. (2019). Managed retreat through voluntary buyouts of flood-prone properties. *Science Advances, 5*(10), 1–9. https://doi.org/10.1126/sciadv.aax8995.

Maldonado, J. K., Shearer, C., Bronen, R., Peterson, K., & Lazrus, H. (2013). The impact of climate change on tribal communities in the US: Displacement, relocation, and human rights. In *Climate change and indigenous peoples in the United States* (pp. 93–106). Cham: Springer.

Maly, E., & Ishikawa, E. (2013). Land acquisition and buyouts as disaster mitigation after Hurricane Sandy in the United States. In *Vol. 2013. Proceedings of international symposium on City Planning* (pp. 1–18).

Marino, E. (2018). Adaptation privilege and voluntary buyouts: Perspectives on ethnocentrism in sea level rise relocation and retreat policies in the US. *Global Environmental Change, 49*, 10–13.

Mobley, W., Atoba, K., & Highfield, W. (2020). Uncertainty in flood mitigation practices: Assessing the economic benefits of property acquisition and elevation in flood prone communities. *Sustainability, 12*(5), 2098. https://doi.org/10.3390/su12052098.

Newman, G. D., Bowman, A. O., Jung Lee, R., & Kim, B. (2016). A current inventory of vacant urban land in America. *Journal of Urban Design, 21*(3), 302–319.

Newman, G. D., et al. (2019). Integrating a resilience scorecard and landscape performance tools into a Geodesign process. *Landscape Research, 45*(1), 63–80.

NIBS (2019) Natural hazard mitigation saves: An independent study to assess the future savings from mitigation activities. National Institute of Building Sciences, Washington, DC 68. https://cdn.ymaws.com/www.nibs.org/resource/resmgr/reports/mitigation_saves_2019/mitigationsaves2019report.pdf Accessed 05/23/2020. Accessed 07/23/2020.

Rijkswaterstaat (2019). Room for the River program completed. Retrieved from https://www.rijkswaterstaat.nl/nieuws/2019/03/ruimte-voor-de-rivierprogramma-afgerond.aspx. Accessed 01/08/2021.

Siders, A. R. (2019a). Managed retreat in the United States. *One Earth, 1*(2), 216–225.

Siders, A. R. (2019b). Social justice implications of US managed retreat buyout programs. *Climatic Change, 152*(2), 239–257.

Steinitz, C. (2012). *A framework for geodesign: Changing geography by design*. Redlands California: Esri Press.

van Alphen, S. (2020). Room for the river: Innovation, or tradition? The case of the Noordwaard. In C. Hein (Ed.), *Adaptive strategies for water heritage*. p. 309.

Watson, K. B., Ricketts, T., Galford, G., Polasky, S., & O'Niel-Dunne, J. (2016). Quantifying flood mitigation services: The economic value of Otter Creek wetlands and floodplains to Middlebury, VT. *Ecological Economics, 130*, 16–24.

Wendland, T. (2020) Water Ways: Dutch Cities are letting water. In: New Orleans Public Radio. Retrieved from https://pulitzercenter.org/reporting/water-ways-dutch-cities-are-letting-water Accessed 01/08/2020.

Woltjer, J., & Al, N. (2007). Integrating water management and spatial planning: Strategies based on the Dutch experience. *Journal of the American Planning Association*, *73*(2), 211–222. https://doi.org/10.1080/01944360708976154.

Yu, S., Brand, A. D., & Berke, P. (2020). Making room for the river: Applying a plan integration for resilience scorecard to a network of plans in Nijmegen, the Netherlands. *Journal of the American Planning Association*, *86*(4), 417–430.

Zavar, E. (2015). Residential perspectives: The value of Floodplain-buyout open space. *Geographical Review*, *105*(1), 78–95.

Zavar, E., & Hagelman, R. R., III. (2016). Land use change on US floodplain buyout sites, 1990–2000. *Disaster Prevention and Management*, *25*(3), 360–374.

CHAPTER 23

Wetlands as an ecological function for flood reduction

Wesley E. Highfield
Institute for a Disaster Resilient Texas, Texas A&M University, College Station, TX, United States
Department of Marine and Coastal Environmental Science, Texas A&M University, Galveston Campus, Galveston, TX, United States

Introduction

The role of wetlands as a natural feature that reduces the adverse effects of floods has shifted from an anecdotal concept to a mainstream consideration to address flood risk reduction. Naturally occurring wetlands are among the many ecological characteristics that encompass nature-based flood defenses (NBFD). For decades, wetlands have been commonly cited for their ability to reduce flooding in both riverine and coastal environments. Despite the acceptance of the "wetland-flooding" relationship, there are variations across results that are worthy of a more thorough discussion concerning this nuanced relationship. Variations in results of wetland-flooding relationships arise out of methodological approaches, the measurement of "flooding," and the type of wetland that is the research subject.

Overall, naturally occurring wetlands have been highlighted for their role in reducing flood velocities, flood peaks, and providing areas for storing precipitation-based floodwaters (Acreman & Holden, 2013). Various studies have shown that naturally occurring wetlands provide flood attenuation by maintaining a properly functioning water cycle (Lewis, 2001; Mitch & Gosselink, 2000). Both anecdotal and empirical research suggest that wetlands may reduce or slow flooding. In one of the most comprehensive literature reviews to date, Bullock and Acreman (2003) noted that 23 out of 28 studies on wetlands and flooding found that "floodplain wetlands reduce or delay floods" (p. 366). The following sections address early, water-balance based studies, more recent numerical- and simulation-based studies, and finally observation-based studies.

Early comparative research

Initial research on the role of wetlands in reducing flooding examined the differences between drained and natural wetlands in riverine environments. These studies showed that nondrained peat bogs reduce low return period flood flow and overall storm flows when compared to drained counterparts (Daniel, 1981; Heikuranen, 1976; Verry &

Boelter, 1978). For example, Novitski (1979) examined four different types of wetlands and found that each had a statistically negative effect on flood flows. The US Army Corps of Engineers (1972) conducted one of the earliest studies on the flood-reduction role of wetlands; they calculated that the flood-reduction function of 3800 ha of floodplain storage on the Charles River, Massachusetts saved US $17 million worth of downstream flood damage each year. Later, Novitski (1985) discovered that basins with as little as 5% lake and wetlands may result in 40%–60% lower flood peaks.

Simulation-based research

Research results based on numeric- and simulation-based models show similar but more variable results. These modeling approaches also begin to shift from inland freshwater (palustrine) dominated wetlands and riverine-based floods to more mixed or estuarine wetlands and coastal flood events. For example, Ammon, Wayne, and Hearney (1981) modeled the effects of wetlands on water quantity for the Chandler Marsh in South Florida. Results indicated that flood peak attenuation was greater with larger areas of marsh. The authors concluded that Chandler Slough Marsh increases stormwater detention times, facilitates runoff into subsurface regimes, and is fairly effective as a water quantity control device. Ogawa and Male (1986) also analyzed a simulation model to evaluate the protection of wetlands as a flood mitigation strategy. Based on four scenarios of downstream wetland encroachment, ranging from 25% to 100% alteration, these researchers found that increased encroachment resulted in statistically significant increases in stream peak flow.

Using a hybrid ADCIRC simulation/regression analysis, Barbier, Georgiou, Enchelmeyer, and Reed (2013) found that wetland continuity and vegetation roughness in Louisiana was effective in reducing surge-induced damage. The authors found that a 0.1 increase in wetland continuity reduces property damages for the average affected area analyzed in southeast Louisiana by $99–$133; and a 0.001 increase in vegetation roughness decreases damages by $24–$43. These figures are equivalent to a mere 3–5 and 1–2 properties per storm, respectively, within an average study area of 1780 properties per study area. Other research using ADCIRC has led to varying findings. Wamsley, Cialone, Smith, Atkinson, and Rosati (2010) concluded that wetlands in Louisiana may reduce storm surge heights, but with variation across the landscape and storm characteristics. Resio and Westerink (2008) found that wetlands may actually increase storm surge heights under wind-driven storm conditions, again illustrating that the role of wetlands may vary spatially. Using Hurricane Rita as an example, surge heights were increased over some Louisiana marshes during this slowly progressing storm. The impact of coastal wetlands is clearly contingent on multiple factors, including vegetation type and density, landscape structure, storm characteristics, and the presence of human interventions, such as levees (Hu, Chen, & Wang, 2015). At the minimum, many experts in the field question whether coastal wetlands would have any dampening effect on extreme storm surges, such as those produced by Hurricanes Katrina and Ike.

Observational research

Another form of research based on direct observation, rather than simulation, also supports the idea that naturally occurring wetlands can reduce flooding events. For example, a constructed wetland experiment along the Des Plaines River in Illinois found that a marsh of 5.7 acres could retain the natural runoff of a 410-acre watershed. The same study estimated that only 13 million acres of wetlands (3% of the upper Mississippi watershed) would have been needed to prevent the catastrophic flood of 1993 (Godschalk, Beatley, Berke, Brower, & Kaiser, 1999). Another empirical research method involves measuring streamflow data from stream gauge stations. Using this approach, Johnston, Detenbeck, and Niemi (1990) found that *even small wetland losses* within watersheds could significantly affect flooding over time.

Examining wetland alteration using Section 404 wetland permits

It is often remarked that wetlands reduce floods and the empirical literature generally supports this statement, especially with respect to riverine-based floods. It should follow then, that a loss of wetlands or their function increases flood events, both in frequency and magnitude. Tying wetland alteration to the scale and impact of flooding has been historically problematic due to a lack of data across broad geographic scales. However, in the early 2000s, Brody and colleagues conducted multiple analyses using geocoded Section 404 permits in Texas and Florida (two of the most flood-afflicted states in the nation). Section 404 of the Clean Water Act charges the US Army Corps of Engineers (USACE) with the responsibility of issuing permits for the "discharge of dredged or fill material" into "waters of the United States." The conditions under which various types of permits are issued vary by the type of activity, the impact of the activity, and the district or region where the activity will be located. USACE currently issues four types of Section 404 permits: Individual permits, Letters of Permission, General permits, and Nationwide permits. Individual permits are the basic and original form of authorization used by the USACE. Activities that entail more than minimal impacts require an individual permit. The first alternate form of authorization used by the USACE for certain prescribed situations is the Letter of Permission. Letters of Permission may be used where, in the opinion of the district engineer, the proposed work would be minor, not have significant individual or cumulative impact on environmental values, and should encounter no appreciable opposition. General permits are an attempt to streamline the permit process for common activities (Downing, Winer, & Wood, 2003). General permits are issued when "activities are substantially similar in nature and cause only minimal individual and cumulative impacts" (33 C. F. R. §352.2). Finally, Nationwide permits are a special type of general permit. They are a key means by which the Corps operates its regulatory program and simplifies its administrative activities. Activities covered under Nationwide permits can go forward without further Corps approval as long as the conditions set forth in the Nationwide permit category of work are met. Nationwide permits are issued for specific activities that are deemed to have "no more than minimal adverse effects on the aquatic environment, both individually and cumulatively" (Issuance of Nationwide Permits; Notice, 2019).

These studies investigating the effects of Section 404 permits on water quantity using measures of streamflow and estimated flood damage are summarized below in Table 1. This body of research, with little variation, found Section 404 permits to have positive and statistically significant effects on flooding and flood impacts. The results hold despite varying units of analysis, changing study periods, and different forms of measuring both flooding and the number of Section 404 permits. Most importantly, these results continue to hold after statistically controlling for additional climatic, hydrologic, socioeconomic, and policy-related variables.

As mentioned above, streamflow measurements collected from United States Geological Survey (USGS) gauging stations provided a sound basis to examine the effects of Section 404 permits. The first of several studies spanned coastal Texas and all of Florida, utilizing hydrologic units established by the USGS (Brody et al., 2007). A total of 85 hydrologic units were incorporated into the analysis, 39 in Texas and 46 in Florida. Average monthly streamflows were calculated for every gauge in each hydrologic unit over a 12-year study period from 1991 to 2002. Counts of "exceedances," or the number of months that a gauge surpassed the study period average, were calculated and averaged within each unit. Results showed that approved Individual and General Section 404 permits were positive and statistically significant drivers of flow exceedances. These relationships held even after controlling for a host of additional variables, including hydrologic unit area, slope, total length of streams in each unit, impervious surface area, the number of dams, population density, and median household income (Brody et al., 2007).

A complementary study using streamflow measurements focused solely on Texas. In this analysis, all USGS stream gauge locations within the 49 county USACE Galveston District were selected from 1996 to 2003 (Highfield, 2012). Using the recorded maximum daily streamflow data for stream gauges, Highfield created a dependent variable, peak annual flow (PAF), log transformed to approximate a normal distribution. Instead of using established hydrologic units, subbasins were delineated around each of the 47 stream gauge locations. Temporal considerations were accounted for by using a cross-sectional time series (CSTS) approach to analysis.

Across four CSTS models, separate models for each permit type—Individual, Letter of Permission, General, and Nationwide—were all positive and significant with respect to increasing peak annual flow. Once again, issued wetland alteration permits proved

Table 1 Summary of studies investigating the relationship between Section 404 permits and flood metrics.

Dependent variable	Section 404 permit measurement	Direction (significance)	Study area (study period)	Citation
Counts of exceedances over study period average	Permits by type	Individual (Positive*) General (Positive**) Nationwide (Positive) Letter (Negative**)	Coastal watersheds in Texas and Florida (85)	Brody, Highfield, Ryu, and Spanel-Weber (2007)
Flood damage (dollar amount)	Permits by type in/out of floodplains	Individual (Positive**) Nationwide (Positive) All permits out (Negative**)	Florida (1997–2002)	Highfield and Brody (2006)
Flood damage (dollar amount)	Total cumulative permits	Positive** ($P < .01$)	Florida (1997–2001)	Brody, Zahran, Maghelal, Grover, and Highfield (2007)
High damage event, dichotomous	Total cumulative permits	Positive* ($P < .05$)	Florida (1997–2001)	Brody et al. (2007)
Flood damage (dollar amount)	Total cumulative permits	Positive** ($P < .01$)	Coastal Texas (1997–2007)	Brody et al. (2007)
Maximum daily streamflow	Total cumulative permits by type	Individual (Positive**) General (Positive**) Nationwide (Positive**) Letter (Negative**)	Coastal Texas sub-basins (1996–2003)	Highfield (2012)

*$P < .05$.
**$P < .01$.
Adapted from Brody, S. D., Highfield, W. E., Kang, J. E. (2011). Rising waters: Causes and consequences of flooding in the United States. Cambridge, UK: Cambridge University Press.

significant factors in increased peak flow even after controlling for variables, including subbasin area, shape, and slope, natural and impervious surface land cover, and soil permeability. Not only did permits have a significant positive effect on peak annual flow but it was found that larger areas of wetlands in each subbasin actually worked to reduce peak annual flow. More specifically, increasing areas of palustrine scrub/shrub wetlands significantly reduced flooding. These results not only confirm the role of wetlands in reducing peak streamflow found in previous research they also simultaneously demonstrate the effects of altering or removing wetlands on increasing peak flows and potential damages from flooding.

While streamflow is an objective measurement of water quantity and potential flooding, it does not address visible impacts that often drive local policy decisions. The second wave of analysis examined the impacts of wetland alteration directly on property damage from flooding events. First, the research team examined the issued Section 404 permits in FEMA delineated 100-year floodplains at the county level in Florida (Highfield & Brody, 2006). Results showed that alteration of wetlands inside the floodplain led to significantly higher amounts of flood damage when controlling for precipitation, median structure improvement value, and population density. Perhaps the most important result of this research in terms of Section 404 activity and flooding was the role of Individual permits. When looking at standardized regression coefficients, Individual permits (0.48) had the highest impact on explaining flood damage—even higher than the amount of precipitation (0.41). This finding demonstrates the potential for naturally occurring wetlands to mitigate the adverse impacts of floods.

The researchers also conducted a related analysis using the gross number of Section 404 permits measured cumulatively from 1997 to 2001 in both coastal Texas and Florida. In Florida, flood damage at the county level (the smallest available spatial unit for measuring property damage at the time) was predicted for individual events by cumulative counts of wetland permits, along with a diversity of geophysical, socio-economic, and planning-related variables (Brody et al., 2007). Despite numerous statistical controls, Section 404 permits remained positive and significant predictors of reported flood damage. In fact, wetland alteration permits had a stronger statistical effect (in terms of standardized regression coefficients) than impervious surface area, dams, floodplain area, flood duration, stream density, and housing value density. Not only did increasing numbers of permits have a strong relationship with damaging floods they also had a positive effect on "high damage" flood events. That is, Section 404 permits had positive and significant impacts on floods that exceeded the aggregate median value of $50,000, even after controlling for the same groups of variables described above (Brody et al., 2007).

In the aggregate, the price of a permit in terms of corresponding flood damages to homes and businesses is quite high. In fact, based on the statistical model calculated in this study, each issued permit in Florida from 1997 to 2001 resulted in almost $1000 in added property damage per flood. When considering the number of permits issued, that equals over $402,465 per flood or about $30,426,354 per year in added damage for the entire state. Strikingly similar relationships hold when conducting this analysis in coastal Texas. Nearly the same set of statistical control variables was used to analyze

SHELDUS[a]-derived flood damage data for 37 counties in the coastal region (Brody, Zahran, Highfield, Grover, & Vedlitz, 2008). Wetland loss, as measured through permits, was statistically a stronger factor in predicting flood damage than the amount of precipitation the day of the flood event, floodplain area, area of impervious surface, number of dams, FEMA CRS rating, and median household income within each county in the sample. According to the model calculated in this study, a single permit to alter a naturally occurring wetland equals an average of $211.88 in added property damage per flood. This figure does not sound like that much, but consider the fact that thousands of permits are issued in this area each year and that number is increasing over time. Just a 5-year glance at the flood problem in coastal Texas (1997–2001) shows that wetland permits equated to over $38,000 in added property damage per flood.

Across all of the research regarding Section 404, several commonalities and important implications come to light. First, aggregate wetland permits are consistently positive and significant in explaining both streamflow and flood damage estimates—in both coastal Texas and Florida. This finding confirms that not only do wetlands reduce floods but Section 404 permits serve as indicators of wetland loss. Second, when permits are treated and analyzed separately, Individual permits always have the greatest effect on flooding. This is an expected result since they are likely to represent the largest wetland losses, but important because it confirms the role of Section 404 permits as an indicator of wetland loss and as having a positive effect on flooding and associated flood damages. The evidence above clearly illustrates the need to incorporate wetland protection and preservation as a key tool in effective flood mitigation. Whatever specific statistical model is analyzed, the result is always the same: that the alteration of naturally occurring wetlands in Texas and Florida has significant positive relationships with the amount of flood events and associated property damage.

Recent advances in identifying the type and shape of wetlands in reducing flood loss

The most recent wave of studies on the role of wetlands in reducing flood loss takes advantage of high-quality remote sensing and parcel-level flood damage data. Improvements in computational power also allow for analyses at much larger scales and time frames. For example, observational analysis of all counties and parishes bordering the Gulf of Mexico (GOM) showed that the loss of an acre of naturally occurring wetlands from 2001 to 2005 increased property damage caused by flooding by an average of $7,457,549,

---

[a] SHELDUS is a county-level hazard data set for the US and covers natural hazards such thunderstorms, hurricanes, floods, wildfires, and tornados as well as perils such as flash floods, heavy rainfall, etc. The database contains information on the date of an event, affected location (county and state) and the direct losses caused by the event (property and crop losses, injuries, and fatalities) from 1960 to present. See https://cemhs.asu.edu/sheldus for more information.

which amounts to approximately $1.5 million per year across the study area (Brody, Peacock, & Gunn, 2012). These studies also further support the importance of *palustrine and upland wetlands* for flood control.

An expanded GOM study examining land cover across 2692 watersheds for over 372,000 insured flood losses claimed under the National Flood Insurance Program per year from 2001 to 2008 indicated that a percent increase in palustrine wetlands is the equivalent to, on average, a $13,975 reduction in insured flood losses per year, per watershed (Brody, Highfield, & Blessing, 2015). Palustrine wetlands, with the ability to absorb, store, and slowly release water, should be considered the most effective land cover for mitigating the effects of rainfall-based flooding. Every percent increase in this type of wetland along the GOM is the equivalent to, on average, a $7580.00 reduction in insured flood losses per watershed, per year. A follow-up analysis of the same study area from 2008 to 2014 found that large, expansive, and continuous patches of naturally occurring open spaces most effectively reduce losses from flood events. In fact, a percent increase in the size of the largest wetland patch corresponds to a 3.3% reduction in damage, or $1801. Among all land cover types assessed, palustrine wetlands have the largest effect in reducing the impacts of flood events across the Gulf of Mexico.

One of the hallmarks of a strong finding is if it can be corroborated at different scales, study areas, and time periods. This was the case with another study using remote sensing-derived landcover data to examine the impacts of adjacent land use and land cover (LULC) on flood damage recorded on parcels within a single coastal watershed in southeast Texas. Brody, Blessing, Sebastian, and Bedient (2014) analyzed empirical statistical models to identify the influence of different LULCs surrounding over 7900 properties claiming insured flood losses from 1999 to 2009. Results indicate that specific types of surrounding LULCs impact observed flood losses and provide guidance on how neighborhoods can be developed more resiliently over the long term. While estuarine wetlands located primarily along the coastline did not have a significant effect in this study, freshwater palustrine wetlands surrounding a property significantly attenuate flood loss ($P < .001$). In fact, this wetland type has the strongest flood damage reduction effect among all LULCs in the model.

Finally, in somewhat contrary to the findings to the previous research presented above, Highfield, Brody, and Shepard (2018) examined solely estuarine wetlands to assess their effect on insured flood damages during Hurricane Ike in 2008. After controlling for a host of parcel-level structural characteristics combined with flood risk and geographic measures, estuarine wetlands had complex effects on surge-induced flood damage. This research found that larger marsh areas significantly decreased structural damage. Perhaps more importantly, the results indicated that structures that were behind, or sheltered, by estuarine wetlands experienced less damage than their unsheltered counterparts, but only within a distance of 500 ft.

Comparing the Dutch experience

What sets the above research in the United States apart from its Dutch counterparts is historical flood experiences combined with a culture of adaptation. In the case of the United States, the above research was focused on demonstrating that wetlands, both fresh and saltwater, have effects on flood and flood-damage reduction. Until recently, these questions in the United States have been asked in a binary manner, seeking to evaluate a single nonstructural solution to flood mitigation. Following Hurricanes Katrina (2005) and Ike (2008) efforts have been made to integrate structural and nonstructural mitigation (Kothuis & Voorendt, 2018; Lopez, 2009; van Berchum et al., 2019). Yet these recent efforts are still in contrast to the Dutch experience, where questions surrounding wetlands and flooding are addressed in the context of both structural and nonstructural working in concert. In addition, the Dutch focus is not on a singular natural amenity (i.e., wetlands), but a broader, system-based approach (Zevenbergen, Rijke, Van Herk, & Bloemen, 2015). This approach applies to both freshwater/riverine floods and saltwater/surge-based floods.

For example, following precipitation-driven floods in 1993 and 1995, the well-known Room for the River projects transitioned from historic structural engineering approaches to flood mitigation to a more integrated system of flood risk reduction through structural and nonstructural techniques. Considered by many to be an exemplary example of integrated basin management and planning (Zevenbergen, Rijke, Van Herk, Ludy, & Ashley, 2013), Room for the River takes a comprehensive view of natural features, wetlands included, as part of a larger system of riverine flood mitigation and risk reduction. In the Netherlands context, research and implementation have also been undertaken with respect to saltwater wetlands and storm surge. Continuing with the theme of wetlands as a part of a larger system of flood defenses, Vuik, Borsje, Willemsen, and Jonkman (2019) found that salt marshes were effective in reducing wave heights on existing coastal dikes. Other research has demonstrated the cost-effectiveness of salt marsh relative to dike heightening (van Loon-Steensma, 2015). Notably, the results here do not suggest *replacing* existing flood defense structures with salt marsh as a one-size-fits-all, but to supplement those structures.

Conclusions

In summary, almost 20 years of peer-reviewed research conducted in the United States at multiple scales, time periods, and study areas converge on the same result: that freshwater wetlands, even when disconnected from main stem river channels or their tributaries, significantly reduce observed flood losses. The evidence in the United States for estuarine or saltwater, wetlands is more complex. This is due at least in part to the vast difference between precipitation-based riverine floods and tropical cyclone-induced storm surge, which presents differences in timing, concentration, and volume of water. Comparing the United States results to those in the Netherlands is difficult. First, due to the long Dutch history of structural flood risk reductions in both fresh and saltwater settings, the historic alteration of the natural environment makes comparisons difficult. Second, the Dutch approach of coupling structural flood mitigation with an early willingness to adopt the inclusion of nature-based infrastructure relative to their US counterparts changes the focus from single-wetland studies to more integrated basin management. While this may be in part to a historic difference in settlement and flood history, the culture of adaptation has produced management approaches that stray from dichotomous thinking about flood mitigation techniques to system-based approaches. Finally, despite these differences, it seems apparent that preservation of wetlands can and do serve as a flood mitigation tool, but one that should be treated as part of an integrated approach to flood risk management and not a single, silver bullet treatment.

References

Acreman, M., & Holden, J. (2013). How wetlands affect floods. *Wetlands*, *33*(5), 773–786.
Ammon, D. C., Wayne, H. C., & Hearney, J. P. (1981). Wetlands' use for water management in Florida. *Journal of Water Resources Planning and Management*, *107*, 315–327.
Barbier, E. B., Georgiou, I. Y., Enchelmeyer, B., & Reed, D. J. (2013). The value of wetlands in protecting southeast Louisiana from hurricane storm surges. *PLoS ONE*, *8*(3), e58715.
Brody, S. D., Blessing, R., Sebastian, A., & Bedient, P. (2014). Examining the impact of land use/land cover characteristics on flood losses. *Journal of Environmental Planning and Management*, *57*(9), 1252–1265.
Brody, S. D., Highfield, W., & Blessing, R. (2015). An Empirical Analysis of the Effects of Land Use/Land Cover on Flood Losses along the Gulf of Mexico Coast from 1999 to 2009. *Journal of the American Water Resources Association (JAWRA)*, *51*(6), 1556–1567.
Brody, S. D., Highfield, W. E., Ryu, H. C., & Spanel-Weber, L. (2007). Examining the relationship between wetland alteration and watershed flooding in Texas and Florida. Natural Hazards, 40(2), 413–428.
Brody, S. D., Peacock, W., & Gunn, J. (2012). Ecological indicators of resiliency and flooding along the Gulf of Mexico. *Ecological Indicators*, *18*, 493–500.
Brody, S. D., Zahran, S., Highfield, W. E., Grover, H., & Vedlitz, A. (2008). Identifying the impact of the built environment on flood damage in Texas. *Disasters*, *32*(1), 1–18.
Brody, S. D., Zahran, S., Maghelal, P., Grover, H., & Highfield, W. (2007). The rising costs of floods: examining the impact of planning and development decisions on property damage in Florida. *Journal of the American Planning Association*, *73*(3), 330–345.
Bullock, A., & Acreman, M. (2003). The role of wetlands in the hydrological cycle. *Hydrology and Earth System Sciences Discussions*, *7*(3), 358–389.
Daniel, C. (1981). Hydrology, geology, and soils of pocosins: A comparison of natural and altered systems. In C. J. Richardson (Ed.), *Pocosins: A conference on alternate uses of the coastal plain freshwater wetlands of North Carolina* (pp. 69–108). Stroudsburg, PA: Hutchinson Ross Publishing Company.
Downing, D. M., Winer, C., & Wood, L. D. (2003). Navigating through Clean Water Act jurisdiction: A legal review. *Wetlands*, *23*(3), 475–493.
Godschalk, D. R., Beatley, T., Berke, P., Brower, D., & Kaiser, E. J. (1999). *Natural hazard mitigation: Recasting disaster policy and planning*. Washington, DC: Island Press.
Heikuranen, L. (1976). Comparison between runoff condition on a virgin peatland and a forest drainage area. In *Proceedings of the Fifth International Peat Congress* (pp. 76–86).

Highfield, W. E. (2012). Section 404 permitting in coastal Texas: A longitudinal analysis of the relationship between peak streamflow and wetland alteration. *Environmental Management, 49*(4), 892–901.

Highfield, W. E., & Brody, S. D. (2006). The price of permits: measuring the economic impacts of wetland development on flood damages in Florida. *Natural Hazards Review, 7*(3), 23–30.

Highfield, W. E., Brody, S. D., & Shepard, C. (2018). The effects of estuarine wetlands on flood losses associated with storm surge. Ocean & Coastal Management, 157, 50–55.

Hu, K., Chen, Q., & Wang, H. (2015). A numerical study of vegetation impact on reducing storm surge by wetlands in a semi-enclosed estuary. *Coastal Engineering, 95,* 66–76.

Issuance of Nationwide Permits; Notice. (2019). *67 Federal Register.*

Johnston, C. A., Detenbeck, N. E., & Niemi, G. J. (1990). The cumulative effect of wetlands on stream water quality and quantity. A landscape approach. *Biogeochemistry, 10*(2), 105–141.

Kothuis, B., & Voorendt, M. (2018). *Multifunctional flood defenses* (pp. 44–46). My Liveable City.

Lewis, W. M. (2001). *Wetlands explained: Wetland science, policy, and politics in America.* New York, NY: Oxford University Press.

Lopez, J. A. (2009). The multiple lines of defense strategy to sustain coastal Louisiana. *Journal of Coastal Research, 10054,* 186–197.

Mitch, W. J., & Gosselink, J. G. (2000). *Wetlands* (3rd ed.). New York, NY: John Wiley & Sons.

Novitski, R. P. (1979). Hydrologic characteristics of Wisconsin's wetlands and their influence on floods, streamflow and sediment. In P. E. Greeson, J. R. Clark, & J. E. Clark (Eds.), *Wetland functions and values: The state of our understanding* (pp. 377–388). Minneapolis, MN: American Water Resources Association.

Novitski, R. P. (1985). The effects of lakes and wetlands on flood flows and base flows in selected Northern and Eastern States. In *Proceedings of the conference on wetlands of the chesapeake* (pp. 143–154). Easton, Maryland: Environmental Law Institute.

Ogawa, H., & Male, J. W. (1986). Simulating the flood mitigation role of wetlands. *Journal of Water Resources Planning and Management, 112*(1), 114–128.

Resio, D. T., & Westerink, J. J. (2008). Hurricanes and the physics of surges. *Physics Today, 61*(9), 33–38.

US Army Corps of Engineers. (1972). *An overview of major wetland functions and values US Fish and Wildlife Service.* FWS/OBS-84/18.

van Berchum, E. C., Mobley, W., Jonkman, S. N., Timmermans, J. S., Kwakkel, J. H., & Brody, S. D. (2019). Evaluation of flood risk reduction strategies through combinations of interventions. *Journal of Flood Risk Management, 12*(S2), e12506.

van Loon-Steensma, J. M. (2015). Salt marshes to adapt the flood defences along the Dutch Wadden Sea coast. *Mitigation and Adaptation Strategies for Global Change, 20*(6), 929–948.

Verry, E. S., & Boelter, D. H. (1978). Peatland hydrology. In P. E. Greeson, J. R. Clark, & J. E. Clark (Eds.), *Wetland functions and values: The state of our understanding* (pp. 389–402). Minneapolis: American Waterworks Association.

Vuik, V., Borsje, B. W., Willemsen, P. W., & Jonkman, S. N. (2019). Salt marshes for flood risk reduction: Quantifying long-term effectiveness and life-cycle costs. *Ocean & Coastal Management, 171,* 96–110.

Wamsley, T. V., Cialone, M. A., Smith, J. M., Atkinson, J. H., & Rosati, J. D. (2010). The potential of wetlands in reducing storm surge. *Ocean Engineering, 37*(1), 59–68.

Zevenbergen, C., Rijke, J., Van Herk, S., & Bloemen, P. J. T. M. (2015). Room for the river: A stepping stone in adaptive delta management. *International Journal of Water Governance, 3*(1), 121–140.

Zevenbergen, C., Rijke, J., Van Herk, S., Ludy, J., & Ashley, R. (2013). Room for the river: International relevance. *Water Governance, 2,* 24–31.

Further reading

Brody, S. D., Highfield, W. E., & Kang, J. E. (2011). *Rising waters: The causes and consequences of flooding in the United States.* Cambridge University Press.

Hallegatte, S., Green, C., Nicholls, R. J., & Corfee-Morlot, J. (2013). Future flood losses in major coastal cities. Nature Climate Change, 3(9), 802–806.

Reja, M. Y., Brody, S. D., Highfield, W. E., & Newman, G. D. (2017). Hurricane recovery and ecological resilience: Measuring the impacts of wetland alteration post Hurricane Ike on the upper TX coast. Environmental Management, 60(6), 1116–1126.

Watson, K. B., Ricketts, T., Galford, G., Polasky, S., & O'Niel-Dunne, J. (2016). Quantifying flood mitigation services: The economic value of Otter Creek wetlands and floodplains to Middlebury, VT. Ecological Economics, 130, 16–24.

CHAPTER 24

Designing and building flood proof houses

Anne Loes Nillesen
Urban and Rural Climate Adaptation, Defacto Urbanism, Rotterdam, The Netherlands

Introduction

Flood proof houses in the Netherlands

The Netherlands is largely positioned under sea level. The "fight against floods" is an important part of the spatial and economic development of the Netherlands and still visible in the Dutch landscape. From early development onward we see that houses are constructed at higher, more flood proof places such as natural sand riches or constructed mounts. After the construction of the elaborate dike ring system, the probability of flooding reduced, as did the need for flood proof buildings. In Dutch unembanked areas building is restricted, however, some experimental flood proof houses are constructed and developed, showcasing and a wide variety of flood proof housing types (Ministry of Transport, Public Works and Water Management, 2005). Regardless of the low probability, potential flooding can have severe consequences (PBL, 2014). Recently, the awareness of the value of flood risk management systems comprising both probability and consequence reduction measures, the so-called "multilayer safety approach" (Gersonius, Veerbeek, Subhan, & Zevenbergen, 2011) increased. This led to renewed interest and explorations regarding the potential of flood proof houses in embanked areas as a second layer of safety.

Different types of flood proof houses

There are different types of flood proof houses (Nillesen & Singelenberg, 2011). An important distinction is whether a house is dry-proof, and is intended to prevent flooding of the interior, or wetproof, intended to prevent or reduce damage in case of a flood. The elevated mount house is historically the most common flood proof housing type in the Netherlands. The functionality of elevating the building is that the floodwater does not reach, it is similar to the pole housing type. Another principle to prevent flooding is the floating house that is positioned in water and can move along with varying water levels, including flood events. The amphibious house is a housing type that during normal circumstances is positioned on land but can float in case of a flood. The floodable but dry-proof buildings have a flood proof facade to prevent water from entering, the floodable wetproof buildings are constructed in a way that minimizes the impact of floodwater, for instance, by using waterproof materials and elevated sockets.

Different types of application areas

Which type of flood proof house is most suitable for a certain area depends on the local flood characteristics. In the Netherlands, there is a big difference between the embanked areas in which the flood risk has a low probability varying from 1:30 to 1:30,000 years (LIWO, 2020) and the areas that are not protected by levees or other flood protection measures. The rivers and flood plains (and the often-connected sand and gravel extraction lakes) are characterized by fluctuations of several meters in water levels due to seasonal changes in discharge loads. The lakes and channels that are dammed usually have a controlled water level that often varies seasonally, these water levels often counter the natural fluctuation since water is stored in summer and drained in winter. The larger water bodies (main rivers, channels, and lakes) can have a strong gradient of some meters during storm events. In Lake IJssel, the country's largest lake, ice can be pushed up by wind in winter.

Flood proof housing types

An overview of different flood proof housing types and their characteristics was made by Nillesen and Singelenberg (2011). In this section, different housing types and their characteristics or special demands are presented, as well as some examples.

Mount houses

The Mount houses are positioned on elevated constructed mounts. In the historic flood-prone areas, we see many century-old mounts, such as under houses along the now dammed Markermeer (Fig. 1A). The constructions on these mounts can be individual

(A)

(B)

Fig. 1 (A) A small settlement on a mount along the Markermeer lake. (B) A historic built housing ribbon on the dike (which in foreseeable time has to be elevated) along the river Lek in the Alblasserwaard-Vijfheerenlanden area. *((A) Photo by Defacto, Rotterdam. (B) Photo by Defacto, Rotterdam.)*

buildings, such as farms or churches, as well as small settlements. The advantage of the mount is that important properties such as cattle can also be protected against floods. Of course, to be effective, the mount needs to be high enough: climate change or differences in water management levels can result in higher flood levels, while the mounts, once built, are often difficult to further elevate.

We also see many houses constructed as ribbons on levees (Fig. 1B). Some levees originated on already elevated sand ridges, others were newly constructed. Though the ribbons provide a high place above flood level, due to climate change and the need for higher protection levels many levees had and have to be elevated at some point, resulting in many dike houses being demolished. For newly constructed dike houses it is therefore important to be prepared for the potential future need for dike reinforcements, for instance, by preactively reinforcing the levee (and creating excess dimensions) or by constructing houses of which the pole foundation can be screwed up (an example of this is houses constructed on a levee in Papendrecht).

The mount principle is still a very interesting model for flood proofing. It is, for instance, applied in the Rotterdam floodplains where land allocation heights are applied, indicating the level to which the plot needs to be elevated before constructing on it (Municipality of Rotterdam, 2013). A famous contemporary example is Hamburg HafenCity, an unembanked former harbor terrain in Germany, now developed as an urban residential and commercial area (Fig. 2A, B). In HafenCity it was decided to not embank the terrain (so the views to the river would be maintained and to prevent costs for reinforcing the quay edges) but to elevate the central part of the terrain. This ensures that there is always a flood proof core that connects this unembanked area to the embanked part of the city. Buildings are either constructed high and dry on this flood proof core or developed in a flood proof way. The transition between the elevated part and the unembanked quays is formed by buildings positioned on the edge between the different levels, built with a flood proof core, and by stepped public spaces that connect both levels. The buildings positioned on the edge of the higher and lower zone often have a flood-proof plinth: either by dry proofing the building with flood doors or by assigning wetproof use—such as parking—to the lower level. All buildings are connected to the elevated core either directly or through an elevated public walkway. The different terrain levels and the reoccurring seasonal flooding of the lower level contribute to the experience of the force and dynamic of the river. With these types of mount development, it is important to include some excess dimensioning anticipating climate change and sea level rise since the cost of further elevating an already constructed development would be substantial.

Pole houses

Pole houses are based on the same principle as mount houses; they are elevated to prevent them from flooding. The pole structure makes it possible to construct them in areas that

Fig. 2 Modern version of mount settlements: flood proof development Hamburg HafenCity along the river Elbe. (A) Building with elevated ground floor, Hamburg HafenCity, Sandtorhafen. (B) Hamburg HafenCity, Flood-proof doors at ground level. *((A) Photo by Defacto Urbanism, Rotterdam. (B) Photo by Defacto Urbanism, Rotterdam.)*

inundate regularly without reducing the water storage capacity, or in places where the possible flood height is severe and the force of the water might erode a constructed mount. They have the same disadvantage as the mount buildings; since the height is fixed, they can flood during unexpected extremes. Other than the mounts, the pole buildings are often easier to adjust and elevate since only the pole structure needs to be replaced. Although pole houses are all around the world applied in flood plains and coastal zones (as is for instance the case in Galveston, where they are called "stilt houses," Fig. 3B) they are relatively rare in the Netherlands. Still, pole houses are applied more frequently in the past decade since the entering of new development standards that prescribe water neutral development, which results in the need for water storage to compensate for development (The Ministry of Transport, Public Works and Water Management & The Ministry of Housing, Spatial Planning and the Environment, 2006). The use of pole houses makes it possible to build the water storage area. These pole houses, often applied in polder areas, can address changing water levels due to extreme rainfall events or small flood events.

(A)   (B)

Fig. 3 (A) Pole houses in the Netherlands polder area at Nesselande, Rotterdam. (B) Pole houses in the Galveston USA coastal zone. *((A) Photo by Defacto Urbanism, Rotterdam. (B) Photo by Defacto Urbanism, Rotterdam.)*

However, in case of a major breach of a primary levee, flood levels will potentially be more severe: as an illustration, in case of a dike breach, the house in Fig. 3A can face flood levels up to 2.8 m (Klimaateffectatlas, 2021).

Floating houses

Floating houses and buildings originate from the houseboat typology. Along many canals in the Amsterdam historic city center canals still many houseboats are in use. Floating houses can move along with changing water levels, which also allows them to function in case of unexpected climate change or flood levels (in contrast to the mount and pole houses). The floating houses often have flexible walkways and pipe connections to the quay. Though in theory a floating house can be transported and relocated, the houses are often positioned at a fixed leased berth or a so-called owned "water-plot."

    Floating houses are often constructed on a waterproof concrete base that is positioned under water. This means the water needs to have sufficient depth to float the house (plus a minimum water depth under the structure to prevent deteriorating of the water quality). In canals, the floating structures can affect the flow capacity needed to drain sufficient water. In Amsterdam for instance, houseboats are becoming a bottleneck in draining (Sikma et al., 2011), especially under climate change inflicted increased amounts of rainwater.

Leveling a floating house is an important comfort aspect. Some houses have a water balancing system in which water can be redistributed to compensate for a weight disbalance due to construction errors or heavy interior elements (such as bookshelves or pianos). Floating houses are impacted by the movement of the water. Whether this is a pro or con is very personal; some believe this is the charm of a floating home; others get seasick and do not prefer floating homes. The stability of a floating house is determined by its mass and its point of gravity. In general, this means that higher and lighter a building, the more unstable it is. However, safe construction of large floating buildings is also possible, since the stability is related to balanced proportions and not to the height itself. In the Netherlands, there are for instance floating hotels and prisons. These bigger floating buildings are often positioned in deeper rivers or channels.

Amphibious houses

A special type of floating house is the amphibious house, which is still mostly applied experimentally. These houses are based on the ground but can float if needed. The floating body is positioned on land but is constructed in the same way as it is in a floating house. Since amphibious houses float incidental, the proportions are often not adjusted to the floating situation: gliding poles are often used to provide stability and prevent tilting. Like with floating houses, the pipe connections and entrance walkways need to be flexible in order to not break off when the building is lifted. The stability is covered by the gliding poles. The amphibious houses can also be constructed from lightweight solid floating devices such as foam and composites. Although current amphibious houses, often positioned in flood plains, are mostly constructed with a concrete waterproof base, lightweight construction versions could be interesting to apply in polders with both flood risk and subsidence challenges (Fig. 4).

Applying consequence reduction measures

In embanked areas in the Netherlands, the probability of a flood occurring is often low, resulting in additional investments for flood proofing of buildings becoming obsolete. Nevertheless, the consequence of a flood (damage and fatalities) can be severe. Therefore, we see a renewed emphasis on exploring suitable options for flood consequence reduction in embanked areas. Since the primary flood risk investments are targeted at probability reductions (constructing levees, pumps, and barriers) big investments in flood proofing all buildings (which have a relatively short lifespan compared to Dutch failure probabilities) does not make sense from a cost-benefit perspective. However, if by small additional or targeted investments in flood proofing of buildings the potential damage or

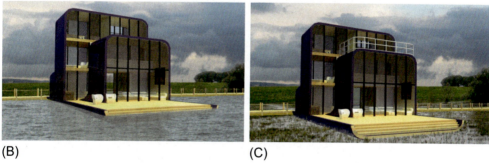

Fig. 4 (A) Amphibious houses in Maasbommel along the river Maas, The Netherlands. (B and C). A design for lightweight composite amphibious houses, that could be positioned in polders with severe subsidence challenges, The Netherlands. *((A) Photo by Defacto, Rotterdam. (B and C) Image and design by Defacto, Rotterdam.)*

fatalities can be substantially reduced or multiple climate change effects can be addressed, these options are interesting to explore. For instance, based on this principle, the Dutch capital Amsterdam drafted a map that indicates areas with a potential for the flood that includes consequence reduction measures in new developments (Omgevingsvisie Amsterdam 2050, 2021). Within the consequence reduction approach, there are several interesting options for flood proof building in embanked areas, like flood proof plinths, zoning, and shelters.

Flood proof plinth
Several areas have a shallow inundation depth (up to circa 20 cm) in most flood scenarios (LIWO, 2020). In some locations, these inundation depths can also be caused by extreme rainfall events (Klimaateffectatlas, 2021). Therefore, introducing small building plinths and thresholds at entrances of parking garages, and elevating sidewalks might be worthwhile investments that can also protect against the much more frequent risk of rainwater inundation. Other small measures, such as elevating electricity sockets and electrical boxes, can be achieved without additional costs in new buildings, and can substantially reduce damage or recovery costs.

Zoning
Applying building regulations and zoning plans can potentially reduce flood fatalities. For instance, houses with only one floor in areas with high potential inundation depths (of 2 m or more) increase the risk of fatalities in case of a flood substantially (Jonkman, 2007; Kolen, Vermeulen, Terpstra, & Kerstholt, 2015). By appointing commercial functions to the ground floor in floodable areas with deep inundation, or by making sure the houses there have access to a second story, the risk of fatalities could be reduced. For vulnerable users such as elderly people or children, these measures should already be considered for lower potential inundation depths (MRA, 2020). Another zoning tool would be building restrictions in area's that can flood very deep; some polders in the Netherlands even have a potential inundation depth of up to 5 m. Making flood risk (and high or extreme potential inundation depths) a factor in designating locations for housing development or for critical functions such as hospitals or electricity suppliers could reduce the potential damage in case of a flood.

Shelters
In areas that are especially vulnerable to inundation and have a large share of households that do not have access to a dry second floor (and therefore cannot apply the so-called 'vertical evacuation'), a nearby public shelter could increase the self-reliance in case of a flood (Kolen et al., 2015). This would especially be valuable in areas where there is not sufficient time to evacuate people out of the area before the floodwaters arrive. The municipality of Dordrecht is now constructing a municipal building that can also function as a shelter for the city in case of a flood (Municipality Dordrecht, 2009).

Technical practicalities of flood proof houses
Although flood proof houses do not have to differ so much from "normal" houses, the flood proof houses have some characteristics that one should be aware of when

constructing them. Some of them, like the leveling of floating houses, the necessity for flexible connections of moving structures, and the different possibilities to adapt to climate change were already addressed in the previous sections.

Some aspects can lead to severe additional costs for flood proof houses. For instance, in river and coastal zones with strong currents or high flow velocity, buildings need to be constructed in a way that they can withstand these forces. This also applies when floating objects such as tree trunks and freight ships can potentially collide with the building. Another (easier to address) practical aspect is maintenance. Houses that are positioned in the water have restrictions on the use of materials that can pollute the water (such as basic paints or cleansers). With that, surfaces and public spaces that flood frequently should be able to withstand the water (this is sometimes a challenge in the case of salt water) and be able to easily cleaned after a flood event. There are also aspects from a planning perspective that require special attention. For instance, fire route regulations can be challenging to apply on walk boards that form the official access route for floating houses. And finally, as it is not possible on the water to construct fences or hedges on the property lines, the design principles for ensuring privacy are different than when planning a neighborhood on land.

Conclusions

Whether or not to construct (and invest in) flood proof housing depends a lot on the flood frequency. The higher the flood frequency, the more chance that a building will be subject to flooding during its lifetime and the more can be invested to prevent this damage. For the Netherlands, consisting mostly of embanked polders and areas protected by dikes, dams and dunes, this means flood proof houses are usually only found in unembanked areas. The flood proof housing developments in these areas are in high demand: they do not only have the technical and financial advantage of withstanding floods but also the experience of and closeness to the water are highly appreciated by its users.

In embanked areas, flood proof houses (and the additional investment they require) are mainly applied at locations where this type of construction opens opportunities for developing areas that would otherwise remain unbuilt, such as lakes and canals. Also, building flood-proof housing in the polder water storages areas can provide so-called "multipurpose" areas for development that may benefit the community.

In embanked areas with a potential shallow flood level as well as a stormwater flooding risk (with a higher frequency), there are feasible opportunities to apply flood risk measures that can address both types of flooding. In addition, there are areas where applying flood proof buildings as part of a multilevel safety strategy could be worthwhile. Because of the Dutch probability-based flood risk approach (*see also Chapter 12 in this volume, by Matthijs Kok*) concepts for the second layer of consequence reduction are not yet matured. The expectation is that by including potential flood depths (even with a lower probability) as a relevant factor in urban planning and development, in some cases interesting options for integrated designs that address flood risk can be achieved without disproportionate investments.

Reflection

The Dutch dike ring system is a very economic one, making additional investment to flood proof individual buildings (that in a densified country as the Netherlands would result in high individual costs) economically unfeasible. However, it also creates a lock-in: when looking at long-term perspectives like those for sea level rise (Deltaprogramme, 2020), we see that the current flood protection system, based mainly on structural measures, is pressured, potentially resulting in the necessity to be extended further and close off more river branches from the sea. An approach that includes more consequence reduction measures might be more flexible regarding future climate adaptation strategies.

Although the same concept of the potential of targeting flood risk by probability reduction or consequence reduction measures applies everywhere, the balance between probability and consequence reduction measures in the USA situation is almost the opposite from the Dutch one. Where in the Netherlands the focus is on nationally organized probability reduction measures and consequences, and reduction and evacuation strategies are almost absent, in the USA the focus is on evacuation, individual consequence reduction, and risk zoning linked to insurance. This is for instance clearly visible along the Galveston coast where the—compared to the Netherlands—high probability of coastal flooding is combined with strict insurance guidelines resulting in many individual home owners building stilt houses. Similarly, the enormous private investments in flood proofing the Texas Medical Center after suffering substantial damages from hurricane Ike are exemplary for the US tendency towards individual flood proofing measures. For both countries, a combined approach that includes multiple layers of flood risk management could be beneficial; in the Netherlands improved awareness of flood risk combined with the traditional flood prevention orientation can result in flood proof integrated design and prevent further lock-in developments. For Houston and the Galveston Bay area, a coastal barrier combined with flood resilient urban design and the current individual property flood proofing measures could help to better protect vulnerable inhabitants from flooding.

A regional integral flood risk management strategy that includes the coastal, Galveston Bay and bayou water management systems, and explores where the economic and societal cost benefits would result in a trade-off for probability and consequence reduction measures in different areas could be a great asset for Houston. However, this would require a level of government coordination and involvement on a regional scale that does not seem to fit the current Houston policy context.

However, especially along Houston's bayous which have frequently faced extreme discharges and damages, a combined approach that includes a citywide bayou integral vision and resulting zoning as a framework for flood proof buildings and developments, could be a valuable addition. With this, it is still possible to meet the Texan culturally embedded need for individual interpretation of urban and architectural design. The combined approach would require an integrated (spatial and water management) framework and vision that systematically defines criteria for required storage and flow capacity of the bayous and provides guidelines for individual developments within and alongside the bayou. The guidelines do not have to prescribe solutions but should prescribe standards; thus requiring (new) developments to be able to withstand certain flood water levels, which still offers the opportunity to be translated into many different individual design solutions.

References

Deltaprogramme (2020). Available online https://english.deltaprogramma.nl/delta-programme.
Gersonius, B., Veerbeek, W., Subhan, A., & Zevenbergen, C. (2011). Toward a more flood resilient urban environment: The Dutch multi-level safety approach to flood risk management. In K. Otto-Zimmermann (Ed.), *Resilient cities: Cities and adaptation to climate change – Proceedings of the global forum 2010, local sustainability*. https://doi.org/10.1007/978-94-007-0785-6_28 (© Springer Science+Business Media B.V.).
Jonkman, S. N. (2007). *Loss of life estimation in flood risk assessment – Theory and applications*. PhD thesis Delft University of Technology.
Klimaateffectatlas. (2021). *Based on LIWO datasets*. Available online https://www.klimaateffectatlas.nl/nl/kaartverhaal-overstroming (in Dutch).
Kolen, B., Vermeulen, C. J. M., Terpstra, T., & Kerstholt, J. (2015). *Randvoorwaarden verticale evacuatie bij overstromingen. [preconditions for vertical evacuation in the event of flooding]*. Lelystad: HKV-lijn-in-water, commissioned by WODC (In Dutch, with English summary).
LIWO. (2020). *Landelijk Informatiesysteem water en Overstromingen [national information system water and floods]*. Available online www.helpdeskwater.nl/onderwerpen/applicaties-modellen/applicaties-per/watermanagement/watermanagement/liwo/.
Ministry of Transport, Public Works and Water Management. (2005). *15 experimenten met bouwen in het rivierbed*. The Hague: Ministeries van V&W, Ministerie van VROM (in Dutch).
MRA-Metropoolregio Amsterdam Klimaatbestendig. (2020). *Concept basisveiligheidsniveau klimaatbestendige nieuwbouw*. Available online https://www.metropoolregioamsterdam.nl/programma/klimaatadaptatie/ (in Dutch).
Municipality Dordrecht. (2009). *Masterplan Stadswerven, nieuwe stedelijkheid voor Dordrecht. [Masterplan Urban Yards, new urbanism for Dordrecht]*. Available online https://arch-lokaal.nl/wp-content/uploads/2014/11/2-Masterplan-Stadswerven-1-februari-2009.pdf (in Dutch).
Municipality of Rotterdam. (2013). *Herijking van Waterplan 2 Rotterdam*. Available online https://www.rotterdam.nl/wonen-leven/waterplan-2/ (in Dutch).
Nillesen, A. L., & Singelenberg, J. (2011). *Amphibious housing in the Netherlands: Architecture and urbanism on the water/waterwonen in Nederland: Architectuur en Stedenbouw op het water*. Rotterdam: Netherlands Architecture Institute (NAi) Publishers, ISBN:9789056627805.
Omgevingsvisie Amsterdam 2050. (2021). Available online https://amsterdam2050.nl (in Dutch).
PBL. (2014). *Kleine kansen – grote gevolgen. Slachtoffers en maatschappelijke ontwrichting als focus voor het waterveiligheidsbeleid. Report #1031*. The Hague: Planbureau voor de Leefomgeving (in Dutch).
Sikma, H., Niessen, Q., Roozendaal, J., Bloemkolk, H., Cleassen, G., & Wolter, A. (2011). *Uitgangspunten verbeteren waterafvoerroute naar gemaal Zeeburg*.
The Ministry of Transport, Public Works and Water Management & The Ministry of Housing, Spatial Planning and the Environment. (2006). *Beleidslijn grootte rivieren*. The Hague: Ministeries van V&W, Ministerie van VROM (in Dutch).

CHAPTER 25

Risk communication tools: Bridging the gap between knowledge and action for flood risk reduction

Samuel Brody[a,b] and William Mobley[a]
[a]Institute for a Disaster Resilient Texas, Texas A&M University, College Station, TX, United States
[b]Department of Marine and Coastal Environmental Science, Texas A&M University, Galveston Campus, Galveston, TX, United States

One of the most important, yet underemphasized aspects of a successful flood management program is the ability to effectively communicate risk. When residents are informed about what to expect during a storm event and how to reduce its adverse impacts, they can minimize property losses and disruption of daily activities. There is a wide range of household mitigation activities and adjustments, from purchasing insurance to elevating structures, that can be implemented in the face of mounting risk. Conveying flood risk information in a way that is accurate, interpretable, and actionable at the household level is especially vital in flood-prone regions such as the upper Texas coast and the Netherlands. This chapter focuses on one aspect of risk communication: the use of data analytics, web services, and mapping software to help residents better understand the risk to flooding and the potential mitigation options that can reduce impacts over the long term. With the advent of high-resolution data and more ubiquitous web dashboards, the technical communication of risk and possible impacts will increasingly become an important issue in both countries.

The role of risk perception

Many researchers have analyzed risk perceptions as a major predictor of the adoption of various kinds of hazard adjustments (see, e.g., Kellens, Terpstra, & De Maeyer, 2013; Lindell & Hwang, 2008). In general, flood risk perception can be understood as the perceived danger of personal consequences owing to inundation. It is one of the strongest and most-studied factors influencing hazard adjustment, especially the purchase of flood insurance. It is important to note that the influence of perceived risk is often bound-up with other factors, including experience, access to information, self-efficacy, previous mitigation, etc., and can be difficult to disentangle (Bubeck, Botzen, & Aerts, 2012).

In the United States, an overall lack of commitment by decision makers to systematically communicate flood risk to the general public has resulted in distorted perceptions, confusion over the nature of risk, and a general lack of action among households. When information and outreach programs do exist, they are primarily targeted to areas within the FEMA-defined 100-year floodplain, despite the fact that in places such as Houston, TX approximately 50% of flood insurance claims are outside this boundary (Blessing, Sebastian, & Brody, 2017). The deleterious effect of geographically limited, overly technical, or nonexistent information on flood risk has clearly taken a toll in terms of awareness. Results from a 2015 survey of residents in League City and Friendswood just south of Houston indicated that over 41% of respondents "did not know" whether they were in or out of the FEMA 100- or 500-year floodplain (the majority of the sample was outside of both). Furthermore, approximately 42% of those surveyed living outside the 100-year floodplain "did not know" how far their residence was situated away from this boundary, whether it is less than a quarter of a mile or over 3 miles (Brody, Lee, & Highfield, 2017). Geographic location within a flood-prone landscape is critical to determining risk of structural inundation. Even an eighth of a mile can mean the difference between water in a home or just on the street.

Flood risk awareness is similarly problematic in the Netherlands, possibly for different reasons. In a household survey, e.g., Botzen, Aerts, and Van Den Bergh (2009) found that in general, perceptions of flood risk are "low" and that increasing citizens' understanding about the causes of flooding may increase flood risk awareness. Terpstra and Gutteling (2008) found a similar lack of concern among Dutch households in part because the vast majority of respondents believed the government is primarily responsible for protecting them against flood damage. Regardless of the motivating factors, research in both countries overwhelmingly finds that communication of flood risk is lacking, despite large amounts of data and government expertise.

Tools that enhance communication of risk

The rise in availability of geospatial data, web services, and high-performance computing provide an opportunity to effectively communicate flood risk like never before. The public increasingly relies on the Internet for key information and guidance in making household decisions. Web-based decision support tools thus provide a critical pathway between technical knowledge and household mitigation. This communication conduit is comprised of the following four dimensions: (1) data analytics, (2) data visualization, (3) risk communication, and (4) learning.

Data Analytics: The increasing ubiquity of spatial data rectified at smaller scales, combined with complex physics and statistical-based models, is enabling researchers to more accurately predict the risk of and impacts from flooding events (Demir & Krajewski, 2013). For example, a new generation of 2D HEC-RAS and spatially distributed hydraulic models can measure and map inundation originating from both stream channels and overland flow, providing a much more complete representation of expected impacts (Demir, Yildirim, Sermet, & Sit, 2018; Macchione, Costabile, Costanzo, & De Santis, 2019). Also, statistical machine-learning models running in a high-performance computing environment can take advantage of parcel-level flood loss data to render risk boundaries at a level of real-world accuracy hydraulic models never could attain (see Chapter 7). Combining these inundation models with data on structures, socioeconomic characteristics, and other measures allows fine-tuned predictions for expected losses, disruption of services, and population displacement—all critical information for informing decision makers and the public on how to prepare for and mitigate against disaster events.

Data visualization: Analytical results, however rigorous, are only partially useful if they cannot be visualized across flood-prone landscapes. Picturing the impacts of various flood scenarios on specific properties and assets is a powerful motivator for enacting change. The advent of web-GIS and related programs provides a prime opportunity to spatially visualize the consequences of living in flood-prone areas. Since 2011, for example, there has been a proliferation of coastal web atlases worldwide (subsequently captured by the International Coastal Atlas Network) in an attempt to better convey the risks of living on the coast (LaVoi et al., 2011; O'Dea, Dwyer, Cummins, & Wright, 2011). Most recently, use of 3D visualizations and augmented reality are being incorporated into GIS web dashboards (Haynes, Hehl-Lange, & Lange, 2018; Kumar, Ledoux, & Stoter, 2018) to provide an immersive user experience.

Communication: Visualizations promoting flood-resilient behaviors must be communicated in a way that makes products usable, interpretable, and viable for stimulating behavioral or policy change. Producing complicated visualizations based on the most cutting-edge statistical models will be useless if they cannot be effectively conveyed. This means taking the time to understand which formats, color schemes, and level of complexity are most effective at getting the message across to target stakeholders. It also means providing an interactive web experience that maintains user interest while at the same time messaging about complex information. Communicating flood-risk visualizations over the Internet does not have to be a self-discovery experience alone. Instead, combining web tools with more traditional communication techniques, such as in-person workshops and training sessions, can enhance the interactive experience and increase the effectiveness of flood risk communication initiatives.

Learning: Learning, as a result of effective visualization and communication of flood risk, is a stepping-stone for facilitating positive change. There are two types of learning that can take place. First, the data and models predicting flood risk are constantly improving. The evolution of flood models from 1D to 2D is a clear example of these model improvements. Running parallel to these advancements are the use of Machine Learning and Artificial Intelligence (AI) techniques to better represent aspects of flood risk (see Chapter 7). Using empirical data, machine learning algorithms can accurately represent flood risk more effectively than physics-based models. In addition, machine learning models are readily improved when new data is incorporated without the need to recalibrate parameters. The second type of learning stems from local resident knowledge and experience, which further increases the accuracy of modeled predictions. Data obtained from user input, citizen sensors, surveys, interviews, etc. can be folded into technical analyses to create a product that is more receptive to local needs, while more time consuming, in-person, reciprocal learning environments, such as community-based charrettes and workshops, will result in more enduring policy and behavioral change around flood risk.

Examples of data-driven web communication tools

The real-estate transaction process presents an ideal opportunity to communicate flood risk and increase awareness of household mitigation options. Purchasing a home or property is usually the largest investment any individual or family will make. As mentioned above, residents in the Houston area and the Netherlands are largely unaware of their geographic risk, despite being located in one of the most flood-prone coastal landscapes in the country. In response to this lack of awareness and the fact that, according to the National Association of Realtors, 74% of home buyers use the Internet as part of their search, researchers developed a prototype called BuyersBewhere.com. Buyers Be-Where is an on-line system to help prospective home buyers, rentors, and sellers better understand their risk (flooding and nine other hazards) relative to other properties in the area. Anyone with an Internet connection can enter a street address or click on a parcel map to receive a graphic and statistical risk assessment for a specific property. Both individual hazard risk scores and a composite rating is relayed through an easy to use and interpreted graphic interface. Users can also click on a color-coded risk score and receive tips on different household mitigation alternatives that will reduce adverse impacts to long-term investments (Fig. 1).

Fig. 1 Buyers be-where risk assessment.

The BuyersBewhere.com system seeks to effectively communicate risk by combining scientific data, models, and geographic visualization around what is a critical decision-making venue. The initial prototype was created for Harris and Galveston counties in Texas and is now poised to improve its analytical capabilities and expand across the state. Machine-learning statistical models will increasingly be integrated into the scoring system to improve the accuracy of risk predictions. The web-platform will also transform from a solely self-discovery risk assessment tool to one that also obtains user information and experience to improve models on the back end.

A comparable web hazard risk communication tool in the Netherlands is called Risicokaart (meaning Risk Card or Risk Map). This web portal (https://www.risicokaart.nl/) provides a geographic and statistical assessment of multiple natural hazards, including floods, fires, earthquakes, and toxic releases. Flood risk can be assessed down to the parcel level based on different return periods, populations exposed, critical facilities, and other built environment factors (Fig. 2). The Risk Map is comprised of a public viewer and government version that helps these entities perform their (statutory) duties more effectively in the field of Spatial Planning (RO) and Permits, Supervision and Enforcement (VTH), respectively.

Buyers Be-where and Risicokaart are great examples of web systems that provide residents hazard information for a single geographic scale. But, there are several issues that neither website addresses successfully. First, residents are affected by natural hazards at a variety of scales. A person's home may flood or she/he may be prevented from reaching their workplace miles away due to flooded roads. Presenting information clearly and at multiple scales will help residents better prepare for large-scale flood events. Second, as

Fig. 2 Risicokaart flood risk assessment.

previously discussed scientists and engineers have designed a variety of natural hazard focused models, making it difficult for users to decide on which one to rely. Without using a standard approach for identifying and communicating the best available data, residents could find themselves obtaining conflicting information. Third, the advent of web2.0 gave web users social media-based opportunities to post thoughts and communicate with friends and colleagues. Currently, the natural hazard-based web applications fail to provide an outlet for residents to discuss their experiences or pose questions regarding their risk profiles. However, these websites could provide residents and decision makers a platform to improve natural hazard literacy through online dialog. If the next generation of hazard web applications can account for these short comings through continued development, initial solutions like BuyersBeWhere.com and ThinkHazard! and their successors can provide opportunities to:

- Develop an online risk assessment tool to provide accurate information at the parcel level.
- Establish an analytical platform that integrates advanced scientific data and models specific to a geographic region.
- Assist decision makers in communicating hazard risk to locally impacted residents, facilitating enhanced risk perceptions and mitigation behaviors.
- Build and test an easily accessible, interpretable, and informative web tool to help communities better understand and prepare for floods and other hazards.
- Create a hazard mitigation tool that serves as a model on how to effectively fuse scientifically derived models, visualization, and online accessibility.

Conclusion: Challenges and opportunities

Additional development of easily accessible, interactive web tools is needed in both countries as individuals are forced to take on more responsibility for protecting themselves against increasing flood impacts. These types of platforms provide a prime opportunity to communicate complex flood information in a way that will increase risk perception, change household behaviors, and reduce impacts of flood events in the future. However, there are various challenges to developing accurate web-based risk communication tools and guidelines that must be followed to ensure their long-term effectiveness. First, integrating the most accurate and appropriate data is paramount to ensure accuracy, especially at a parcel level. Many models are rendered at a spatial scale that is not appropriate to apply to individual parcels without generating large errors that could misinform homeowners or renters. Second, capturing changing contextual conditions on a regular basis is necessary to keep tools relevant. Constantly fluctuating environmental, socioeconomic, and built environment local characteristics can make a tool obsolete in a matter of months. In particular, the effects of new buildings, impervious surfaces, and other seemingly nuanced physical changes can have large impacts in flood-prone areas. Third, a human-centered approach to designing flood risk communications tools will better ensure their usability. This approach requires user input early and often throughout tool development, including the use of surveys, focus groups, workshops, and other methods of stakeholder input. Finally, even after being implemented, ongoing assessment should be done on the usability and interpretability of the tool to ensure it will affect long-lasting behavioral change.

References

Blessing, R., Sebastian, A., & Brody, S. D. (2017). Flood risk delineation in the United States: How much loss are we capturing? *Natural Hazards Review*, *18*(3), 04017002.

Botzen, W. J. W., Aerts, J. C. J. H., & Van Den Bergh, J. C. J. M. (2009). Dependence of flood risk perceptions on socioeconomic and objective risk factors. *Water Resources Research*, *45*(10).

Brody, S. D., Lee, Y., & Highfield, W. E. (2017). Household adjustment to flood risk: A survey of coastal residents in Texas and Florida, United States. *Disasters*, *41*(3), 566–586.

Bubeck, P., Botzen, W. J., & Aerts, J. C. (2012). A review of risk perceptions and other factors that influence flood mitigation behavior. *Risk Analysis*, *32*(9), 1481–1495.

Demir, I., & Krajewski, W. F. (2013). Towards an integrated flood information system: centralized data access, analysis, and visualization. *Environmental Modelling & Software*, *50*, 77–84.

Demir, I., Yildirim, E., Sermet, Y., & Sit, M. A. (2018). FLOODSS: Iowa flood information system as a generalized flood cyberinfrastructure. *International Journal of River Basin Management*, *16*(3), 393–400.

Haynes, P., Hehl-Lange, S., & Lange, E. (2018). Mobile augmented reality for flood visualisation. *Environmental Modelling & Software*, *109*, 380–389.

Kellens, W., Terpstra, T., & De Maeyer, P. (2013). Perception and communication of flood risks: A systematic review of empirical research. *Risk Analysis: An International Journal*, *33*(1), 24–49.

Kumar, K., Ledoux, H., & Stoter, J. (2018). Dynamic 3D visualization of floods: Case of the Netherlands. In *International archives of the photogrammetry, remote sensing & spatial information sciences*.

LaVoi, T., Murphy, J., Sataloff, G., Longhorn, R., Meiner, A., Uhel, R. J., et al. (2011). Coastal atlases in the context of spatial data infrastructures. In *Coastal informatics: Web atlas design and implementation* (pp. 239–255). IGI Global.

Lindell, M. K., & Hwang, S. N. (2008). Households' perceived personal risk and responses in a multihazard environment. *Risk Analysis: An International Journal*, *28*(2), 539–556.

Macchione, F., Costabile, P., Costanzo, C., & De Santis, R. (2019). Moving to 3-D flood hazard maps for enhancing risk communication. *Environmental Modelling & Software*, *111*, 510–522.

O'Dea, E. K., Dwyer, E., Cummins, V., & Wright, D. J. (2011). Potentials and limitations of coastal web atlases. *Journal of Coastal Conservation*, *15*(4), 607–627.

Terpstra, T., & Gutteling, J. M. (2008). Households' perceived responsibilities in flood risk management in the Netherlands. *International Journal of Water Resources Development*, *24*(4), 555–565.

SECTION V

Immersive place-based learning through convergence approach

CHAPTER 26

How to design a successful international integrative research and education program

Yoonjeong Lee[a,b] and Baukje Bee Kothuis[c,d]
[a]Institute for a Disaster Resilient Texas, Texas A&M University, College Station, TX, United States
[b]Department of Marine and Coastal Environmental Science, Texas A&M University, Galveston Campus, Galveston, TX, United States
[c]Department of Hydraulic Engineering and Flood Risk, Faculty of Civil Engineering and Geosciences, Delft University of Technology, Delft, The Netherlands
[d]Netherlands Business Support Office, Houston, TX, United States

Introduction

The integration of multiple disciplines in research and education is becoming an important subject in higher education, combined with an increasing emphasis on multidisciplinary approaches and international collaborations in research projects. In a traditional university setting, a discipline-based organizational structure makes it hard for students to have a chance to learn how to communicate, collaborate, and interact with other students or faculty from different disciplines. In addition, students are not adequately provided with opportunities for international research or learning experiences due to lack of funding or nonexistent systemic programs. To remedy this gap, funding agencies and universities extended efforts to provide students and faculty with international multidisciplinary research and education experiences. However, there are not many existing studies that discuss how to design such programs to ensure the successful learning outcomes of participating students. This chapter explains the program design process, as well as reflections and lessons, acquired from developing and implementing 4 years of student research and education that involves international research trips and interdisciplinary collaboration. Having a set of guidelines for a program design based on concrete experiences would benefit researchers and faculty who intend to create a similar program in the future.

The Partnership for International Research and Education (PIRE) is a National Science Foundation (NSF)-wide program that supports high-quality international multidisciplinary projects to facilitate the development of a diverse and globally committed scientific and engineering workforce. The NSF PIRE Coastal Flood Risk Reduction Program (CFRRP) was initiated under the PIRE in 2015 as an international integrative research and education project to address flood risk. This multiyear binational program includes a synergetic set of faculty, researchers, experts, and students with different academic backgrounds, such as engineering, hydrology, landscape architecture, economics, and planning. The program aims to contribute to flood risk reduction in the United States, especially along the Upper Texas coast by benchmarking successful cases in the Netherlands, a country known for effective flood mitigation practices. It has become clear that the rising costs of flooding are not just a function of changing weather patterns or a problem that can be solved with technical solutions alone. Flood risks and associated losses can only be understood and ultimately reduced through integrated research across multiple disciplines, cultures, and international borders. The approach of this program, therefore, involves combining physical and social science data, research methods, and analytical techniques to form a more comprehensive understanding of flood risks.

An integral part of the CFRRP is the educational component where teams of students from different academic and cultural backgrounds conduct problem- and place-based research with the guidance of a multidisciplinary team of faculty mentors. The primary goal of the education component of the program is to provide "transformational learning experiences" for participating students by creating "authentic learning environments" that both support and benefit from the research components.[a]

This chapter describes the design and implementation of the educational aspects of the CFRRP. First, a brief review of theories and concepts applied to the program design is addressed, followed by a detailed description of the design process including a discussion on how to incorporate convergence into the program components. The chapter concludes with reflections and lessons learned that could help other researchers or educators in higher education who intend to create an international integrative research education program.

Background theories and concepts of the program design

The design of the education component of the program embraces convergence, an expanded concept of interdisciplinary research (NRC, 2014). The NSF identified "convergent research" as one of the main paths the organization should pursue. Convergence in research is defined as the deep integration of knowledge, tools, expertise, and ways of thinking and communication from multiple fields to form a comprehensive framework for addressing scientific and societal challenges (Herr et al., 2019; Kitney et al., 2017).

With the inevitable increasing interactions among various academic disciplines in research projects dealing with societal challenges, convergence has become an opportunity for the education communities. The integration of crosscutting core knowledge and concepts requires a perspective shift from "what is taught" to "how is it taught" (Kitney et al., 2017).

---

[a] For more details about the background theories and concepts of transformative education, authentic learning, and problem- and place-based learning, see Chapter 5.30.

The NSF PIRE CFRRP is an international research education program in which students conduct case study research focusing on flood-related problems. Due to the increasing complexity of flood impacts and associated losses, integrative efforts to tackle the problem across disciplines is essential for both faculty and students participating in the project. To create a convergent setting that facilitates a holistic approach to the problem in a research education program, integration across disciplines should be incorporated into educational and training strategies as well (Herr et al., 2019).

Many factors affect the success of integrative research education, and the most important ingredient is the physical incorporation of classroom and field by forming a multidisciplinary team that focuses on a particular overarching research question and problem solving. This approach helps students overcome fragmented perceptions and realize connections among different discipline domains as well as across participants' experiences gained in various academic and cultural contexts (Booth, 2011; Clapton et al., 2008; Kulasegaram, Martimianakis, Mylopoulos, Whitehead, & Woods, 2013). Jamison, Kolmos, and Holgaard (2014) described the educational effects of the integrative setting in engineering education. They stated that the integrative approach allows students to grasp the connections between the theoretical and practical aspects of engineering work, broadening their cultural and social understanding to more effectively deal with real-life problems. In this way, integrative convergent education trains students to grow into professionals with enhanced communication skills through interacting and building a network with other students and faculty from various disciplines (Herr et al., 2019).

Another pivotal factor for convergent research and education is the diversity that takes multiple forms. The National Research Council (2014) suggests two types of diversity that are needed to secure innovation in convergent research: functional and identity. Functional diversity refers to the range of approaches to problem solving; identity diversity is the variance in demographic, cultural, and ethnic characteristics within a research team. Indeed, previous studies showed that groups of individuals with various perspectives generate more innovative solutions to complex problems than do groups comprised of individuals with similar perspectives (Hong & Page, 2004). In addition, vertical diversity in education levels (e.g., bachelor's, master's, and PhD), as well as horizontal diversity (variations in disciplines), are also pivotal features in achieving the goal of convergent education. Bolli, Renold, and Wörter (2018) found that vertical educational diversity significantly enhances innovation, particularly in the invention phase of a new product. This is because different levels of education accompanied by different insights and experiences of participants might create various interpretations of problems which can lead to a wider spectrum of possible solutions. Further, the diversity in convergence is generally more inclusive than in interdisciplinary research because the cross-fertilization of knowledge, ideas, and experiences occur not only among different academic disciplines but also with stakeholders, partners, and policymakers outside academia (NRC, 2014).

While physical integration and diversity are the factors needed at the initial stage of the program design, creating an open and inclusive learning environment that allows students to freely represent their thoughts on issues is essential at the implementation phase. A high level of integration requires one to move beyond their own expert language to be able to communicate across disciplines and build a common understanding of the shared problem solving and goals (NRC, 2014). This is where transformative learning occurs through immersive discourses guided by faculty and researchers from various disciplines.

Tables 1–3 show the program timeline along with expected educational effects and organizational tasks objectives of each phase in chronological order: pretrip (Table 1), on-trip (Table 2), and posttrip (Table 3). These tables have been adopted and modified from Lee, Kothuis, Sebastian, and Brody (2019).

Program design incorporating convergence
Integrative case study research design

Every year, the first and most important step of the program design process is to configure the case studies that will be investigated by the cohort of students. The overarching research theme of these case studies is flood risk reduction (problem-based) and the cases are located in the Netherlands (place-based). Fig. 1 shows the map of the case study areas of the 2019 program (case study locations change every year).

How to present the case study areas with relevant local flood-related issues is crucial for integrative group formation because students who apply for the program are required to write a case study research proposal based on the information provided. One of the most important tasks of the program managers and faculty mentors is to set up the case study presentation properly so it can attract students from various disciplines and academic levels.

A brief description of the regional context addressing local problems and issues concerning the flood risk of the area is provided with an example of a set of research questions (see e.g., Fig. 2). Suggested research questions are developed based on the local context embracing various disciplines. To the students who intend to apply for the program, this array of multidisciplinary research questions acts as a guide providing a hint for how to come up with a research question(s) for application based on their major or research focus. For example, an undergraduate student majoring in urban planning or architecture could choose "Focus area 1: City of Vlissingen" and "RQ1: What is a flood resilient building code in Vlissingen and how can this be applied in the United States?" and a civil engineering PhD student whose research focus is infrastructure resilience could write a research proposal for application around "Focus area 2: Vlissingen Port" and "RQ4: What are the impacts of climate change or extreme weather on the area, especially infrastructure networks?" or formulate their own research question.

Table 1 Pretrip.

Phase	Timeline	Phases	Educational aspects	Organizational aspects
Pretrip Application and orientation	Sep–Nov	Application Phase #1	• Writing skills: LOI • Knowledge increase: Get acquainted with study areas and local flood risk related issues	• Design case studies • Organize mentor teams • Publish application information online • Select 25 students and notify them to submit a full proposal
	Dec–Jan	Application Phase #2	• Writing skills: full research proposal • Knowledge increase: Obtain a better understanding of a specific flood risk related issue in a specific case study area	• Select 16 students and announce selection • Organize accommodations and other logistic arrangements
	Feb–Mar	Application Phase #3	• Writing skills: revise and update the proposal based on the comments from the review committee • Knowledge increase: international field work travel preparation	• Scoping trip to case study locations for arranging fieldtrips, lectures, and expert meetings • Prepare the orientation • Travel arrangements for students and faculty
	Mar	Orientation Phase #1	• Presentation skills: first presentation to a multidisciplinary group (mentors and students) • Collaboration skills: meet with peers and discuss different ways of approaching the topic within a multidisciplinary team	• Orientation day • Review the updated proposals and guide students in how to do a literature review

Continued

Table 1 Pretrip—cont'd

Phase	Timeline	Phases	Educational aspects	Organizational aspects
	Apr–May	Orientation Phase #2	• Knowledge increase: General overview of the flood risk issues and cultural differences in both countries • Knowledge increase: individual literature review • Collaboration skills: virtual meetings with the team members, share the research proposal development process and combine individual literature reviews into one paper for the case study team	• Finalize day-to-day schedule of the research trip • Give final comments on the updated proposals

At this stage, a team of multidisciplinary faculty mentors and researchers provide input on the proposed case study descriptions and research questions to provide students with proper guidance throughout the program. An integrative case study design is a part of the foundational work that enables students to implement interdisciplinary research activities through problem- and place-based learning. Every year a program announcement with a description of case study areas and research questions is published online and advertised through campus emails, social media, and presentations in classes to recruit students.[b]

Student recruitment and application review for securing diversity

Student recruitment emphasizes attracting underrepresented groups in science and engineering and integrating all relevant disciplines to form a diverse multidisciplinary cohort of students. So far, the program has been successful in both functional (various disciplines) and identity (demographic, cultural, and ethnic characteristics) diversity, which resulted in a 72% underrepresented minority (Black, Hispanic women) participation in this STEM-intensive program between 2016 and 2020. Additionally, the program drew on the extensive experience in undergraduate and graduate student recruitment and retention offered by a historically Black college and university (HBCU) collaborating institution.

[b] http://www.tamug.edu/ctbs/PIRE/application.html.

Table 2 On-trip.

Phase	Timeline	Phases	Educational aspects

On-trip	May	Group Field visits (2–3 days)	• Knowledge increase: Firsthand experience of the Dutch flood risk approach (see and hear)
			• place-based learning:
			   - Field visits to all the case study locations
			   - Meet with local experts and stakeholders (lectures/discussion)

		Research meeting #1 and #2 (half day each)	• Presentation skills: second presentation to a multidisciplinary audience (mentors and students) • Problem-based learning: Various approaches to a problem (flood risk reduction) • Transformative learning: multidisciplinary perspectives on multiple cases
		Research time #1 and 2 (4 days each)	• Place-based research: Individual field visit to a case study location • Authentic learning: individual meetings and discussions with local experts and stakeholders • Knowledge increase: deep understanding of the issues and the study area from meetings with local stakeholders and experts and desk research on individual research case study • Writing skills: start writing a research paper
		Bi-national multidisciplinary design studio (full day)	• Knowledge increase: lectures by local experts and stakeholders, description of the area • Place-and-problem-based research: field visit to the location, discussions with local experts and stakeholders • Collaboration skills: interaction within an international multidisciplinary team for the assignment • Presentation skills: group presentation of the final outcome to the jury consisting of academic experts and local stakeholders
		Research meeting #3 (half day)	• Presentation skills: fourth and final presentation to audience with various backgrounds (PIRE mentors and cohort students plus local experts and stakeholders, host university faculty, and students)

Table 3 Posttrip.

Phase	Timeline	Phases	Educational aspects
Posttrip	Jul–Aug	Finalize	• Knowledge increase: finalize desk research • Writing skills: Write a research paper • Presentation skills: Produce a poster
	September 1		Students submit paper and poster

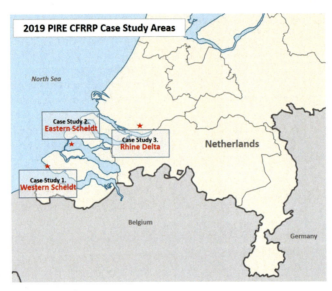

Fig. 1 2019 program case study areas.

Each year, participants are chosen from all applicants through a rigorous two-step selection process. The two steps include submission of a preproposal in the form of a letter of intent (LOI) and a full proposal by a narrowed-down group of candidates in the form of a more detailed research plan. In the LOI, students are asked to describe their project preference, their preliminary research question(s), how their background and research interests contribute to the overall place-based flood resilience research, and how they expect this program to influence their current research or future career. Applicants are also required to submit their resume or CV along with the LOI as supporting evidence of their background and research focus.

The faculty mentors review all applications based on the following assessment rubric. The reviewers score each subcriteria from 0, meaning "not at all," to 5, meaning "definitely yes."

> **Case Study #1: Western Scheldt**
>
> The Western Scheldt region is an intertidal area vulnerable to storm sure and riverine flooding. While a critical navigational channel to Antwerp, Belgium, this area experiences declining development, making it an interesting case for examining the role of restoring natural functions for flood mitigation. Due to its special geographic location and integral functions, regular dredging is required to secure safe navigation to the port of Antwerp and it has caused several environmental and intergovernmental issues that need attention. Vlissingen is a coastal city located in Western Scheldt region, and well-known for its comprehensive flood risk reduction strategies. With its critical location between the Scheldt and the North Sea, Vlissingen has played a pivotal role as a harbor for centuries.
>
> | Focus areas | Examples of local issues and topics |
> | --- | --- |
> | 1. City of Vlissingen | - Multipurpose flood defenses
- Land use plan for flood risk reduction
- Dike in dune constructions
- Flood resilient building code
- Flood resilient housing policy for socio-economically vulnerable population |
> | 2. Vlissingen port | • Vulnerable infrastructure protection:
- Nuclear power plant protection (critical infra)
- Above ground Storage Tanks (AST) protection |
> | 3. Ship channel | • Environmental and social-political issues:
- Dredging
- Intergovernmental management/collaboration |
> | 4. Coastal inlet | - Morphology of multiple channel tidal inlet systems
- Oyster reef protection |
>
> **Examples of Research Questions**
> **RQ 1:** What is a flood resilient building code in Vlissingen and how can this be applied in the U.S?
> **RQ 2:** How can land use plans incoporate flood risk reduction and are they effective in reducing losses?
> **RQ 3:** How can a local governments produce housing policies that promote the flood safety of socioecoimically vulnerable populations?
> **RQ 4:** What are the impacts of climate change or extreme weather on the area, especially infrastructure networks?
> **RQ 5:** How do the Netherlands and Texas differ in dredging methods and practices?
> **RQ 6:** How can multiple local governments or agencies collaborate to secure the ship chanel safety/ecosystm protection?
> **RQ 7:** How do the Netherlands and Texas differ in flood risk perception and how does this influence their mitigation approach?

Fig. 2 Case study area description with examples of multidisciplinary research questions.

(1) Quality of the LOI/research plan
   – *Research question*: Is/are the research question(s) in the LOI/research plan well reasoned and relevant to the issues of the preferred case?
   – *Motivation*: Does the LOI/research plan describe the motivation for the preferred case study and is the motivation a likely basis for a good research project?
(2) Scientific approach
   – *Scientific contribution*: Is the proposed research project a significant contribution to the field of study?

- *Feasibility*: Will the proposed research project be feasible as part of the preferred case; can the necessary data be collected and the report written within the research time frame?
- *Disciplinary contribution*: Is the proposed research project a potential contribution from a multidisciplinary perspective to the group?

(3) Applicant's background
- *Cultural diversity*: Does the applicant contribute to the overall diversity of the program or this year's class?
- *Academic diversity*: Will the background and research interests of the candidate contribute to the preferred case study?
- *Potential for success*: Does the applicant's educational background, previous coursework, and/or GPA indicate that the applicant will succeed?

(4) Quality of presentation
- *Presentation*: Does the presentation of the LOI/research plan satisfactorily show overall quality? Are all fields in the application form filled in correctly? Is the text of the LOI/research plan well written?

After the review of the LOIs, around 30 undergraduate and graduate students are selected to submit a 900–1000-word research plan and two reference letters from their academic advisors. The same rubric is used for the second-round review of the research plans, but in the Scientific Approach category, reviewers focus more on whether the following factors are identified: (1) specifics on methodology to answer the proposed research questions, (2) data sets needed and how they are to be used, (3) possible sources for those data, and (4) opportunities to apply the expected results in the United States for flood resilience. The final review results in the selection of 16 students per year (8 undergraduate and 8 graduate) to participate in the program.

It is important to note that the application review process is systemically devised to secure multidisciplinarity and diversity among participating students, not only academically but also demographically.

Open and inclusive learning environment
Group research meetings and presentations
Throughout the program, students are placed in situations where they have to present their research plan, process, and results to a group of people with diverse cultural and academic backgrounds. The main audience is their cohorts and faculty mentors, but Dutch experts and stakeholders also attend these meetings to give students feedback and local knowledge on their case study research. While they prepare for the meetings and presentations, students learn how to make their research comprehensible for a diverse audience and also learn how other people look at the same problem from different perspectives. This is where transformative learning occurs through a frame of reference moving away from ethnocentrism and dualistic epistemologies and

acquiring a new perspective with which to approach the problem (Bell, Gibson, Tarrant, Perry III, & Stoner, 2016).

During these group meetings and presentations, students go through comprehensive immersive discourses on their research, guided by faculty mentors, researchers, and Dutch experts. Open discourse is important not only for convergent research and education by providing a variety of intellectual viewpoints on a certain issue but is also an essential feature of authentic learning environments that the curriculum aims to provide throughout the program because it helps students construct hypotheses to test (Duignan, 2012). In addition, the debates and brainstorming that occur in the discourse encourage students to talk, think, and explore their research topics in a stereoscopic way. Faculty mentors play a role here as guides rather than just "information providers," leading students to the path but not directly indicating the destination. In addition, the mentors focus on creating an open and inclusive atmosphere helping students feel free to talk about their research or comment on somebody else's work.

Multidisciplinary design workshop
The design workshop is a 1-day long integrative research activity held at the actual project location of an innovative flood protection project along the Dutch coast.[c] Participating students from US institutions collaborate with a group of 10–15 multidisciplinary students from several Dutch universities. The workshop starts with talks by local stakeholders and experts. Then, the students are split into multidisciplinary binational teams consisting of 5–6 members and collaborate on envisioning a design to address the complex flood-related local problems. The workshop provides an authentic learning environment given existing flood issues in the local area and a collaborative problem-solving process. Two extra learning aspects are included in the workshop: (1) students are introduced to a Dutch-specific design-concept called Building-with-Nature, which they must incorporate into their design. This change of perspective stimulates transformative learning (Mezirow, 1997); (2) students learn about the different roles of experts representing diverse disciplines and get a chance to conduct a self-assessment to discover which role fits them best. During the workshop, they practice various roles and experience culturally different interpretations of these roles.

The on-site execution of the workshop offers students a more conducive learning environment. In addition, allowing Dutch and US students to collaborate in producing an actual design enables them to acknowledge cultural differences in approaching problems. On top of that, this collaboration facilitates the development of a potential international professional network for both the United States and Dutch students.

---

[c] See also Chapter 27, "Measuring the educational effects of problem- and place-based research education programs: The student survey".

The design workshop is an effective educational tool to help student participants learn about and experience convergent research. Not only do US students get to interact with international students and faculty, but they also communicate with stakeholders, including local residents, authorities, and private sectors, for problem solving.

Lessons learned

To date, four of the seven planned research trips have been completed and each year, significant improvements are made to different parts of the program based on lessons learned from previous years.

First, there was a need for a major improvement in the application and review system. For the first 2 years of the program, there was no rubric for the evaluation of applications. The committee assessed the applications based on their own judgment without clear rating criteria. On top of that, the committee had to review, all at once, a 30-page application packet of each applicant that included a research proposal and other supplemental documents, which resulted in an overwhelming review process that made it hard for the committee to evaluate the applications from various aspects. To address these issues, the application was changed to a two-step process, which allowed the reviewers to spend more time on each application. In addition, we developed and refined a review system equipped with clear criteria that specifically reflects the multidisciplinarity and diversity of the applications and applicants on top of measuring the systemic quality of the research proposal. This modification of the application and review system played a pivotal role in increasing the number of underrepresented participants and the inclusion of students from a wider variety of disciplines.

Second, the case studies, which change each year, have become more integrative. We found that students from certain disciplines tended to concentrate on specific case studies, which hinders the convergent research and transformative learning effects among students. To avoid this problem, the description of the case study areas is presented more inclusively and more broadly so it can cover a wider range of local issues and research topics that can be approached by various disciplines.

Third, the program struggled with some students who lost interest in finishing their research work after coming back from the Netherlands. The final product required is the submission of a report and a poster, which students work on for 3 months after the research trip. Since this program is not an actual course for which they get credit or has other consequences, some (but not many) students were reluctant to finalize their work. To fix this problem, additional resources have been allocated to implement a "carrot-and-stick" strategy. The stick is to hold an annual student research symposium where students are required to present their work to receive a certificate of program completion. The carrot is to fund six top-performing students to attend an international conference to present their work. We hope this strategy will encourage students to engage even more in the later phase of the program and improve the quality of their final work.

Conclusions

Convergent research and education through transformative and authentic learning can be successfully implemented if it is carefully designed by incorporating the integration of students from various disciplines, ensuring diversity, and creating an open and inclusive learning atmosphere. Well-structured international integrative research and education programs could significantly benefit students by providing career-transforming insights, resources, and connections that go far beyond regular classroom teaching. This chapter describes the design process of the NSF PIRE CFRR Program as a vehicle for offering transformative education to students through problem- and place-based learning in authentic learning environments. Built on our 4 years of experience, we have been continuously improving and modifying the contents of the program to provide the best learning opportunities to participating students. The program has demonstrated how to break down existing disciplinary silos and develop the skillsets needed to address growing flood risks through convergent research and education. Upon completion of the program, participating students will have obtained a deep understanding of necessary connections between different disciplines to address major societal challenges, which will make them valuable future research leaders and practitioners with a global perspective on problem solving.

Acknowledgments

This work was supported by the US National Science Foundation Partnership for International Research and Education (Grant no. OISE-1545837). The authors are grateful for the unwavering support and commitment provided by our partner institutions: Texas A&M University, Texas A&M University at Galveston, Rice University, Jackson State University, Delft University of Technology and Vrije Universiteit Amsterdam.

References

Bell, H. L., Gibson, H. J., Tarrant, M. A., Perry, L. G., III, & Stoner, L. (2016). Transformational learning through study abroad: US students' reflections on learning about sustainability in the South Pacific. *Leisure Studies, 35*(4), 389–405.

Bolli, T., Renold, U., & Wörter, M. (2018). Vertical educational diversity and innovation performance. *Economics of Innovation and New Technology, 27*(2), 107–131.

Booth, A. (2011). 'Wide-awake learning' Integrative learning and humanities education. *Arts and Humanities in Higher Education, 10*(1), 47–65.

Clapton, G., Cree, V. E., Allan, M., Edwards, R., Forbes, R., Irwin, M., et al. (2008). Thinking 'outside the box': A new approach to integration of learning for practice. *Social Work Education, 27*(3), 334–340.

Duignan, P. (2012). *Educational leadership: Together creating ethical learning environments.* Cambridge University Press.

Herr, D. J., Akbar, B., BrummetF, J., Flores, S., Gordon, A., Gray, B., et al. (2019). Convergence education—An international perspective. *Journal of Nanoparticle Research, 21*(11), 1–6.

Hong, L., & Page, S. E. (2004). Groups of diverse problem solvers can outperform groups of high-ability problem solvers. *Proceedings of the National Academy of Sciences, 101*(46), 16385–16389.

Jamison, A., Kolmos, A., & Holgaard, J. E. (2014). Hybrid learning: An integrative approach to engineering education. *Journal of Engineering Education, 103*(2), 253–273.

Kitney, R., Brummet, J. L., BCS, N., Gordon, M. A., Gray, U. D. B., & EFMA, N. (2017). Global perspectives in convergence education (workshop report). In *National Science Foundation*. Retrieved from www.nsf.gov/nano/ConvergenceEducation.

Kulasegaram, K. M., Martimianakis, M. A., Mylopoulos, M., Whitehead, C. R., & Woods, N. N. (2013). Cognition before curriculum: Rethinking the integration of basic science and clinical learning. *Academic Medicine, 88*(10), 1578–1585.

Lee, Y., Kothuis, B. B., Sebastian, A., & Brody, S. (2019). Design of transformative education and authentic learning projects: Experiences and lessons learned from an international multidisciplinary research and education program on flood risk reduction. In *2019 ASEE annual conference & exposition*.

Mezirow, J. (1997). Transformative learning: Theory to practice. *New Directions for Adult and Continuing Education, 1997*(74), 5–12.

National Research Council. (2014). *Convergence: Facilitating transdisciplinary integration of life sciences, physical sciences, engineering, and beyond*. National Academies Press.

CHAPTER 27

Measuring the educational effects of problem- and place-based research education programs: The student survey

Yoonjeong Lee[a,b] and Baukje Bee Kothuis[c,d]
[a]Institute for a Disaster Resilient Texas, Texas A&M University, College Station, TX, United States
[b]Department of Marine and Coastal Environmental Science, Texas A&M University, Galveston Campus, Galveston, TX, United States
[c]Department of Hydraulic Engineering and Flood Risk, Faculty of Civil Engineering and Geosciences, Delft University of Technology, Delft, The Netherlands
[d]Netherlands Business Support Office, Houston, TX, United States

Introduction

The evaluation the effectiveness of a research education program is vital to ensure its success. Although some valuable feedback can be obtained through discussions with participating students and faculty mentors, it is crucial to empirically evaluate the impact of the program to clarify and document the contribution that it is making to the achievement of educational goals (Fien, Scott, and Tilbury, 2001). This study examined the educational and learning effects of the program based on the results of a survey of 56 participating students.

The main purpose of the survey was to address the following questions:
(1) Does the 2-week research trip to the Netherlands significantly increase students' general knowledge of flood risk?
(2) Does the multidisciplinary approach of the program have a significant impact on changing and diversifying students' perspectives?

A total of 56[a] students (21 undergraduate and 35 graduates) responded to the questionnaire comprising of both structured and open-ended questions. The survey is designed to quantify how well the program is achieving its goal and consists of 23 items asking about students' flood risk perception, knowledge, and feedback on the program itself. Six items were constructed to measure the degree of flood-related knowledge of the students and how diverse perspectives lead to different problem-solving approaches. Using a pretest-posttest design, student learning over the 2-week research trip in the Netherlands was tracked and analyzed quantitatively to see whether the program led to significant differences in knowledge level and perspectives of the students.

---

[a] A total of 58 students have participated in the program but the data of two students were dropped because they were not able to complete the program.

The results show that the 2-week long problem- and place-based research education program does increase the level of knowledge of the participating students in general. The results also indicate the students acquired a perspective transformation.

The following section briefly reviews literature that contributed to the fundamental concepts of the program: transformative learning and authentic learning. Next, we describe the method used in the study including survey data description and analysis. Then, we report on the statistical results of paired t-tests assessing changes in knowledge and perspectives of the students. This chapter concludes with implications of our findings in terms of improving research education programs for authentic and transformative learning experiences at the university level.

Transformative and authentic learning and education
Transformative learning and education

Transformative learning (TL) can be defined as a process of change in a frame of reference, achieved when the transformation occurs as a result of experience or by acquiring a new perspective (Mezirow, 1997, 2003; Strange & Gibson, 2017). In Mezirow's Transformative Learning Theory (Mezirow, 1997), "frames of reference" is defined as the structure of assumptions that constitutes a person's cognitive habits and points of view. These fixed assumptions form a set of codes that can be influenced by the cultural, social, and educational environment of individuals. TL aims to encourage learners to question and transform these assumptions, the ways they see and think about the problem, and enable them to deepen their understanding of that particular topic (UNESCO, 2017).

In higher education, TL is considered as a primary objective and important outcome that enables students to think critically and analytically, while enhancing communication and collaboration skills, and global understanding (Calleja, 2014; Nichols, Choudhary, and Standring, 2020; Strange & Gibson, 2017). Furthermore, students are encouraged to create new meanings of a problem from an altered view and identity. The importance of TL is widely recognized in various areas and United Nations Educational, Scientific, and Cultural Organization (UNESCO) designated TL as a key element of learning approaches for the 2030 sustainable development agenda (Harder, Dike, Firoozmand, Des Bouvrie, and Masika, 2021; UNESCO, 2017). The pivotal role of TL is getting more attention as the importance of multidisciplinary approaches to address complex natural and societal challenges grows. In this sense, TL should be properly incorporated when designing an international multidisciplinary research and education program to provide students with an opportunity to recognize various perspectives in approaching a complex problem such as flooding.

When implementing TL, students should be presented with a challenging problem in an unfamiliar environment with other students they can relate with in the same process and educators or lecturers should challenge them to discuss with their peers (Christie,

Carey, Robertson, and Grainger, 2015). This approach may evoke a process of seeing, acting, or thinking outside their comfort zone (Perry III, 2011). Since a shift of perspective can be achieved by changing culturally entrenched meaning structures, even short-term foreign experiences—combined with strong academic content—can lead to transformation (Bell, Gibson, Tarrant, Perry III, and Stoner, 2016). Furthermore, a direct teacher intervention may occur to ensure changing frames of reference to help students develop insights about different perspectives and to encourage participation in critical dialectical discourse. These attributes are crucial for constructing a new perspective (Mezirow, 2003).

Authentic problem- and place-based learning and education

Authentic learning is widely recognized as an effective approach based on real-world problems that are closely related to a specific field (Herrington, Reeves, & Oliver, 2014). This approach enables students to constantly experience and engage with interdisciplinary problems in a real-world situation, using active learning pedagogies such as fieldwork, group work, and rigorous discourse, in a way that emphasizes the complexities of tackling challenging problems (Cross & Congreve, 2020). Debate is also considered an essential feature of authentic learning environments because it enables students to construct hypotheses, test them against what they think is true, and view knowledge and information from multiple perspectives.

In authentic learning environments, the role of the teacher changes from information provider and test maker to learning guide and problem presenter by demonstrating "care" or "passion" for a subject and by motivating students to care as well. Because students regulate this learning process themselves, they are encouraged to think, discover, and become more reflective practitioners (Cross & Congreve, 2020; Duignan, 2012). This kind of learning, presented as an iterative discovery process around an authentic task, enables students to conceptualize the nature of a complex problem and allows students to develop solution-focused thinking, problem-solving skills, and confidence in their own learning ability.

One of the best ways to implement authentic learning is the use of problem-based learning (PBL) techniques. In PBL, students are presented with a challenging problem and required to analyze and find solutions to the problem (Zamroni, Hambali, & Taufiq, 2020). In this process, a team of students shares the problem with the aim of solving it collectively, leading to a greater level of responsibility, competence, and learning outcomes (Donnelly, 2006; Friedman & Deek, 2002). A key point of PBL is linking theoretical knowledge to practical application by working together in mixed disciplinary groups in which students decide for themselves what to learn.

To combine authentic learning and PBL, a problem can be presented in a case study setting that invites students to investigate, analyze, and solve problems in collaborative groups (Cockrell, Caplow, and Donaldson, 2000). Some notable features of such case studies make them an ideal strategy to facilitate authentic learning. First, a case is based on a real situation or event that forces students to think through problems they may encounter in the workplace. Second, the case study is designed and developed through careful research and study involving local experts and stakeholders. Third, and most importantly, a case provides learning opportunities at different levels for both those involved in designing the case and those who may be involved with the case (Wallace, 2001).

Another key application of authentic learning is the use of place-based learning. Place-based education is based on the principles of authentic learning and applies them to a particular spatial environment, for example, a floodplain or vulnerable community. Collaborative learning is then tailored to the local context in which students can experience a specific problem first-hand, how it affects their own lives, and the actions needed to address the problem. In these situations, students have the ability to produce rather than consume, teachers act as guides rather than just instructors, and groups work together to develop a set of strategies to address a real problem (Smith, 2000).

Methods

This study uses a pretest-posttest research design to measure and explain the changes in knowledge and perspectives of 56 participating students of the program from 2016 to 2019. Comparing pre- and postintervention student survey responses is the most common evaluative approach for an education program assessment (Carleton-Hug & Hug, 2010; Stern, Powell, and Hill, 2014). The data were collected using student surveys, conducted before and after the 2-week research trip in the Netherlands. A pretrip questionnaire was given to the students on the plane going to the Netherlands, where no Internet access was available; the posttrip survey was administered right after the last group meeting of the research trip and students were instructed not to use any resources while they answered the questionnaire. The total of 56 students consisted of 21 undergraduate, 12 Master's, and 23 PhD students, with an age ranging from 20 to 52 years old. Respondents came from five different US institutions and diverse disciplines, such as engineering (25), social science (12), natural science (9), and architecture/urban planning (10). The students had a variety of educational and professional backgrounds.

The educational objectives of the research trip in the Netherlands are to provide students with:

(1) A comprehensive and integrative research and education experience to produce a diverse generation of researchers and practitioners equipped to solve societal challenges of increasing flood hazards.

(2) A deep understanding of the necessary connections between different disciplines and the ability to approach an issue with a varied and holistic perspective.

In general, the effectiveness of environmental education programs can be measured by changes in students' knowledge, skills, awareness, attitudes, and behavior (Stern, Powell, and Hill, 2014). For this study, change in students' perspectives on flood issues was measured along with increases in baseline flooding knowledge. We measured each student's knowledge level by including quiz type of questions related to general and specific flood issues in the pretrip and posttrip surveys and a scoring rubric was created to grade the responses. The score of baseline knowledge was calculated by aggregating the points given to the answers according to the rubric. The questions asked students if they can list or address any examples of flood mitigation strategies, the main differences of flood risk mitigation between the United States and the Netherlands, and the major drivers of flood risk in two countries. Each mitigation example was given up to three points if a student had provided specific information (name and location of the mitigation example). For the question about the differences in risk mitigation approach between two countries, the answer was given up to four points if a student had clearly stated and explained the differences using proper examples. Additionally, students were asked to suggest flood risk mitigation ideas for each case study area in both countries according to its flood risk drivers. Not only was the knowledge measured, but also the comprehensiveness of students' perspectives was assessed using the mitigation ideas they had suggested. We measured the degree to which their approaches to solving flood problems were integrative based on the number and scope of mitigation responses: if a respondent proposed more than three mitigation strategies that included both nonstructural and structural or both engineering and nonengineering approach (e.g., Flood risk drivers: storm surge and land use pattern; Mitigation ideas: building a storm surge barrier and adopting a zoning system), the answer would get the highest possible score (see Appendix for the questionnaire used for analysis and employed rubric to score them).

Participation in the program requires a case study research proposal addressing a specific issue related to flood risk reduction. Students are advised to revise and update their research plan by completing a literature review during the pretrip period. Thus, it was expected that students develop a baseline knowledge level prior to traveling to the Netherlands.

As the same questions were asked before and after the research trip to the Netherlands, we used a paired *t*-test of means to assess the change in the average score of students' responses between the pretrip and posttrip survey. A paired *t*-test is also called a repeated-measures *t*-test since it is used to measure one group of people at two different points in time in order to determine the mean difference between the two sets of observations (Acock, 2008).

Results

Improved knowledge. Based on the results reported in Table 1, students' knowledge level showed significant change comparing the means of the pretrip (pretest) and posttrip (posttest) scores. The total average score of the flood-related knowledge of the students changed from 42.90[b] to 51.61, a statistically significant 16.88% increase ($P<0.01$). This result provides an initial indication of the educational effects of the 2-week long research trip to the Netherlands on individual levels of flood-related knowledge. The results of paired t-test of each component of the knowledge questions follow.

Major flood risk drivers of case study areas: Students were asked to select two major drivers (among "rainfall," "storm surge," "subsidence," "sea level rise," "land use patterns," "social vulnerability," and "climate change") of flood risk in case study areas (Houston-Galveston metropolitan area and the Netherlands). This question was intended to measure students' basic understanding of the "problem" (floods) in the "place" (case study areas) that the program is based on. We expected the posttrip score would be higher than pretrip, however, the analysis did not show any significant change. There was a slight increase in the average score of the drivers of the Netherlands but not statistically significant.

Differences in flood risk mitigation approach between the United States and the Netherlands: The second component was an open-ended question that asked students if they could address any differences between the two countries when it comes to flood risk mitigation

Table 1 The paired t-tests of pretrip and posttrip survey.

Variable	Pre-test (Mean)	Post-test (Mean)	t-value	P-value
Total average score	42.95	51.61	−3.63	0.0006
Knowledge				
Flood risk drivers (US)	45.76	41.52	1.21	0.23
Flood risk drivers (Dutch)	30.58	33.06	−0.67	0.51
Differences in mitigation approach (US and Dutch)	61.16	68.30	−2.37	0.02
Flood mitigation examples	53.57	51.61	0.69	0.49
Flood mitigation ideas	39.05	51.24	−3.69	0.0005
Perspective				
Integrative mitigation ideas	18.17	34.95	−4.70	0.0000
Self-assessed knowledge				
Knowledge level	4.21	5.34	−7.80	0.0000

[b] Highest possible score for the knowledge level was 69 for the 2016 and 2017 cohort, and 62 for 2018 and 2019 cohort. Each respondent's score was divided by the highest possible score and multiplied by 100 for generalization.

approach. The results showed 11.67% increase ($P<0.05$) in the score, which indicates that the research and learning activities in the Netherlands for 2 weeks were informative and enabled students to realize that two different countries would take a different approach to mitigate flood risk.

Mitigation examples: Another open-ended question asked students to list five examples of flood risk mitigation strategies and the result showed a statistically significant increase in the score of this item. This question did not restrict areas or types of mitigation measures, so students were able to freely list any mitigation examples no matter where they are located or how they mitigate the flood risk. However, surprisingly, there was no statistically significant knowledge change on this component.

Mitigation ideas according to the major drivers in the area: The last component of the questions attempted to measure students' ability to connect the problems of places (flood risk drivers of each case study area) and possible solutions (flood mitigation ideas). The score showed a clear increase of 32.14% ($P<0.001$) on this question, indicating the positive educational effects of the problem- and place-based research education program.

Changed perspectives: Comprehensive and integrative problem-solving approach. Flood mitigation ideas addressed by students were graded to see if students' perspectives had changed to more integrative or perspective after participating in the research trip to the Netherlands. A response with more than three mitigation ideas that include both structural and nonstructural or both engineering and nonengineering approaches received the highest possible score. The average score for this item has increased by 93.15% ($P<0.001$).

Other findings: Self-reported knowledge level and the most helpful program features. In addition to the questions about the knowledge and perspective, students were asked to self-measure their current knowledge level on the flood-related topics and answer from 1, meaning "very low," to 7, meaning "very high." The result shows a statistically significant 27% ($P<0.001$) increase in the score, meaning students feel that they have a better understanding of the topic after the program, no matter how much their actual knowledge has increased. Also, we asked students to rank the program components ("field trips," "connecting to experts," "literature review," "lectures," "breakfast meetings," "individual research time," and "free time") based on how much they can contribute (pretrip)/how much they actually contributed (posttrip) to their knowledge about flood risk reduction. The majority of the students selected "field trips" and "connecting to experts" for the most helpful components before and after the trip. In total, 17 students ranked the field trips the first, and 16 students ranked the second for the pretrip survey, and the number slightly increased after the trip to 18, and 19 students ranked the field trips first and the second. Regarding connecting to experts, 25 students ranked it the first and 16 students the second. After the trip, it also shows a negligible increase to 27 and 18 students. This descriptive analysis implies that the program has met the students' expectations of the place-based and authentic learning approach.

Discussion

All survey responses, except ones asking about major flood risk drivers and the examples of flood mitigation, showed a clear improvement after the research trip. The most notable change was found in mitigation ideas, suggesting a substantial increase in flood-related knowledge. It is important to note that this question measured not only students' understanding of major drivers of flood risk in each case study area, but also their ability to come up with mitigation ideas to alleviate the food risk. It requires students to have contextual knowledge of the place and the ability to think comprehensively to present the solutions accordingly. This result implies that the transformative and immersive characteristics of the program helped students to approach a problem with a more comprehensive and integrative perspective that is beyond their own academic discipline by communicating closely with people with diverse backgrounds: other students, faculty, Dutch experts, and Dutch local stakeholders.

Another noteworthy finding is from the results of descriptive analysis. Students reported that the "field trips" and "connecting to experts" were the most helpful components of the program, which indicates the distinguished effects of a place-based and authentic learning approach in comparison with a traditional in-class approach in which learning opportunities like visiting actual places and talking with Dutch experts in person cannot be offered.

By completing a 2-week long problem- and place-based program, students became more knowledgeable and their approach to the problem became more integrative as well. This result shows a clear contrast with the result of the item that asked students to list general flood mitigation examples without any limitation of location or types. Although the results indicate that the students learned thoroughly about local specific flood issues and to suggest its associated possible solutions, there was no statistically significant improved knowledge on the universal flood mitigation measures. This might imply that when it comes to designing a problem- and place-based program, it is important to incorporate a way to guide students to see a problem not only in a horizontally diverse perspective—multidisciplinary approach, but also in a vertically diverse perspective—in a different scale: local, regional, national, and global level.

Conclusions

The results of this study suggest that a problem- and place-based research education program offering immersive and transformative training could significantly improve students' knowledge and help students to approach a problem with a more integrative and holistic perspective in only 2 weeks if the program is properly designed. Findings support that the activities of the 2-week long research trip such as field trips, lectures, and a series of immersive discourses with other students from a variety of disciplines, led by a multidisciplinary group of faculty mentors could offer substantial knowledge increases and an opportunity to learn how to approach a problem with different perspectives. Another notable finding that the analysis offers concerns the students' self-assessed level of knowledge on flood risk. The result of the pretest-posttest analysis shows that students think their knowledge has increased significantly after the 2-week long research trip to the Netherlands. This finding might be considered more important than the actual knowledge score increase because it indicates that in 2 weeks, students have gained not only knowledge but also confidence in having a broad base of experience, which gives them more room for academic growth in the future.

Future studies should include an untreated control group (traditional in-class, lecture and reading focused learning group) for analysis to have a better understanding of the educational effects of this program, as the actual differences between the study group and the control group can thus be assessed. On top of that, a follow-up study after a certain time that has elapsed can be conducted by tracking down students' career paths and see if the program has any substantial impacts on their capability to deal with real-world problems in a professional or academic setting. Furthermore, within this program, supporting faculty members gathered multiple times to discuss the effectiveness of the diverse program items and methods. This was done in group meetings during and after the research trip to the Netherlands. To be able to further evaluate and standardize this valuable information, future studies could include a survey for supporting faculty mentors to inquire about their experience and insights into students' learning progress and processes, methods and other relevant issues concerning the effectiveness of the program.

Appendix: Scoring Rubric for NSF PIRE CFRRP Student Survey[c]

Question 3. What are the primary drivers of flood risk in the Houston-Galveston region? (1: least important ~7: most important)   [2 points total]

Criteria:
- Allow 2 points if the student places "Rainfall" and "Land Use Patterns" in 5, 6, or 7 (most important).
- Allow 1 point if only one of them (Rainfall or Land Use Patterns) is placed in 5, 6, or 7.

No credit if "Rainfall" or "Land Use Patterns" is placed in 1, 2, or 3.

---

[c] This rubric was created using an example rubric provided by New York State Alternative Assessment in Science Project (NYSED, n.d.). Retrieved from https://pals.sri.com/tasks/9-12/Testdrug/rubric.html.

Question 3-1. Do you think there are any others? [2 points total]
   Criteria:
- Allow 2 points if the student states two or more other drivers.
- Allow 1 point if the student states only one other driver.
- No credit if the student does not answer or states something completely irrelevant.

Question 4. What are the primary drivers of flood risk in the Netherlands? [2 points total]
   Criteria:
- Allow 2 points if the student places "Rainfall" and "Storm Surge" in 5, 6, or 7 (most important).
- Allow 1 point if only one of them (Rainfall or Storm Surge) is placed in 5, 6, or 7.
- No credit if "Rainfall" or "Storm Surge" is placed in 1, 2, or 3.

Question 4-1. Do you think there are any others? [2 points total]
   Criteria:
- Allow 2 points if the student states two or more other drivers.
- Allow 1 point if the student states only one other driver.
- No credit if the student does not answer or states something completely irrelevant.

Question 5. What in your opinions are the main differences between Dutch and American flood risk mitigation? [4 points total]
   Criteria:
- Allow 4 points for clearly stating and describing the differences between two countries with examples or detailed explanations.
- Allow 3 points for clearly stating and describing the difference between two countries without examples or detailed explanations.
- Allow 2 points for simply listing examples of different approaches between two countries without explanations.
- Allow 1 point if the student vaguely or unclearly states differences between two countries (or guessing).
- No credit if the student does not answer or states something completely irrelevant.

Question 6. List 5 examples of mitigation strategies that are innovative and where they have been applied (not limited to the United States or the Netherlands). [15 points total—3 per example]
   Criteria:
- Allow 3 points if the student states a specific mitigation example and its location (e.g., Maeslant barrier, Rotterdam).
- Allow 2 points if the student states a somewhat general example and its location (e.g., storm surge barrier, the Netherlands).

- Allow 1 point if the student states an example but missing its location (e.g., Maeslant barrier/storm surge barrier).
- No credit if the student does not answer or states something completely irrelevant.

Question 7. As part of NSF PIRE CFRRP, we have identified 5 or 6 case study areas: 2 or 3 in the United States and 2, 3, or 4 in the Netherlands. Based on your *existing* knowledge of each area, please list what you view as the primary driver of flood risk. List two to three ideas for mitigation in each area.   [25–30 points total—5 point per case]

   Criteria:
- Allow 5 points if the student states two or more primary drivers and two or more mitigation ideas that can mitigate the stated drivers—the drivers and mitigation ideas are matched (e.g., Drivers: storm surge; Mitigation idea: building a storm surge barrier/ Drivers: land use pattern; Mitigation idea: adopting a zoning system).
- Allow 4 points if the student states:
  - Two or more primary drivers and two or more mitigation ideas but they are randomly listed—the drivers and mitigation ideas are not matched (e.g., Drivers: land use pattern; Mitigation idea: building a storm surge barrier); **or**
  - Two or more drivers with only one matching mitigation idea.
- Allow 3 points if the student states one primary driver and two or more mitigation ideas that can mitigate the stated drivers—the drivers and mitigation ideas are matched.
- Allow 2 points if the student states:
  - One or more primary driver and one or more mitigation ideas but they are randomly listed—the driver and mitigation idea are not matched;
    or
  - One primary driver with only one matching mitigation idea
- Allow 1 point if the student states either primary drivers or mitigation ideas, but not both.
- No credit if the student does not answer or states something completely irrelevant.

<Mitigation ideas>   [10–12 points total]
- Allow 2 points if the student states three or more mitigation ideas that include both nonstructural and structural or both engineering and plan/policy approaches.
- Allow 1 point if the student state two mitigation ideas that include both nonstructural and structural or both engineering and plan/policy approaches.
- No credit if the mitigation idea(s) are solely nonstructural, structural, engineering or plan/policy-based, or no mitigation ideas are mentioned.

Highest possible score   62 points (2019/2018); 69 points (2017/2016)

References

Acock, A. C. (2008). *A gentle introduction to Stata*. Stata Press.

Bell, H. L., Gibson, H. J., Tarrant, M. A., Perry, L. G., III, & Stoner, L. (2016). Transformational learning through study abroad: US students' reflections on learning about sustainability in the South Pacific. *Leisure Studies, 35*(4), 389–405.

Calleja, C. (2014). Jack Mezirow's conceptualisation of adult transformative learning: A review. *Journal of Adult and Continuing Education, 20*(1), 117–136.

Carleton-Hug, A., & Hug, J. W. (2010). Challenges and opportunities for evaluating environmental education programs. *Evaluation and Program Planning, 33*(2), 159–164.

Christie, M., Carey, M., Robertson, A., & Grainger, P. (2015). Putting transformative learning theory into practice. *Australian Journal of Adult Learning, 55*(1), 9–30.

Cockrell, K. S., Caplow, J. A. H., & Donaldson, J. F. (2000). A context for learning: Collaborative groups in the problem-based learning environment. *The Review of Higher Education, 23*(3), 347–363.

Cross, I. D., & Congreve, A. (2020). Teaching (super) wicked problems: Authentic learning about climate change. *Journal of Geography in Higher Education*, 1–26.

Donnelly, R. (2006). Blended problem-based learning for teacher education: Lessons learnt. *Learning, Media and Technology, 31*(2), 93–116.

Duignan, P. (2012). *Educational leadership: Together creating ethical learning environments*. Cambridge University Press.

Fien, J., Scott, W., & Tilbury, D. (2001). Education and conservation: Lessons from an evaluation. *Environmental Education Research, 7*(4), 379–395.

Friedman, R. S., & Deek, F. P. (2002). Problem-based learning and problem-solving tools: Synthesis and direction for distributed education environments. *Journal of Interactive Learning Research, 13*(3), 239–257.

Harder, M. K., Dike, F. O., Firoozmand, F., Des Bouvrie, N., & Masika, R. J. (2021). Are those really transformative learning outcomes? Validating the relevance of a reliable process. *Journal of Cleaner Production, 285*, 125343.

Herrington, J., Reeves, T. C., & Oliver, R. (2014). Authentic learning environments. In *Handbook of research on educational communications and technology* (pp. 401–412). Springer.

Mezirow, J. (1997). Transformative learning: Theory to practice. *New Directions for Adult and Continuing Education, 1997*(74), 5–12.

Mezirow, J. (2003). Transformative learning as discourse. *Journal of Transformative Education, 1*(1), 58–63.

New York State Alternative Assessment in Science Project (NYSED). (n.d.). *Performance assessment links in science*. Retrieved July 13, 2020, from: https://pals.sri.com/tasks/9-12/Testdrug/rubric.html.

Nichols, M., Choudhary, N., & Standring, D. (2020). Exploring transformative learning in vocational online and distance education. *Journal of Open Flexible and Distance Learning, 24*(2), 43–55.

Perry, L. G., III. (2011). *A naturalistic inquiry of service-learning in New Zealand university classrooms: determining and illuminating the influence on student engagement: A thesis presented to the Faculty of the College of Education, University of Canterbury, in partial fulfillment of the requirements for the degree of Doctor of Philosophy*. University of Canterbury.

Smith, P. J. (2000). Flexible delivery and apprentice training: Preferences, problems and challenges. *Journal of Vocational Education and Training, 52*(3), 483–503.

Stern, M. J., Powell, R. B., & Hill, D. (2014). Environmental education program evaluation in the new millennium: What do we measure and what have we learned? *Environmental Education Research, 20*(5), 581–611.

Strange, H., & Gibson, H. J. (2017). An investigation of experiential and transformative learning in study abroad programs. *Frontiers: The Interdisciplinary Journal of Study Abroad, 29*(1), 85–100.

UNESCO. (2017). Division for Inclusion, Peace and Sustainable Development, Education Sector *Education for sustainable development goals: Learning objectives*.

Wallace, J. (2001). *Introduction: Science teaching cases as learning opportunities*. Springer.

Zamroni, E., Hambali, I. M., & Taufiq, A. (2020). Does problem base learning effective to improve decision making skills student? In *6th international conference on education and technology (ICET 2020)* (pp. 139–144). Atlantis Press.

CHAPTER 28

A specific transdisciplinary co-design workshop model to teach a multiple perspective problem approach for integrated nature-based design

Jill H. Slinger[a,b] and Baukje Bee Kothuis[c,d]
[a]Faculty of Technology, Policy and Management, Delft University of Technology, Delft, Netherlands
[b]Institute for Water Research, Rhodes University, Makhanda, South Africa
[c]Department of Hydraulic Engineering and Flood Risk, Faculty of Civil Engineering and Geosciences, Delft University of Technology, Delft, The Netherlands
[d]Netherlands Business Support Office, Houston, TX, United States

Introduction

Interdisciplinary, place-based learning by international groups of students formed an integral component of the Partnerships for International Research and Education Coastal Flood Risk Reduction (PIRE-CFRR) program "Integrated, multiscale approaches for understanding how to reduce vulnerability to damaging events." The program aimed to create "authentic learning environments" that supported and benefitted from ongoing research efforts related to flood risk management. The challenges lay in designing such environments to accommodate the place-based and contextual nature of flood risk management, to integrate across multiple disciplinary fields, and to complement the diverse educational backgrounds and programs from which the staff and students in PIRE-CFRR were drawn. Moreover, the program sought to learn collectively about new approaches to flood risk reduction through innovative nature-based infrastructure design. Such nature-based solutions are characterized by disciplinary integration, including multiple perspectives in the determination of design requirements, and long-term time frames that balance the limitations of the Earth's natural systems and the socio-technical systems created by humans (Klaassen, Kothuis, & Slinger, 2021). The infrastructural artifacts reflect these characteristics in their form (Slinger & Vreugdenhil, 2020) and are sometimes designed to disappear over time, e.g., the Sand Engine in South Holland (Bontje & Slinger, 2017; Stive et al., 2013). The novelty of the nature-based solution concept presented an additional challenge to the design of an appropriate learning environment.

We report on two transdisciplinary workshops undertaken within the PIRE-CFRR program to teach a multiple perspective problem approach for integrated nature-based design, and examine their efficacy. The first workshop in May 2016 focused on the ebb-tidal delta offshore of the southwestern corner of Texel, an erosion hotspot on the Dutch coast, whereas the second workshop in June 2017 focused on the anticipated failure of the Hondsbossche Pettermer Sea Dike to continue to meet Dutch flood protection standards in the future (Fig. 1). The potential to apply nature-based solutions in managing these flood risks was a hot topic in the Netherlands at the time, making these cases attractive choices for teaching (see Pruyt, Slinger, van Daalen, Yucel, & Thissen, 2009).

Both workshops were convened and facilitated by the authors. The effects of these learning interventions are reported and analyzed in terms of (i) the co-design workshop process, (ii) the substantive outcomes, and (iii) evidence of learning at the individual level. The (shared) changes in understanding of (engineering) roles in a design team are reported in a publication by Klaassen et al. (2021). This chapter will not focus on the analysis of shifts in individual design roles nor on the design process followed by each of the student teams. Instead, the sequence of activities comprising the workshop process, the design outcomes of the transdisciplinary workshop, and the efficacy of the workshop method for achieving learning outcomes form the focus of the study.

First, we provide a theoretical background on problem-based learning and authentic learning pedagogies, on the design of participatory workshops within a policy process, and on nature-based solutions in hydraulic engineering (Section "Theoretical background"). After a brief description of the method (Section "Method"), we then describe the workshops in terms of the co-design process followed and their knowledge content (Section "Results"). The effects of the workshops are then evaluated in terms of the learning outcomes (Section "Learning outcomes") and the chapter concludes with a reflection.

Fig. 1 The sandy southwestern coast of the island of Texel separated by a narrow channel from the ebb-tidal delta *(left)*, and the old Hondsbossche Pettermer Sea Dike and groynes *(right)*.

Theoretical background

Standard teaching practices in traditional classroom environments focus on transferring formalized knowledge (textbooks, exercises) from the expert (the teacher) to novices (the students) and have long been criticized as lacking (i) the authentic problem contexts essential for effective learning (Schmidt, 1993) and (ii) the collaborative environment in which students can learn together by exploration (Duignan, 2012). Alternative forms of education have been developed and applied to address these issues. For instance, problem-based learning (Barrows, 1985, 1992) is a pedagogical approach in which students are challenged to solve an open-ended problem. The learning activity centers on realistic complex case study material associated with a particular local setting. The local setting can be a specific geographical area or biogeophysical environment or can encompass a specific social setting such as a community. The problem-based learning process does not focus on solving the case study problem via a preexisting or defined solution but encourages the development of skills such as critical appraisal, problem structuring, literature review, creative design, and iterative reflection and synthesis. The process involves working in small groups of learners. Students collaboratively identify what they know, what they need to know, and how to develop new knowledge to resolve the case study problem. The role of the teacher is envisaged as supporting, guiding, and monitoring the learning process. Problem-based learning originated in the medical sciences but has gained ground in the fields of design studies, engineering, and the natural sciences (Nicaise, Gibney, & Crane, 2000).

In a parallel development, the authentic learning pedagogy concentrated on teaching students to undertake complex and realistic tasks through situated cognition. The aim is for students to develop problem-solving skills and robust knowledge that transfers to real-world practice in a particular field of study (Herrington, Reeves, & Oliver, 2014). The design of an authentic learning environment in which the student is placed centrally and the teacher acts to facilitate learning became a core focus of authentic learning endeavors. Learning environments are the physical or virtual settings in which learning takes place. Their design is not simply a matter of following a recipe (Bransford, Brown, & Cocking, 2000). Instead, it requires crafting to weave the activities, the (collaborative) interactions, and the tasks into a set of conditions that resemble the real-world situation sufficiently for the learning goals to be achieved (Boettcher, 2007).

Indeed, Yadav, Subedi, Lundeberg, and Bunting (2011) and Warren, Dondlinger, McLeod, and Bigenho (2012) established that involving students in authentic and meaningful work enhances their engagement and performance. By combining problem-based learning and authentic learning approaches, students can be offered opportunities to produce rather than solely consume knowledge, teachers can act as learning facilitators rather than simply as instructors, and groups can work together to develop designs and strategies to address an actual problem. In such situations, students regulate their learning process internally and are encouraged to become more reflective practitioners (Duignan, 2012; Slinger, Kwakkel, & van der Niet, 2008).

The emphasis placed on teaching students to become reflective practitioners aligns with the work by Schon (2011) in the field of policy analysis. Indeed, in their book on new developments in public policy analysis, Thissen and Walker (2013) emphasize the necessity for iterative and reflective processes in decision-making on complex problems. McEvoy et al. (McEvoy, 2019; McEvoy, van de Ven, Blind, & Slinger, 2018; McEvoy, van de Ven, Brolsma, & Slinger, 2020) draw on policy analysis work by Thissen and Twaalfhoven (2001) in distinguishing process and content in evaluating the efficacy of participatory planning workshops for the design and selection of flood mitigation measures in urban environments. They conceptualize these workshops as "policy analytic activities" or interventions nested within ongoing planning processes and explore how process and content choices within the workshops affect the overall process. A significant finding is that the effect of a half-day workshop in which participants learned about the different perspectives of representatives from other departments within a city authority could be traced one and a half years later. This emphasizes the learning impact of participatory workshops and signifies their potential value as procedural and substantive learning environments within the context of flood risk management.

The field of flood risk management has undergone significant developments in the last decade. Most notable is the insurgence of new concepts such as "Building with Nature" (EcoShape, 2020; Waterman, 2010), "Working with Nature" (PIANC, 2011), and "Engineering with Nature" (Bridges, Banks, & Chasten, 2016). Building with Nature specifically seeks to use natural materials and interactions in the design, realization, operation, and maintenance of hydraulic infrastructures (Waterman, 2010), striving for more ecosystem-based hydraulic engineering, while acknowledging social complexity (Slinger & Vreugdenhil, 2020). New types of nature-based hydraulic infrastructure have resulted. For instance, in the coastal area of North Holland, the Hondsbossche Dunes now offer protection from flooding where previously the oldest Dutch stone dike was located (RWS, 2015). On the Wadden Sea coast tidal marshes on the dike foreshore aid in protecting the hinterland from flooding, while permeable bamboo fences retain sediment and promote mangrove forest regrowth, preventing part of the Indonesian coast from eroding further (EcoShape, 2020). Such innovations in flood protection infrastructure design require the integration of knowledge from the fields of ecology and geomorphology with planning and engineering. They also require a broad consideration of the perspectives of multiple actors whose lives the infrastructure will affect over its whole lifecycle (Slinger & Vreugdenhil, 2020). Clearly, new methods for teaching the transdisciplinary, collaborative design skills necessary to develop integrated nature-based solutions for flood risk reduction are required.

Method

Two learning interventions were designed in the form of transdisciplinary co-design workshops—"Building with Nature" Living Labs—between international groups of students and faculty members drawn from three Texan universities and four Dutch universities. The Texan universities included Texas A&M, Rice University, and Jackson State University, while the involved Dutch universities included the Delft University of Technology, the University of Twente, Wageningen University, and the Vrije Universiteit Amsterdam.

The first workshop took place on May 31, 2016 and was attended by a total of 20 master and doctoral students, 10 from Texas, and 10 from the Netherlands. The Texan participants spanned a wide range of disciplines associated with flood risk reduction including civil, chemical, and environmental engineering, urban and regional planning, geography and environmental science. Students from a similarly wide range of disciplinary backgrounds were recruited to attend by Dutch faculty, namely civil engineering, architecture, policy analysis, and environmental science. Key to the workshop design is the combined problem-based learning and authentic learning environment pedagogy. Accordingly, four additional local experts were invited to share their deep situated knowledge with participants in formal presentations and to act together with the faculty members as "service desks" for knowledge sharing during the entire workshop. The local experts were drawn from public authorities, such as the water authority Heemraadschap Hollandsnoorderkwartier (HHNK), from nongovernmental organizations, and from knowledge institutes located in the Netherlands such as UNESCO-IHE and Deltares. The first workshop spanned a day and took place in a large open space within the Science Museum on the campus of the Delft University of Technology. The schedule of activities comprising the workshop is listed in Box 1.

The second workshop took place on June 9, 2017, in Petten at the Beach Pavillion Zee&Zo near the Hondsbossche Dunes on the coast of North-Holland. This meant that the participating 26 students (16 from Texas and 10 from the Netherlands) from a wide range of disciplinary backgrounds were able to experience the actual location of an innovative nature-based flood defense for themselves. Four experts drawn from the water board Heemraadschap Hollandsnoorderkwartier (HHNK), and from a local citizens initiative, as well as from knowledge institutes, provided situated knowledge in the form of presentations and "service desk" advice, together with faculty members. The schedule of activities (Box 1) of the second workshop spanned a full day.

Data used in this analysis comprise (i) detailed "shooting scripts" prepared by the organizers detailing the choices made regarding the activities and their intended outputs, (ii) the presentations by local and disciplinary experts, (iii) photographs and notes on the designs made by the students, (iv) notes taken during the plenary feedback sessions, (v) proceedings of the workshops, and (vi) questionnaires completed by the Texan students on their return journey as part of the PIRE-CFRR exchange program.

> **Box 1: The eight activities making up the transdisciplinary co-design workshop method**
>
> The **transdisciplinary co-design workshop method** comprises a specific sequence of eight activities:
> - Activity 1. Getting acquainted with each other
> - Activity 2. Getting acquainted with the problem context
> - Activity 3. Identifying key stakeholders and characterizing the biogeophysical system
> - Activity 4. Acquiring the nature-based infrastructure design concept and the design assignment
> - Activity 5. Collaboratively designing nature-based infrastructure
> - Activity 6. Communicating the nature-based infrastructure designs
> - Activity 7. Reflecting on learning
> - Activity 8. Receiving expert feedback on the nature-based infrastructure design.
>
> Throughout such a workshop, diverse disciplinary and situated, experience-based knowledge is offered to the small groups of students undertaking the design challenge. The effort is directed to ensuring an open, friendly atmosphere in which experts can easily be consulted and where documentary and visual information is freely accessible. Each student manages their own process of inquiry and discovery, although this takes place within the context of a small team of students from diverse backgrounds within the larger workshop setting

Results

The co-design process of the workshop

The purpose of the workshops was specified as "teaching integrated nature-based design for a specific coastal and societal context." The contexts were (i) the dynamic area on the southwestern coast of the island of Texel, and (ii) the sandy coast of Petten, the location of the oldest stone sea dike constructed in the Netherlands (Fig. 1), as these areas were either regarded as a potential site for nature-based solutions (Wijnberg, Mulder, Slinger, van der Wegen, & van der Spek, 2015) or were experiencing such interventions (EcoShape, 2020).

Earlier work on game-structuring approaches to complex environmental management problems in coastal communities (Cunningham, Hermans, & Slinger, 2014; Kothuis, Slinger, & Cunningham, 2014; Slinger, Cunningham, Hermans, Linnane, & Palmer, 2014) built an understanding of the problem by first identifying the key players, next gathering information on the system from the situated experience of participants, and then discussing and defining future outcomes. In essence, participants answered the questions (i) Who cares? (ii) Why we care? and (iii) What do we care about? before proceeding to define payoffs, identify future moves, and then negotiate with each other regarding the potential resolution of the problem. The success of this approach in creating an environment within which participants felt safe to discuss deeply contested issues and start to collaborate to address flood risk reduction issues (see Kothuis et al., 2014), led us to include similar steps in the teaching workshops. Clearly, the students were not familiar with the case study environment. This meant that they first needed information on the local bio-geomorphological and social environment and its current use and management. Accordingly, after a formal welcome and short getting acquainted exercise (*Activity 1*), disciplinary experts provided information on the hydro-geomorphology and ecology of the case study. This was followed by information on the local community, their uses of the area and their concerns, provided by a local resident or representative. Information on Dutch flood risk management practices and details applicable to the specific location was provided by a representative from the water board HHNK. At the close of this activity, many perspectives of the case study location and its natural and social dynamics had been communicated to the students. The presentations occurred in a theatre setting with opportunities for questions and discussion in the plenary. However, the experts were present throughout the workshop and were available for bilateral discussion and questions as and when needed. The provision of the contextual information formed *Activity 2* of the workshop method. The information was complemented by preprepared fact sheets designed to give an impression of the problem location, such as colorful photographs, tourist information, types of ecosystem, biodiversity importance, conservation status, employment, and human uses of the area. Aerial images and maps were also available for study as were the annual reports on the position of the Dutch coastline and sand nourishment volumes.

In *Activity 3*, the students are tasked with understanding the system by identifying the key stakeholders (Who cares?) and engaging with the biogeophysical and use components of the problem situation (Why they care?). This was undertaken using brainstorming techniques followed by clustering and grouping to come to a shared understanding of the key actors and the issues at play in the natural environment. The students undertook this task in small groups standing around a number of central tables and were free to consult the available information and the experts present.

Box 2: Design assignments for the 2016 and 2017 workshops

Texel Design Assignment

Design alternative coastal management strategies (or improve the current strategy) for the coast of southwestern Texel using the natural channel-shoal dynamics of the ebb-tidal delta to ensure safety from flooding and serve other functions.

Hondsbossche Pettemer Sea Defense

The Hondsbossche Pettemer Sea Defense no longer meets the required safety standards. Design alternative coastal protection strategies (or improve the current strategy) so as to comply with required safety standards both now and in 2050, taking compatibility with the biophysical, social, and institutional environment into account in your integrated design.

In *Activity 4*, the design assignments were explained (Box 2). The instructions given regarding the design assignments were intended to stimulate creativity and indicate to students that they should not be limited by considerations of whether their solution is financially or institutionally viable. So, for instance, they need not consider legal feasibility but should consider physical feasibility. So, "turning the sea red" was infeasible, but designing a flood risk reduction option that did not conform to the then legally required flood protection level of 1 in 10,000, was feasible. This explanation occurred in an informal plenary setting, with the assignment projected on a central board and the first author, as workshop lead, emphasizing the core thinking of nature-based infrastructure design and reiterating the particular place-based challenge. In the second workshop, the explanation of the design assignment commenced with the students still standing around in groups after completing the previous step. They then adjusted their position in order to be able to hear and see clearly, contributing to the open atmosphere. In both workshops, a task relating to each student's role in a design team formed an additional component of the design assignment and is fully reported in Klaassen et al. (2021).

At the close of *Activity 4*, each student was asked to look on the back of their name badge to find a picture of a bird. They were then tasked with finding all the other students with the same bird, to form their small design team. Each team was handed a pail with a variety of equipment potentially useful for making their design—coloring pens, tacks, colored paper, markers, post-its, and flip-overs. They were encouraged to eat lunch together, provided in picnic form in the first workshop and as a buffet in the second workshop, and to begin to design. Each small group then set about making their own integrated, nature-based solution for the problem situation (*Activity 5*). Students made full use of the space available to them. Some sat outside to develop their design, others spread themselves liberally across the available space, some worked very neatly and quietly, and still others employed their artistic talents in the service of the group. The students were in charge of their process of inquiry, discovery, and design, learning to translate their understanding of the system and the actors into an integrated design with their small group. According to Klaassen et al. (2021), the early, divergent way of looking at the design problem and the search for common ground across the diverse perspectives of the team members, each bringing different disciplinary backgrounds to the design table, supported the realization of integrated nature-based solutions.

In *Activity 6* each of the small groups presented their designs. Participants asked questions of other groups to clarify their understanding and also questioned the reasoning behind specific design choices. The jury, comprising the experts who had presented information earlier in the day supplemented by faculty members with a sound understanding of "Building with Nature," then went into a separate room to deliberate.

The students then reflected together on their roles in the design team and how this had influenced the collaborative design process of their small group (*Activity 7*). Finally, the jury returned to give feedback on the integrated, nature-based infrastructure designs. This formed the final formal component of the workshop (*Activity 8*). In the time remaining before the departure of the bus, most students explored their surroundings and made plans to meet up again.

Overall, the collaborative design workshop focused first on allowing the students to formulate a picture of the real, complex design (solution) space of the problem (Activities 1–3) before tasking them with a design assignment (Activities 4–6), and reflection (Activities 7 and 8). Furthermore, the atmosphere in both workshops shifted from some discomfort at the beginning, to a more relaxed atmosphere during the day, and finally to an informal, open atmosphere in which students and faculty interacted freely and binational network contacts were formed.

The substantive outcomes of the workshop

In 2016, there were four design teams with five students. In 2017 there were four design teams with five students and one design team with six students. The design assignments are specified in Box 2. This is followed by a detail of each of the designs produced in the first workshop (Figs. 2–4) and a summary of the designs produced in the second workshop. The observations of the expert jury are included in the analysis of the range and character of the students' designs.

The design of group 1 was entitled the Wadden Sea Education Research Center (TWERC) and focused on educating people (tourists, locals, authorities) about the dynamic, transient nature of features of the ebb-tidal delta like the "Razende Bol" (Fig. 2). This design shows that the students were able to acquire and use knowledge on the real problem situation in producing their multifunctional, adaptive design. When they were asked whether educating people really contributed to flood defense, they responded that the ebb-tidal delta would over time coalesce with the southwestern corner of Texel. This would mean that sand nourishment would no longer be needed for quite some time.

The design of group 2 envisaged the Razende Bol as a recreational sand engine (Fig. 2). They sought to give free reign to the natural processes of erosion and accretion that lead to the migration of this part of the ebb-tidal delta. They also envisaged constructing a vegetated dune ridge and erecting seasonal structures (demobilized in winter) to support recreational activities and enhance economic value. The Razende Bol would be accessible only by boat. In responding to questions the group explained their motivation as tackling the erosion problems, while simultaneously trying to realize a new

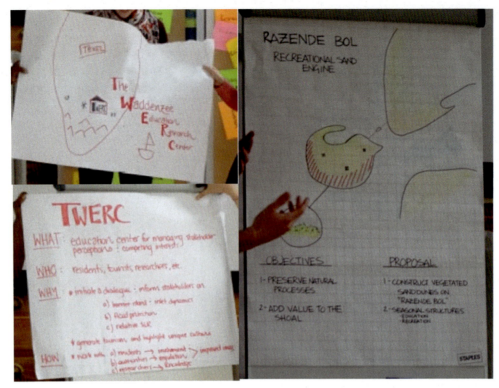

Fig. 2 The integrated nature-based solution envisaged by group 1 *(left)* and group 2 *(right)*. Both of these solutions considered education regarding the transient and dynamic character of the Razende Bol an important aspect of their designs.

recreational function for the area. They used the analogy of the Sand Engine, as here the sand migrates along the coast as well as strengthening the dunes, and recreational use diversifies and intensifies over time. Again, their design reveals that the students were able to acquire and use problem-based knowledge to produce a creative design alternative.

Group 3 took a different approach, focusing on the problem of erosion on southwest Texel and the navigation channel between the Razende Bol and the island of Texel (Fig. 3). There are many uncertainties that this group took into account using scenarios. They explored business as usual—Scenario Zero in which the present erosion hotspot remains, there is sand nourishment each year, and the sand from the Razende Bol slowly silts up the navigational channel. There are no additional risks. Next, they explored the Full Stop scenario in which all intervention is halted. This means the ebb-tidal delta will continue to migrate and Texel will continue to erode. It is the most risky in terms of flooding danger. Then, they considered expanding the current dredging program to include the navigational channel, concluding that the ecological consequences of this would be severe. They advised following the business as usual approach as the problem

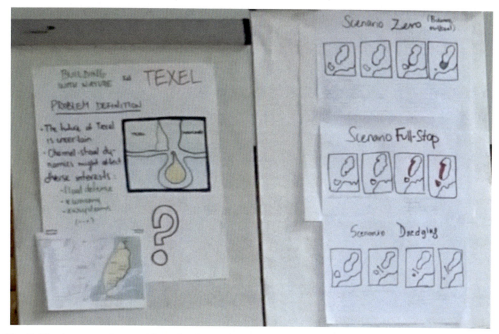

Fig. 3 The integrated nature-based solution envisaged by group 3 takes uncertainty into account. They advised buying time by proceeding as usual and waiting for the Razende Bol to adhere to the island of Texel.

will disappear when the ebb-tidal delta joins to the island, and in the meantime encouraging a dialogue with all stakeholders. The design of this group reflects the integrative character of the authentic, problem-based learning environment by combining both procedural and substantive elements.

Group 4 proposed a staged approach (Fig. 4). First, they envisage using vegetation to stabilize the sand on the northwestern edge of the Razende Bol and on the southwestern edge of Texel. Next, they proposed dredging the channel to induce offshore migration (by disposing sediment on the Texel side to create shallow and steep slopes) and finally using mussel beds to attenuate waves. The mussel beds offer more natural value, serve to trap sediment, and increase flood protection. The plan is that the increased natural value will lead to enhanced use by nature lovers, birdwatchers, etc. In response to questioning, this group acknowledged a high level of uncertainty about the time frames for mussel beds to become established but indicated that this is likely to provide a truly sustainable solution or at least build evidence of what can be achieved using nature-based measures. This design revealed that the students could effectively integrate a wide range of knowledge of the real problem situation into an adaptive "Building with Nature" solution within a 1-day workshop.

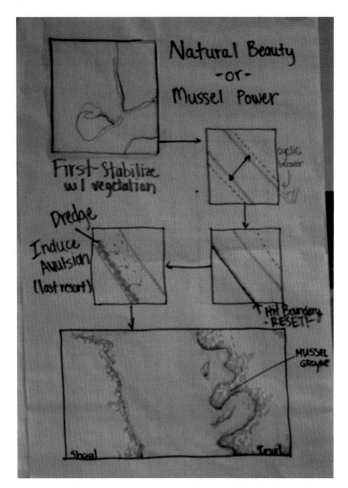

Fig. 4 The integrated nature-based solution envisaged by group 4 employs vegetation, dredging, and finally mussel beds to realize flood protection for the island of Texel.

The expert jury (*Activity 8*) considered that all the designs showed evidence of an understanding of dynamic natural processes and an appreciation for the range of uses and perspectives held by stakeholders. It was slightly surprising that such a strong focus on education and awareness building appeared in two of the groups. However, the serious consideration given to uncertainties by group 3 was acknowledged, while group 4 was awarded the prize (an apple pie) for their integration of dynamic aspects of erosion and sedimentation with the natural feature of mussel beds.

Of all the groups, only group 3 took into account the degree to which their design affected flood protection standards. The absence of elements of conventional engineering and the degree to which the students embraced the concept of using natural dynamics were noteworthy. All designs revealed that the student groups were able to acquire and use diverse problem-based knowledge to achieve integrated nature-based solutions.

The design assignment of 2017 specified that the designs had to meet flood protection standards in the future. This requirement, and the fact that the Texan students had already been to the Sand Engine in South Holland, meant that the wide range of designed solutions for the HPZ case was still narrower than that for the Texel case. The manner of depicting the solutions was more varied, as more diverse crafting materials were available to the students. This included clay, wool, and matchsticks so some groups constructed maquettes of their designs. All of the designs took local interests into account, primarily tourism and recreational interests. There was also an emphasis on dune landscapes and their ecological and flood protection value. All of the designs focused on placing large volumes of sand at the Pettemer coast. The major differences in the designs lay in how quickly and where the sand volumes were placed. One team included the construction of wetlands inland of the position of the old HPZ dike as wetlands have ecological and tourism value. This group was awarded the prize by the expert jury (*Activity 8*). The designs of 2017 consistently exhibited multidisciplinary integration and a strong "Building with Nature" philosophy in that they integrated ecological and engineering knowledge, included multiple perspectives in the design requirements, considered the full lifecycle of the infrastructure, and designed artifacts different from conventional flood defenses (see Slinger & Vreugdenhil, 2020).

In summary, the design outcomes of 2016 and 2017 revealed that the problem-based, authentic learning pedagogy embodied in the 1-day transdisciplinary workshop method provided an effective procedural and substantive environment for acquiring transdisciplinary, nature-based design skills.

Learning outcomes

In the reflection activity (*Activity 7*) during the workshop, only positive remarks were made about the authenticity of the design challenge (see also Klaassen et al., 2021). This indicates that the students recognized and enjoyed the case studies chosen and the way they were able to engage with the material—an important factor in learning and acquiring new skills. However, we were specifically interested in the knowledge acquisition of the Texan students and their opinions on the codesign workshop, particularly, as they previously had limited exposure to nature-based design concepts. In their responses to the confidential survey administered on the return journey, the majority of the Texan students could explain the "Building with Nature" concept after the workshop (Table 1). They were enthusiastic about the idea as captured in the quotes: "I was not aware of this concept until coming to the Netherlands. I am very impressed on their innovative creation and engineering" and "Building with nature works with natural processes instead of against them. Building with nature in the context of flood mitigation includes strategies such as dunes, permeable surfaces etc." In 2017, a student was even able to express their knowledge gap in relation to the concept, namely "I know the goals of engineering with nature and its benefits, but I don't know how its design process differs from traditional design." This anticipates upon the type of learning about design that occurred through the 1-day workshop method.

Table 1 Responses of Texan students to the survey question: "What do you know about the concept 'Building with Nature'?."

What do you know about the concept "Building with Nature"?	
2016	2017
- How the natural processes play the major role in transporting sediment - Build with Nature" mainly makes things look natural, but it is actually a man-made structure for flood mitigation. I am interested in solid core dunes, but they are robbed of the sediment transport processes that natural dunes have, i.e., migration patterns, flood response to climatic regimes, and this bums me out, but I think they work. It is something we should strive for from a process-based understanding. I like the sand engine, actually - That it's awesome! For example, the sand engine utilizes natural processes (currents and waves) to transport sediment along the shoreface in an economical fashion. The Dutch are trying to utilize natural processes to minimize coastal risk from flood impacts - Building using naturally occurring processes - The only thing I know is to use what's in nature to build a conceptual design - The basic idea and process - It's a pretty big deal in the Netherlands, and becoming more so in the rest of the developed (and developing) world. It is a wide-ranging concept that includes multiple kinds of interventions and noninterventions all with careful consideration - I was not aware of this concept until coming to the Netherlands. I am very impressed on their innovative creation and engineering - Sand engine—building for defense with what nature provides	- Building with Nature; applying certain amount of sand to let it do its flooding/coastline protection job - This is a recent concept with both pro's and con's depending on personal opinion, experiences, project requirements - Building with Nature should be a major requirement where applicable for proposed flood infrastructure - Building with nature works with natural processes instead of against them. Building with nature in the context of flood mitigation includes strategies such as: dunes, permeable surfaces, etc. - Let the nature do (most of the work). Integrated design with existing/prior ecosystems - I know the goals of engineering with nature and its benefits, but I do not know how its design process differs from traditional design - This is basically integrating nature aspects in designing solutions for flood protection - Building with Nature is incorporating designs that allow natural processes to work for you instead of against you - I know that is finding ways to use nature or work with it for designs. I still do not know a lot about design processes, but I learned a lot more here - I mostly know about "biophilic" design in the area of architecture. Respecting nature and integrating it into design instead of working against it - Human inputs are involved at the initial stage, and let it build its adaptive capacity by nature without any artificial approach - Building with Nature is working with the natural system to increase the safety and resiliency of a region - Need to consider input from many stakeholders. Need to design multifunctional solutions. High emphasis on spatial quality - Building with nature is letting nature "do the work for you"

The Texan students' responses to the broadly formulated question "What are the main things you have acquired from the design workshop?" (Table 2) focused primarily on the different ways of working and the acquisition or deepening of transdisciplinary collaboration skills. This is exemplified in the following quotes: "I learned how to work with others from another country to solve a problem," and "The biggest skills I gained were about bonding with a team and coming to an agreement of a design; We needed to work together." A number of students indicated explicitly that they worked outside their knowledge boundaries, e.g., "Working outside of your natural ability could be valuable as it helps you bridge knowledge between the different field, however one always fall back to their natural profile." Only one of the 46 students complained that the time was too rushed to learn adequately.

The overwhelming majority of enthusiastic and positive responses to the workshop method, and the integrated nature-based designs that were produced validate this learning intervention as a means of teaching transdisciplinary collaborative skills.

Table 2 Responses of Texan students to the survey question: "What are the main things you have acquired from the design workshop?."

What are the main things you have acquired from the design workshop?	
2016	2017
- Understanding the engineering role in each group. 2. Applying my role in the group was very interesting because I found that the engineer can play any role based on his/her experience. 3. Thinking for innovative solution for building with nature - I learned how Dutch students organize group work and attack problems—I will use this method - I found from the design workshop that Dutch students are (1) extremely practical and organized in problem solving, and (2) are very humble! Their approach to flood risk reduction is refreshing. In the US, we consider living on the coast (with an ocean view) an inherent right ("you can build whatever you want if it's your property" or "it's my loss if my property gets destroyed"). We have to juggle individual property rights and flood risk reduction when designing protection structures. It seems simpler in the Netherlands…as if the entire nation understands flood risk and has bought into living with a modified coast	- Brainstorm; Itemize the goals of design; strategies can be reasonable and creative - I learned how research questions are formulated along with the process required to answer; Working with new people outside my core group allowed me to work with even more views and knowledge (expanded my thoughts) - Working with engineers and assuming a non-natural role - Diversity of issues that need to be addressed when planning; Stakeholders - Collaborative work; Building with nature (had very little knowledge before) - I learned a ton about the Dutch political system and Dutch culture from talking to Dutch students on the bus - The interaction with experts, students in different fields that helped a lot - The biggest skills I gained were about bonding with a team and coming to an agreement of a design; we needed to work together - I learned from the Dutch students on a public knowledge / student level; Simple

Continued

Table 2 Responses of Texan students to the survey question: "What are the main things you have acquired from the design workshop?."—cont'd

What are the main things you have acquired from the design workshop?

2016	2017
- The design workshop validated the fact that Dutch students and American students approach problems differently - I learned how to work with others from another country to solve a problem - I did not gain much in terms of knowledge since the workshop way completely out of my discipline. However, it was a good experience to see the perspectives of different groups and working in a multidisciplinary group - Collaboration between Dutch students was very helpful by understanding their approach - I learned to think more critically and look at a larger range of things - Within short frame of time, I was able to address the problem and work on finding a creative solution - Again, Dutch approach to organizing goals + decide - There are many approaches to solving coastal engineering problems and there are lots of stakeholders to consider - Working outside of your natural ability could be valuable as it helps you bridge knowledge between the different field, however one always fall back to their natural profile - Being able to work with others and try to understand what role I play - I do have experience with multidisciplinary research. The workshop only reinforced my experiences - The design workshop was an interesting experience, and I am glad I had the chance to participate, but I didn't acquire much additional knowledge, as far as I can tell. I've experienced such workshops before - I am a "anonymized." I work on process improvement from mechanical standpoint. I have applied a few strategy of mechanics into civil engineering, especially in whole system design and problem solving - I learned to think more critically and look at a larger range of things	explanations from people my age / education level helped me grasp complex concepts better - The Dutch students were much familiar to the Petten's condition that they helped us get into the main problems and possible strategies efficiently - I enjoyed meeting the Dutch students and sharing our perspectives on flood risk mitigation design and planning - Main skills I acquired was learning how to work in team with multiple backgrounds and experience levels - Felt like I didn't gain much because it was rushed and not explained too well

Concluding remarks

Informed by problem-based and authentic learning pedagogies, the game-structuring approach, and nature-based design concepts, an eight-step transdisciplinary collaborative design workshop method was developed. The effects of the method, in the form of two learning interventions—"Building with Nature" Living Labs—are evaluated in this chapter. Specifically, two workshops were conducted in 2016 and 2017. The participants were international students and faculty members drawn from three Texan universities and four Dutch universities, supplemented by local experts familiar with the problem situations. The authentic problem contexts were provided by (i) the erosion hotspot of the southwest coast of Texel island and (ii) the future noncompliance of the old Hondsbossche Pettermer Sea Dike with Dutch flood defense standards. The effects of the workshops are evaluated in terms of the process of co-design, the substantive outcomes, and the learning outcomes. The collaborative design process created an open and authentic environment in which the students could experience undertaking integrated, nature-based design. Each student managed their own process of inquiry and discovery, although this took place within the context of a small team of students from diverse backgrounds tasked with designing together, within the larger workshop setting. Overall, a wide range of integrated nature-based designs was produced and diverse biogeophysical and social aspects were included in all designs. This indicated that the students were able to work collaboratively to produce novel designs that incorporated multiple perspectives from the problem situations. They were exposed to information beyond their own disciplinary fields and learned to synthesize relevant aspects into a coherent nature-based design by collaborating with other students. The overriding enthusiastic and positive responses of the Texan students to survey questions on their learning experiences affirm the success of the transdisciplinary co-design workshop method as a means of teaching integrated, nature-based infrastructure design.

Acknowledgment

Financial support from the Multi-Actor Systems Research Programme of Delft University of Technology, the Dutch Research Council under grant number 850.13.043, and the Partnerships for International Research and Education Coastal Flood Risk Reduction (PIRE-CFRR) program "Integrated, multi-scale approaches for understanding how to reduce vulnerability to damaging events" is acknowledged.

References

Barrows, H. S. (1985). *How to design a problem-based curriculum for the preclinical years*. New York: Springer Publishing Company.
Barrows, H. S. (1992). *The tutorial process*. Springfield, Illinois: Southern Illinois University School of Medicine.
Boettcher, J. V. (2007). Ten core principles for designing effective learning environments: Insights from brain research and pedagogical theory. *Innovate: Journal of Online Education, 3*. Article 2. Available at: https://nsuworks.nova.edu/innovate/vol3/iss3/2.

Bontje, L., & Slinger, J. H. (2017). A narrative method for learning from innovative coastal projects—Biographies of the Sand Engine. *Ocean and Coastal Management, 142*, 186–197. https://doi.org/10.1016/j.ocecoaman.2017.03.008.

Bransford, J. D., Brown, A. L., & Cocking, R. R. (2000). *How people learn: Brain, mind, experience, and school.* Washington, DC: National Academies Press.

Bridges, T. S., Banks, C. J., & Chasten, M. A. (2016). Engineering with nature. Advancing system resilience and sustainable development. *The Military Engineer, 2016*, 52–54. https://ewn.el.erdc.dren.mil/pub/Pub_1_EWN_TME-JanFeb2016.pdf.

Cunningham, S. C., Hermans, L. M., & Slinger, J. H. (2014). A review and participatory extension of game structuring methods. *EURO Journal on Decision Processes, 2*(3-4), 173–193. http://link.springer.com/article/10.1007/s40070-014-0035-8.

Duignan, P. (2012). *Educational leadership. Together creating ethical learning environments* (2nd ed.). Cambridge, UK: Cambridge University Press, ISBN:9781107637894.

EcoShape. (2020). *Building with nature.* Available online: www.ecoshape.nl. (Accessed 19 December 2020).

Herrington, J., Reeves, T. C., & Oliver, R. (2014). Authentic learning environments. In J. M. Spector, M. D. Merrill, J. Elon, & M. J. Bishop (Eds.), Handbook of research on educational communications and technology (pp. 401–412). Springer.

Klaassen, R., Kothuis, B., & Slinger, J. H. (2021). Engineering roles in building with nature interdisciplinary design—Educational experiences. In J. Bergman, et al. (Eds.), *Building with nature perspectives*.

Kothuis, B. L. M., Slinger, J. H., & Cunningham, S. W. (2014). *Contested issues game structuring approach—CIGAS workshop Houston report, results and reflection: Exploring stakeholder-based joint commitment to action for flood protection decision-making in the Houston Galveston Bay Area.* Amsterdam: Bees Books Publishers, ISBN:9789074767170.

McEvoy, S. (2019). *Planning support tools in urban adaptation.* PhD thesis Delft, Netherlands: Delft University of Technology. Available online: https://repository.tudelft.nl/islandora/object/uuid:48b7649c-5062-4c97-bba7-970fc92d7bbf?collection=research.

McEvoy, S., van de Ven, F. H. M., Blind, M. W., & Slinger, J. H. (2018). Planning support tools and their effects in participatory urban adaptation workshops. *Journal of Environmental Management, 207*, 319–333. https://doi.org/10.1016/j.jenvman.2017.10.041.

McEvoy, S., van de Ven, F. H. M., Brolsma, R., & Slinger, J. H. (2020). Evaluating a planning support system's use and effects in urban adaptation: An exploratory case study from Berlin, Germany. *Sustainability, 12*(1), 173. https://doi.org/10.3390/su12010173.

Nicaise, M., Gibney, T., & Crane, M. (2000). Toward an understanding of authentic learning: Student perceptions of an authentic classroom. *Journal of Science Education and Technology, 9*(1), 79–94. https://doi.org/10.1023/A:1009477008671.

PIANC. (2011). *The World Association for Waterborne Transport Infrastructure.* PIANC Position Paper 'Working with Nature', Available on: https://www.pianc.org/uploads/files/EnviCom/WwN/WwN-Position-Paper-English.pdf. (Accessed 29 May 2020).

Pruyt, E., Slinger, J. H., van Daalen, E., Yucel, G., & Thissen, W. (2009). Hop, step, step and jump towards real-world complexity @ Delft University of Technology. In *Proceedings of the 27th international conference of the System Dynamics Society, Albuquerque, USA* (pp. 1–9).

RWS. (2015). *Rijkswaterstaat annual report 2015. Rijkswaterstaat | cd0616mc74.* Den Haag, The Netherlands: Ministry of Infrastructure and the Environment.

Schmidt, H. G. (1993). Foundations of problem-based learning - some explanatory notes. *Medical Education, 27*(5), 422–432.

Schon, D. (2011). The reflective practitioner. How professionals think in action. *Journal of Policy Analysis and Management, 34*(3). https://doi.org/10.1080/07377366.1986.10401080.

Slinger, J. H., Cunningham, S. C., Hermans, L. M., Linnane, S. M., & Palmer, C. E. (2014). A game structuring approach applied to estuary management in South Africa. *EURO Journal on Decision Processes, 2*(3-4), 341–363. http://link.springer.com/article/10.1007/s40070-014-0036-7.

Slinger, J., Kwakkel, J., & van der Niet, M. (2008). Does learning to reflect make better modelers? In *Proceedings of the 26th international conference of the System Dynamics Society, July 2008, Athens, Greece.*

Slinger, J. H., & Vreugdenhil, H. S. I. (2020). Coastal engineers embrace nature—Characterizing the metamorphosis in hydraulic engineering in terms of four continua. *Water, 12*(9), 2504. https://doi.org/10.3390/w12092504.

Stive, M. J. F., de Schipper, M. A., Luijendijk, A. P., Aarninkhof, S. G. J., van Gelder-Maas, C., van Thiel de Vries, J. S. M., et al. (2013). A new alternative to saving our beaches from sea-level rise: The Sand Engine. *Journal of Coastal Research, 29*, 1001–1008. https://doi.org/10.2112/JCOASTRES-D-13-00070.1.

Thissen, W. A. H., & Twaalfhoven, P. G. J. (2001). Towards a conceptual structure for evaluating policy analytic activities. *European Journal of Operations Research, 129*, 627–649.

Warren, S. J., Dondlinger, M. J., McLeod, J., & Bigenho, C. (2012). Opening the door: An evaluation of the efficacy of a problem-based learning game. *Computers & Education, 58*, 397–412. https://doi.org/10.1016/j.compedu.2011.08.012.

Thissen, W. A. H., & Walker, W. E. (Eds.), (2013). Public policy analysis: New developments. Springer. https://doi.org/10.1007/978-1-4614-4602-6

Waterman, R. E. (2010). *Integrated coastal policy via Building with Nature*. PhD thesis Delft, The Netherlands: Delft University of Technology. http://resolver.tudelft.nl/uuid:fa9a36f9-7cf8-4893-b0fd-5e5f15492640.

Wijnberg, K., Mulder, J., Slinger, J., van der Wegen, M., and van der Spek, A. (2015). Challenges in developing 'Building with Nature' solutions near tidal inlets. In Presented at the Coastal Sediments '15 conference, Understanding and working with nature, May 11–15, San Diego, CA, USA.

Yadav, A., Subedi, D., Lundeberg, M. A., & Bunting, C. F. (2011). Problem-based learning: Influence on students' learning in an electrical engineering course. *Journal of Engineering Education, 100*(2), 253–280. https://doi.org/10.1002/j.2168-9830.2011.tb00013.x.

CHAPTER 29

Flood risk assessment of storage tanks in the Port of Rotterdam

Sabarethinam Kameshwar
Department of Civil and Environmental Engineering, Louisiana State University, Baton Rouge, LA, United States

Introduction

The Port of Rotterdam, the largest seaport in Europe, houses several ports, oil and gas facilities, and petrochemical industry installations in close proximity to several residential areas, farmlands, and coastal areas that have high recreational and ecological value. Aboveground storage tanks (ASTs) are widely used in these industries to store a variety of substances such as crude oil, petrochemicals, and other hazardous substances. ASTs are susceptible to failure during hurricanes, which can disrupt the oil and gas supply chain. At the same time, hazardous spills due to AST failure can be catastrophic to the surrounding ecosystems and communities. For example, during the infamous Murphy oil spill (shown in Fig. 1), the failure of just one AST in Meraux, LA, during hurricane Katrina released over 25,000 barrels of mixed crude oil into the surrounding environment and rendered over 1700 houses uninhabitable. The susceptibility of ASTs has been evidenced repeatedly by recurring failures during past hurricane and flood events all in several countries (Cozzani, Campedel, Renni, & Krausmann, 2010). Yet still, design codes such as Euro code 14015 (EN, 2005), American Petroleum Institute (API) 620 (API, 2002), and API 650 (API, 2013) do not have provisions for preventing AST failure during hurricane surge events.

In the context of the Netherlands, even though there is a strict risk acceptance threshold of 1 in a million chance of loss of life for an individual person due to hazardous facilities (Beroggi, Abbas, Stoop, & Aebi, 1997), these standards typically do not apply to failure of ASTs since damage during hurricanes and potential spills do not pose a direct threat to life safety for surrounding communities. Furthermore, limited studies have assessed the vulnerability of ASTs in the Netherlands, specifically in the Port of Rotterdam. Existing studies such as Lansen and Jonkman (2010, 2013) use simplistic approaches to assess the hurricane risk to ASTs. At the time of this research in 2016, there was a lack of probabilistic AST performance assessment models for various failure modes such as flotation and surge buckling. As a consequence, the risk of AST failure in the Port of Rotterdam was not well understood.

Fig. 1 Aerial view of Murphy oil spill. *Source: https://archive.epa.gov/katrina/web/html/index-6.html.*

More recently, since 2016, several studies have assessed the hurricane performance of ASTs. Physics-based fragility models using finite element simulations have been developed for assessing the failure probability due to hurricane-induced flotation (Kameshwar & Padgett, 2018b), sliding (Bernier & Padgett, 2019a), wave buckling (Bernier & Padgett, 2019b) and surge buckling (Kameshwar & Padgett, 2018a), and wind buckling failure modes (Kameshwar & Padgett, 2018a). Khakzad and Van Gelder (2017, 2018) and Qin, Zhu, and Khakzad (2020) developed fragility models for ASTs subjected to flood events for flotation, sliding, and buckling failure modes using simplified analytical equations. Using models from Kameshwar and Padgett (2015), Knulst (2017) assessed the risk to ASTs in Vopak's AST farms in the Port of Rotterdam. Even though models for assessing the hurricane performance of ASTs exist, there is limited understanding of the vulnerability of ASTs in the Port of Rotterdam.

A better understanding of risks to ASTs is essential to understand the economic, ecological, and social effects of hurricane-induced AST failures. Furthermore, insights into the hurricane risk of AST failure can be used to inform the Netherland's acceptable risk thresholds and identify potential mitigation strategies to reduce the risk of AST failure. Therefore, the two objectives of my research trip were to (1) quantify the hurricane surge-induced risk to ASTs in the Port of Rotterdam and (2) understand the Netherland's flood risk management philosophy to gain insights on how the risk of AST failure can be mitigated in the Port of Rotterdam and elsewhere. To achieve these objectives, first, in Section "AST inventory analysis," I developed a database containing information on geometry, location, and elevation of ASTs in the Port of Rotterdam. Next, in Section "Storm surge hazard data," I estimated the exposure of ASTs to storm surge for various return period events for current and future sea levels. I combined the information on ASTs from provided in Section "AST inventory analysis" and the hazard information from provided in Section "Storm surge hazard data" to estimate the failure probability of ATSs for various return periods for current and future conditions that include sea-level rise in Section "Vulnerability analysis." In Section "Dutch flood risk management philosophy," I have discussed my interactions with experts, research meetings, and field trips that helped understand the Netherland's flood risk management philosophy. In Section "Impact of PIRE program," I have discussed how I benefitted from PIRE's program components and how it influenced my future research path. Finally, in Section "Summary," conclusions from my research and experiences from the PIRE program are presented.

AST inventory analysis

As a first step toward assessing the vulnerability of ASTs in the Port of Rotterdam, I performed an inventory analysis of ASTs to develop a database with information on the location of ASTs, their size, and their structural characteristics such as the type of roof. Typically, port authorities and AST owners have some or all details on ASTs; however, it is not publicly available. Therefore, I used the approach from Kameshwar and Padgett (2018a) and Bernier, Elliott, Padgett, Kellerman, and Bedient (2016) to perform an inventory analysis, which is briefly described below. First, I used aerial images to identify the location, diameter, and the roof type for all ASTs in the Port of Rotterdam. Once a tank was identified, its diameter was measured manually in Arc-GIS (Beyer, 2004). Additionally, since the type of roof can influence buckling failure of ASTs, each AST's roof type was classified into three categories: open (no roof), fixed, and floating, based on visual inspection. In order to estimate the height of the ASTs, I obtained the digital elevation and surface models from the Actueel Hoogtebestand Nederland (AHN, 2020) database to measure the elevation at the base and the roof of each AST, respectively. The difference in elevation between the roof and the base of ASTs was used as a measure of ASTs' height. Additionally, I used the elevation at the base of the ASTs for vulnerability analysis in Section "Vulnerability analysis."

Fig. 2 shows the location of 3122 ASTs identified in the Port of Rotterdam using the abovementioned methodology. The location of each AST is color coded to show the ground elevation at each AST's location with respect to the Normaal Amsterdams Peil (NAP) datum. Additionally, the figure shows the dikes in the area using thick solid lines along with their annual failure probability. From Fig. 2, it can be seen that the ASTs are spread in three regions: Maasvlakte, Europort, and Botlek. The majority of the ASTs are located in the Botlek region and most of them are protected by dikes and the Maeslant barrier, but these ASTs are situated at low elevation which makes them susceptible to flooding if the flood protection systems fail. Approximately 750 ASTs are in the Maasvlakte and Europort regions, where they are not protected by dikes. But, these ASTs in the unprotected regions are situated at a higher elevation, which could reduce their risk for current sea levels. But, as sea levels rise in the future, their vulnerability will increase.

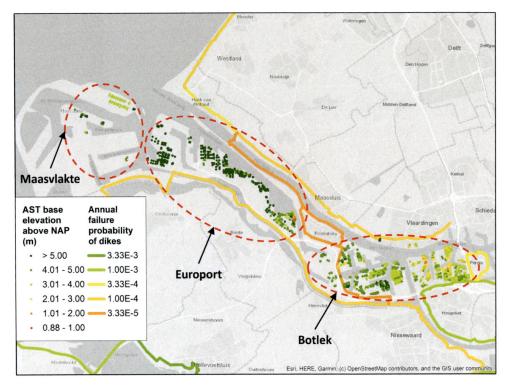

Fig. 2 Location and ground elevation of tanks.

Inventory analysis shows that only 20 ASTs have open roof, 325 ASTs have a floating roof, and the remaining 2777 ASTs have a fixed roof. Further, approximately 64% of the ASTs have a diameter less than 20 m, 30% of the ASTs have a diameter between 20 and 60 m, and 6% of the ASTs have a diameter greater than 60 m. This analysis shows that majority of the ASTs are very small and only a small fraction of ASTs are extremely large. Tank size is important to estimate the spill volumes in the event of a failure. The information of ASTs geometry is essential for vulnerability assessment and will be used in Section "Vulnerability analysis."

Storm surge hazard data

The primary focus of my study was to assess the storm surge performance of ASTs. I did not consider wind load performance since several studies have already assessed the wind load response (Kameshwar & Padgett, 2016; Portela & Godoy, 2005a, 2005b). Therefore, storm surge inundation maps for current and future conditions, considering sea-level rise, for different return periods (100, 300, 1000, 4000, 10,000, and 30,000 years) were obtained from the Port of Rotterdam. The current conditions correspond to 2015 and future conditions correspond to 2050 and 2100. However, this data was only available for the Botlek region.

Fig. 3 shows the storm surge inundation depth above the ground level for a 10,000 year return period event for the current (2015) and a future condition (2100). From Fig. 3A and B, the effect of the dikes on reducing the flood hazard can be clearly seen as the unprotected area in the western side of Botlek has significantly larger water depths compared to the protected areas in the east. Furthermore, the effect of sea-level rise for the future condition can also be clearly seen as the eastern part of Botlek becomes inundated for the event in year 2100, while it remains uninundated for the same return period event in year 2015. It must be noted that the storm surge data obtained from the Port of Rotterdam only provides the range of storm surge inundation, as shown in Fig. 3A and B. While the information on the exact inundation would be result in more accurate risk estimates, the inundation depth ranges in Fig. 3 will be used for vulnerability assessment in the next section. Moreover, the inundation maps in Fig. 3 alone can provide additional valuable insights. For example, for current and future conditions, it can be seen that only a small portion of the unprotected area in Botlek has inundation depth greater than 2.0 m. This observation can be used to infer that other unprotected areas in Maasvlakte and Europort can be expected to have lower inundation depths since these unprotected areas have higher ground elevation.

Vulnerability analysis

This section will primarily discuss the results of vulnerability analysis of ASTs in the Botlek area since the storm surge information is only available for this area and it has over 75% of the ASTs in the Port of Rotterdam. I used the parameterized logistic regression-based fragility model for flotation failure of ASTs presented in Kameshwar and Padgett (2018b) to assess the storm surge-induced flotation failure probability of each AST in the Botlek region. I did not consider buckling failure because storm surge-induced buckling is more likely to happen at high storm surge inundation depths (~4.0–5.0 m) and storm surge inundation depth did not exceed 2.0 m at any of the AST locations even for the 30,000 return period surge level in year 2100. The inputs to the fragility model included: tank height (H), tank diameter (D), fill level inside the ASTs (L), density of stored contents (ρ_s), inundation depth height (S). The range for each parameter used herein, which was also used in Kameshwar and Padgett (2018a), is shown in Table 1. To determine ρ_s, it was assumed that all the ASTs contain crude oil. This conservative assumption was made considering the potential consequence of spills caused by AST failure. The fill level in each AST cannot be determined a priori; therefore, it was assumed to be uniformly distributed between zero to 90% of AST's height; typically 10% capacity is reserved for spill protection. Finally, since the storm surge inundation depth maps (such as the one in Fig. 3) only provide a range for the inundation depth (S), for each AST, storm surge was assumed to be uniformly distributed within the range obtained from Fig. 3. AST geometry was obtained from the inventory analysis.

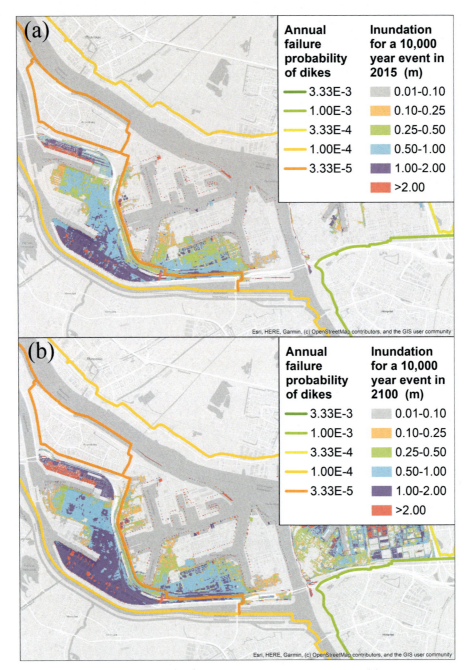

Fig. 3 Storm surge inundation in Botlek for a 10,000 year event (A) in 2015 (B) in 2100.

Table 1 Key parameters and their distributions.

Variable	Value	Distribution
H	Estimated from inventory analysis	Deterministic
D	Estimated from inventory analysis	Deterministic
L	$0–0.9H$	Uniform
ρ_s	$835–893 \, kg/m^3$	Uniform
S	Estimated from hazard maps (e.g., Fig. 3)	Uniform

Uncertainty in variables shown in Table 1 is propagated using Monte Carlo simulation, wherein for each AST, 100,000 simulations are performed. In each simulation, the fragility model was used to determine if the AST fails and determine the spill volume by conservatively assuming that the entire contents of the failed AST will leak in case of failure. This conservative assumption can be justified by the potential social, economic, and environmental consequences of spills. Using this approach, I assessed the performance of ASTs in Botlek for different return period surge levels and future conditions. For example, Fig. 4 shows the failure probability of ASTs in Botlek along with the storm surge inundation depth for the 30,000-year return period level inundation depth for 2015, 2050, and 2100. As can be seen, for current and future conditions, even at the 30,000 year return period inundation level, none of the ASTs have a failure probability greater than 0.3. Nevertheless, further analysis was conducted to determine the expected number of failed ASTs and the potential consequences of AST failure (spill volumes). Fig. 5 shows the expected number of failed ASTs for each of the six return period inundation depths where it can be seen that for current conditions there are no failures for 100- and 300-year inundation depths and a maximum of approximately 20 ASTs are expected to fail even for a 30,000-year event. In contrast, for future conditions, even the 100-year inundation depth is expected to cause 20 AST failures, increasing up to 45 failures for the 30,000-year inundation depth for conditions in 2100. This increase in the number of AST failures for future conditions can be attributed to larger inundation depths for the future conditions caused by a combination of sea-level rise and changes in frequency and intensity of storms. Fig. 6 correspondingly shows the expected spill volumes. Even though for the 100-year event in 2100, about 20 ASTs fail, they do not lead to a significant spill volume because most of these AST failures happen when the ASTs are nearly empty. For an economic perspective, cleanup costs for 1 m^3 oil spill is estimated at $12,000 in 2005 (Gaurd, 2006); a spill of 7000 m^3, which is expected for a 30,000-year event for year 2100, could cost $84 million (in terms of dollar value in 2005) in cleanup alone. Additionally, such spills can have long-term adverse impacts on the surrounding communities and ecosystems. The results provide insights into the expected number of ASTs that can fail for levels of inundation at present and future conditions and the expected spill volumes. In addition to spill volumes, further analysis can also identify areas with most spills.

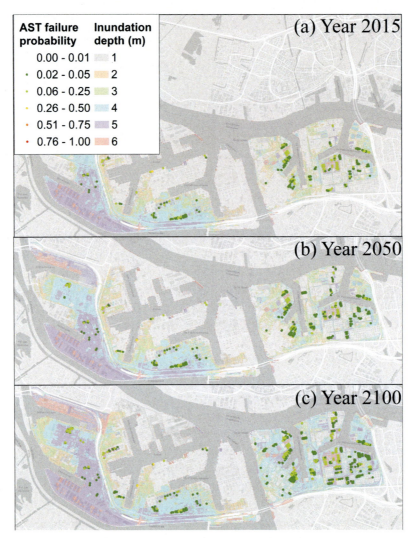

Fig. 4 Failure probability of ASTs in Botlek subjected to 30,000-year return period inundation depth (A) in 2015, (B) in 2050, and (C) in 2100.

All these results provide insights into the risk of AST failure in the Port of Rotterdam, which can inform mitigation measures if required.

Dutch flood risk management philosophy

The PIRE program's components were instrumental in meeting the second objective of understanding the flood risk management philosophy in the Netherlands. Specifically, during my meetings with experts, I learned about the concepts of individual and societal

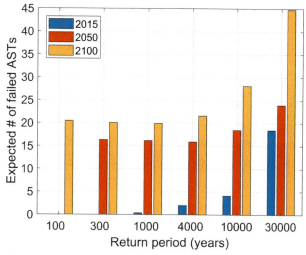

Fig. 5 Expected number of failed ASTs.

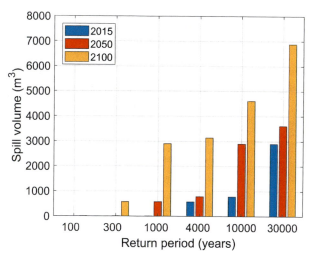

Fig. 6 Expected spill volumes.

risk thresholds, which are widely used in the Netherlands for determining safety standards for flood defense systems. In the context of flood risk management, individual risk pertains to the probability of death of a person due to floods and societal risk determines the probability of many people dying during an extreme event. Specifically, in the context of industrial accidents, such as failure of ASTs, the risk acceptance threshold is set at 1 in a million chance of death of an individual who is always present at the location of interest. This threshold is equivalent to annual radiation expose of 0.4 mSv (*millisieverts*). For individuals exposed to multiple risks, the acceptable threshold can increase up to one in

100,000 annual probability of death. These thresholds for individual risk also consider risk of injury, where it is assumed that each death will be accompanied by 10 injuries. The threshold on individual risk alone cannot prevent accidents that can cause multiple fatalities. Therefore, societal risk thresholds are also included considering the n^2 risk aversion effect, where in a single large event is considered worse than several small events. An annual risk acceptance threshold of $10^{-3}/n^2$ was set up for an establishment, where n is the number of fatalities. For example, for an accident involving 10 fatalities, $n=10$, an acceptable annual risk will be 10^{-5} ($10^{-3}/100$). The Dutch approach to flood risk management might not be applicable directly to the United States due differences in socio, economic, and governance structures. However, the general principle of risk-based mitigation planning can be adapted alongside existing risk mitigation planning methods that are either scenario based or use cost-benefit analysis.

At present, the flood risk mitigation of AST related Natech events, i.e., spills, in the United States is primarily left to the owners of facilities (API, 2013, p. 650), which can be partially attributed to lack of guidelines or directives to precent Natech incidents like the Seveso directive (European Commission, 2021) used in the European Union. Nevertheless, existing flood risk mitigation planning for ASTs in the Houston Ship Channel (HSC) is based on cost-benefit analysis or scenario-based analysis (SSPEED, 2016). However, there is an opportunity to include the Dutch concepts of individual and societal risks in the cost-benefit analysis and holistically extend it to include exposure to harmful substances, the long and short terms effects of exposure, and environmental impacts. Such considerations are necessary due to the presence of several residential communities and fragile ecosystems in close proximity to the HSC.

Site visits to the Maeslant Barrier, the Afsluitdijk, the Oosterscheldekering, and the polders demonstrated how multiple of flood defense systems were integrated to reduce the risk below the individual and the societal flood risk thresholds. Similar approaches, with multiple lines of defense are being considered for flood risk mitigation in the HSC (SSPEED, 2016) and elsewhere such as Louisiana. Furthermore, site visits also provided the historic and the social context that drive flood mitigation planning in the Netherlands, e.g., the North Sea flood of 1953. For storage tanks, interactions with decision makers from the Port of Rotterdam also provided insights in to how the flood risk management philosophy affects the interaction between the facilities at the port and the port management. For example, facilities such as tank farm owners are required by law to share pertinent flood mitigation plans with the port officials. While similar measures exist in the United States where each facility needs to have an emergency management plan but the accessibility to these plans is limited. There is a need for improved communication on Natech risks between the facility owners and community stakeholders. My discussions with experts at the Royal Haskoning further emphasized the need for understanding and identifying different flood induced failure modes that can lead to AST failure and Natech incidents, such as lack of liquid nitrogen for cold storage tanks leading to explosions, a similar incident was observed in HSC in the aftermath of Hurricane Harvey.

Impact of PIRE program

My interactions with experts and students from other disciplines such as urban planning and hydrology provided me a broad perspective of different ways to mitigate risk of AST failure. I believe that an integrated holistic effort can be more effective in tackling flood risk for ASTs and in general as well. My interactions with the general public also provided me new perspective on how an informed public can be an asset for championing flood risk mitigation. These interactions motivated me to focus on collaborative multidisciplinary ways to identify flood risk mitigations strategies for ASTs. So, my later research focused on identifying engineering based mitigation measures to reduce ASTs' flood risk, such as installation of anchors (Kameshwar & Padgett, 2018a) and stiffening rings (Kameshwar & Padgett, 2019). Next, I worked on an interdisciplinary approach to identify risk mitigation strategies for AST failure (Bernier, Kameshwar, Elliott, Padgett, & Bedient, 2018) where social science-based measures were identified along with an engineering-based approach to mitigate risks for all the ASTs in the Houston Ship Chanel. In this study, I also developed a heuristic framework to for developing acceptable spill thresholds for individual ASTs based on a regional level threshold (Bernier et al., 2018). The framework assesses the contribution of an individual AST to the total regional level risk before mitigative actions and determines the required reduction in risk for each AST such that the regional risk threshold is met. In the proposed framework, the required risk reduction for each AST was proportional to its contribution to the overall regional risk.

My interdisciplinary experiences during the PIRE program and my doctoral and postdoctoral training have influenced and shaped my research direction. These experiences have led me to work more collaboratively within my research discipline and also seek interdisciplinary collaborations. These interdisciplinary collaborations have increased my awareness of the interconnectedness of different systems such as social, economic, and infrastructure and the cascading impacts of failures in one system due to extreme events like floods on other systems. Overall, this has led me to believe that a holistic approach combining different disciplines, including engineering, may be best suited to address flood risk. Finally, in addition to help shaping my research direction, the PIRE program also helped me build a network of international researchers focused on flood risk mitigation.

Summary

As a part of the PIRE program my research objectives were to quantify the flood risk to above ground storage tanks (AST) in the Port of Rotterdam and understand the flood risk management approach used in the Netherlands. For the first objective, I first developed a database of ASTs in the Port of Rotterdam containing information on tank locations, dimensions, and roof characteristics. Next, I used a parameterized physics based fragility model along with AST information from the tank database to estimate the probability of flotation failure and expected spill volume for ASTs in the Botlek region of the port for various inundation depths corresponding to current and future conditions, obtained from the Port of Rotterdam. The results highlight the magnitude of the flood risk to ASTs, which could be used to inform mitigation measures, if needed. To understand the process used to manage flood risk, i.e., the second objective, I met with several experts and researchers from the Netherlands. Site visits to places like Maeslant Barrier, the Afsluitdijk, the Oosterscheldekering, and different polders also provided me insights on how the risk management philosophy was implemented. Other program components such as interactions with researchers and students from different disciplines offered me new perspectives on flood risk management, which has influenced my research direction. Overall, the PIRE program was an excellent learning experience and has helped me connect with researchers from different disciplines and counties.

References

AHN. (2020). *Actueel Hoogtebestand Nederland [WWW Document]*. AHN. https://www.ahn.nl/. (Accessed 3 December 2020).

API. (2002). *620: Design and construction of large, welded, low pressure storage tanks*. Wash., DC: API.

API. (2013). *650: Welded steel tanks for oil storage*. Wash., DC: API.

Bernier, C., Elliott, J. R., Padgett, J. E., Kellerman, F., & Bedient, P. B. (2016). Evolution of social vulnerability and risks of chemical spills during storm surge along the Houston Ship Channel. *Natural Hazards Review, 18*.

Bernier, C., Kameshwar, S., Elliott, J. R., Padgett, J. E., & Bedient, P. B. (2018). Mitigation strategies to protect petrochemical infrastructure and nearby communities during storm surge. *Natural Hazards Review, 19*, 04018019.

Bernier, C., & Padgett, J. E. (2019a). *Neural networks for estimating storm surge loads on storage tanks*. https://doi.org/10.22725/ICASP13.015.

Bernier, C., & Padgett, J. E. (2019b). Buckling of aboveground storage tanks subjected to storm surge and wave loads. *Engineering Structures, 197*, 109388.

Beroggi, G. E., Abbas, T. C., Stoop, J. A., & Aebi, M. (1997). *Risk assessment in the Netherlands*.

Beyer, H. L. (2004). *Hawth's analysis tools for ArcGIS*.

Cozzani, V., Campedel, M., Renni, E., & Krausmann, E. (2010). Industrial accidents triggered by flood events: Analysis of past accidents. *Journal of Hazardous Materials, 175*, 501–509.

EN. (2005). *14015: 2004, Specif. Des. Manuf. Site Built Vert. Cylind. Flat—Bottomed Ground Welded Steel Tanks Storage Liq. Ambient Temp. Above*.

European Commission. (2021). *Seveso—Major accident hazards—Environment [WWW Document]*. https://ec.europa.eu/environment/seveso/. (Accessed 29 January 2021).

Gaurd, U. S. C. (2006). *Oil spill liability trust fund hurricane impact*.

Kameshwar, S., & Padgett, J. E. (2015). Fragility assessment of above ground petroleum storage tanks under storm surge. In *12th international conference on applications of statistics and probability in civil engineering, ICASP* (pp. 12–15).

Kameshwar, S., & Padgett, J. E. (2016). Stochastic modeling of geometric imperfections in aboveground storage tanks for probabilistic buckling capacity estimation. *ASCE-ASME Journal of Risk and Uncertainty in Engineering Systems, Part A: Civil Engineering, 2*. https://doi.org/10.1061/ajrua6.0000846.

Kameshwar, S., & Padgett, J. (2018a). Fragility and resilience indicators for portfolio of oil storage tanks subjected to hurricanes. *Journal of Infrastructure Systems*, *24*. https://doi.org/10.1061/(ASCE)IS.1943-555X.0000418, 04018003.

Kameshwar, S., & Padgett, J. E. (2018b). Storm surge fragility assessment of above ground storage tanks. *Structural Safety*, *70*, 48–58. https://doi.org/10.1016/j.strusafe.2017.10.002.

Kameshwar, S., & Padgett, J. E. (2019). Stiffening ring design for prevention of storm-surge buckling in aboveground storage tanks. *Journal of Structural Engineering*, *145*. https://doi.org/10.1061/(ASCE)ST.1943-541X.0002275, 04019002.

Khakzad, N., & Van Gelder, P. (2017). Fragility assessment of chemical storage tanks subject to floods. *Process Safety and Environment Protection*, *111*, 75–84.

Khakzad, N., & Van Gelder, P. (2018). Vulnerability of industrial plants to flood-induced natechs: A Bayesian network approach. *Reliability Engineering and System Safety*, *169*, 403–411.

Knulst, K. (2017). *Flood risk management in the unembanked areas: An optimal approach?*.

Lansen, A. J., & Jonkman, S. N. (2010). *Flood risk in unembanked areas part D: Vulnerability of port infrastructure*. Knowl. Clim. Natl. Res. Programme KfC Rep.

Lansen, A. J., & Jonkman, S. B. (2013). Vulnerability of port infrastructure for the port of Rotterdam. In *Climate adaptation and flood risk in coastal cities* (pp. 73–94).

Portela, G., & Godoy, L. A. (2005a). Wind pressures and buckling of cylindrical steel tanks with a conical roof. *Journal of Constructional Steel Research*, *61*, 786–807. https://doi.org/10.1016/j.jcsr.2004.11.002.

Portela, G., & Godoy, L. A. (2005b). Wind pressures and buckling of cylindrical steel tanks with a dome roof. *Journal of Constructional Steel Research*, *61*, 808–824. https://doi.org/10.1016/j.jcsr.2004.11.001.

Qin, R., Zhu, J., & Khakzad, N. (2020). Multi-hazard failure assessment of atmospheric storage tanks during hurricanes. *Journal of Loss Prevention in the Process Industries*, *104325*. https://doi.org/10.1016/j.jlp.2020.104325.

SSPEED. (2016). *The SSPEED Center 2016 annual report. Severe storm prediction, education, and evacuation from disasters, Houston*.

CHAPTER 30

Experiences on place-based learning and research outcomes from the perspective of a student

Alaina Parker-Belmonte
Professional Landscape Architect, TX, United States

Background

Prior to knowledge of the NSF PIRE program and 2017 application, my first introduction to flood-risk reduction and coastal resiliency occurred during my 2016 fall landscape design studio, which revolved around the master planning and design of an undeveloped site located in League City, TX. Perched between Houston and Galveston, TX, League City is at high inundation risk from both hurricane and local rain events; this risk will be exacerbated as the coastal city doubles in population (City of League City, 2013), experiences an increase of impervious surfaces, and faces threats associated with global warming (Climate Change Impacts, 2021). Undergraduate students were expected to conduct a GIS-layered analysis to map flood risk, project land use, and analyze other attributes to inform the master plan. The studio also focused on learning both structural and nonstructural flood-control mechanisms to integrate strategies within the site design and assist in protecting future residents from flood threats. Specifically, mechanisms focused on how to integrate low-impact development techniques and large-scale green infrastructure to challenge conventional flood control measures. The studio took a siloed approach to understanding and solving problems. This approach restricted students' methods and techniques compared to the design solutions that would have been implemented with interdisciplinary collaboration. Regardless, I recognized the importance of broadening my perspectives and skillsets, to work collaboratively in a team with other disciplines to truly master plan and integrate design solutions that reduce flood risk. This desire to enrich my skillsets led me to apply for the NSF PIRE program and, ultimately, was the reason for selecting the "New Towns Coping with Flood Risk" case study.

Introduction

The "New Towns Coping with Flood Risk" case study revolved around the relatively new towns of Almere and IJburg, located on Lake IJssel (Fig. 1). Both towns emerged within the

Fig. 1 Locator diagram for Almere and IJburg. 52011′46.66″ N and 2044′30.76″ E. Google Earth, December 13, 2015. Retrieved August 8, 2017.

last 40 years and take differing approaches to development and flood-risk reduction. IJburg is a series of artificial archipelagos designed to have finished floor elevations and grades above sea level; upon build-out the new town will feature six new islands at varying dwelling densities (Claus, Dongen, Schaap, Burg, & Huizinga, 2001). In contrast, Almere follows the traditional polder development in the Netherlands and is planned to contain six districts inspired by Ebenezer Howard's Garden City Movement (IP Course ACCE, 2013). Both new towns have the capacity for urban growth, with Almere being one of the fastest growing cities in the Flevoland (IP Course ACCE, 2013). Each town can serve as a case study for the greater Houston-Galveston Metropolitan Area as both new towns have implemented strategic planning and design mechanisms to accommodate population growth demand and climate change effects.

Landscape architects have a significant role in shaping our built environment, and like other professionals in the design industry, are educated and trained to design sites that protect critical ecosystem services while creating aesthetic experiences for users (Calkins, 2012). There are many strategies by which designers can plan for the protection of critical ecosystem services; the following serves as a small snapshot: integration of low-impact development (LID) techniques to reduce peak stormwater runoff, protecting ecosystems integral in flood-risk reduction by preserving critical boundaries and suitable placement of project components, providing project sites that serve mixed-uses, reducing impervious coverage, preserving existing vegetation and tree canopy, etc. For the New Towns case study, I was able to utilize my academic background in landscape architecture and urban planning to expose the contrasting flood-risk reduction strategies and analyze the critical flood-risk infrastructure employed in the new towns of Almere and IJburg. The result of my conducted research was the creation of a "new towns" tool kit that visualizes the design strategies that can be utilized to mitigate flood risk, ultimately to serve as a reference for the design and development of communities in the Houston-Galveston Metropolitan Area.

Program methodology

Over the course of 7 months, spanning from February through August 2017, the NSF PIRE Coastal Flood Risk Reduction Program provided invaluable structure by which students were able to: incrementally plan their research topics and strategies, interact with students from different academic and cultural backgrounds, meet with international and national experts and faculty, conduct interviews with expert panels, and participate in field trips. The cumulation of all the above experiences allowed for students to navigate the complexities of international research and ultimately apply place-based learning to finalize multidisciplinary research reports.

Case study research and results

Research for the NSF Program was conducted in three phases: pretravel, during travel, and posttravel. During all phases, students were constantly tasked with developing research topics tailored to their case study and specific academic skill set.

Pretravel research and planning occurred in the months leading up to May 31. This phase consisted of two pretrip meetings with selected program participants and served as the official introduction to the NSF Coastal Flood Risk Reduction Program and involved faculty, and fellow graduate and undergraduate students. Students were able to present initial research plans and ideas for their field of study. Many of these ideas came from the research plan submitted with the program application. As an undergraduate student with a minimal research background, I began with a multitude of research topics to be vetted as the program progressed. A literature review was required of all participants prior to leaving for the Netherlands. I focused on learning more on the implications of green versus gray infrastructure and how these different flood control strategies impact flood risk in the United States. Ultimately, the literature review revealed that the precedence in the United States is to rely on flood control as the primary strategy against inundation, typically in the form of stand-alone structures (such as seawalls) to eliminate flood risk; often, maintenance and monitoring for these structures are inadequate for long-term flood protection (Williams, 1994). Localized flood-control mechanisms, such as detention ponds designed to accommodate runoff from a 100-year storm, were also unable to properly eliminate flood risk when facing urbanization, increased impervious coverage, increased storm water runoff and peak discharges, and higher occurrences of stream peak flows (O'Driscoll, Clinton, Jefferson, Manda, & McMillan, 2010). Understanding that current flood-risk reduction practices in the United States are inadequate solutions to an evolving urban environment allowed me to set up the comparative background for the research I conducted in the New Towns case study, and ultimately assisted in the evaluation of how Almere and IJburg are successful in flood-risk reduction.

During-travel research occurred from May 31 to June 16, and consisted of a structured schedule intended to provide students with a variety of place-based learning opportunities including: four field trips to case study locations, pit-stops at related museums and other major flood-risk infrastructure sites, seven full days dedicated to research time (individual and team), four meetings with the larger group to discuss process and progress, presentations from expert panels, one international/multidisciplinary design workshop charrette, one progress presentation, and one final presentation of research concept.

- **Field Trips:** Field research consisted of visiting all four case study sites; each visit was guided by municipal representatives and consisted of a tour; a presentation covering regional context, site history, and development over time; flood-risk management strategies; and how the site is currently responding or will respond to increased inundation risk from climate change (Parker, 2018). Students had the opportunity to have guided tours at each case study site that allowed for the observation of critical flood-risk infrastructure in person. For the New Towns case study, this included the intersection of the High and Low Canals at the Blocq van Kuffeler in Almere and the floating houses in IJburg (Parker, 2018). The opportunity to physically be in the Netherlands and learn the qualitative features of the case study sites was incomparable to any research experience that would have occurred if students were only conducting research within the United States.

- **Dedicated Individual and Team Research Time:** During the 7 days dedicated to individual and team research time, students had the opportunity to travel to conduct interview panels with experts, and dedicate time to further develop individual research topics. As a team, the New Towns case study participants traveled to meet and interview with city representatives from the City of Almere, Water Authority Zuiderzeeland, and Amsterdam Rainproof (Parker, 2018). In preparation for the interviews, our team collaborated to discuss what questions were best suited to ask experts about our individual research topics within the meeting time allotted. Subject matter for the expert panels involved urban planning, spatial planning, hydrology, and low-impact development techniques (Parker, 2018). The interviews revealed specific flood infrastructures employed in both Almere and IJburg, how the infrastructure is designed, how the infrastructure functions, and how the infrastructure is resilient to pressures of urbanization and climate change (Parker, 2018). Above all, interviews revealed the three tiers of inundation risk for New Towns Almere and IJburg: (1) dike failure, (2) inundation from waterways, and (3) inundation from extreme rainfall events (Parker, 2018).

The ability to interview expert panels was an invaluable experience for students, as the information gained was direct, concise, and provided students with quality research information that may not have been published or accessible had research been conducted from the United States. In addition, having the opportunity to physically interact and meet with expert panels assisted students with growing their professional network, and taught students how to interact in different cultural settings. Students were able to have direct explanations on what specific flood-risk infrastructure Dutch terminology is inherent in the New Towns, and through the interviews they were able to understand the English term counterparts or become enlightened on how specific terminology is unique solely to the Netherlands. Personally, the interviews assisted me in better understanding the differences and nuances between the traditional polder infrastructure that is employed in Almere, specifically the term *wadi*, in which I struggled to find information digitally on how the *wadi* functioned or what its overall purpose was. After interacting with the expert panels, they were able to directly answer my questions on the *wadi* terminology and even diagram how they function. Ultimately, I was able to then compare the *wadi* to a dry canal or what's referred to as a *swale* in the United States. Practicing these cultural communication skills allowed for myself and other students to better navigate the complexities of international research. This cross-cultural interaction is arguably one of the biggest strengths of place-based learning that cannot be replicated solely by researching within the United States.

- **Student Interactions:** The NSF PIRE Coastal Flood Risk Reduction program was the first time in my undergraduate career that I was able to work with other graduate and undergraduate students from different academic backgrounds and specializations. Students of the New Towns team were able to provide a multiscaled and multilayered analysis approach that documented flood-risk reduction strategies of Almere and IJburg. The New Towns research subject matter involved topics related to adaptive capacity, social vulnerability, urbanization, flood-risk communication, critical flood-risk infrastructure identification, and pumping strategies. Discussing how our research topics related better allowed our team to create the midpoint and final case study presentations for the trip. Personally, this multidisciplinary approach to research provided me with a better understanding of the strengths of my team's academic background and training, and how this approach is necessary to properly reduce and manage flood risk. This understanding was amplified in the 'Texas Meets HPZ' Multidisciplinary Design Workshop, in which PIRE students from the United States were paired with students from TU Delft for a fast paced, collaborative, design charrette. Teams were tasked to design solutions to reduce flood risk for a shoreline town. This experience revealed the benefits of different perspectives on problem solving for flood-risk reduction, and provided the opportunity for students to learn from one another while collaboratively creating a design solution (Fig. 2).

Upon leaving the Netherlands, I was able to leave with a defined research question: *What critical flood-risk reduction infrastructure should be incorporated into new towns? Specifically, what structural and nonstructural mechanisms should be employed?*

Fig. 2 Snapshots of "TX meets HPZ" Multidisciplinary Design Workshop. Image *(left)* shows students from different backgrounds collaborating to create design solutions related to flood-risk reduction. Image *(right)* is an example of one student group's design strategy for flood-risk reduction. *Adapted from Studenten maken ontwerpen voor Petten [Video file]. (2017). Netherlands: Noordkop Centraal. Retrieved November 19, 2020, from https://www.youtube.com/watch?v=-W9uw66hfLU.*

With research questions posed, posttravel research then occurred from June 16 to the end of August, and entailed students finalizing their research topics, conducting further research, and publishing a research report, poster, and related diagrams/figures. The cumulation of knowledge during-travel provided the foundation from which my individual research stemmed, which was the categorization of flood-risk infrastructure of Almere and IJburg into three tiers of inundation risk: (1) dike failure, (2) inland inundation from waterways, and (3) inundation from extreme rainfall events (Parker, 2018). I was then able to analyze information gained from the various expert panels and presentations of Almere and IJburg to categorize critical flood-risk infrastructure employed in the New Towns into the three tiers of inundation risk, and further investigate additional infrastructure that was not described during-travel. Sources such as the Almere comprehensive plan, the IJburg comprehensive plan, design documents describing the conceptualization of the New Towns, and the Dutch Dikes book (Pleijster & Van der Veeken, 2014) served as the foundational sources for my research report. Many documents presented a language barrier challenge, as no English translations were offered; however, I was able to utilize my familiarity of certain Dutch terminology and Google Translate to transcribe sources into English. I found that searching with the terminology and keywords that I learned internationally allowed me to better tailor my research and produce publications that would have been challenging to find had we not traveled internationally.

The New Towns toolkit identifies the mechanisms employed by Almere and IJburg that protect against the three categories of inundation risk. Mechanisms are categorized by: primary, secondary, and tertiary critical flood-risk infrastructure (Fig. 3). Dike failure is the primary inundation risk category and reflects the Zuiderzee Water Authorities main focus to protect health, safety, and welfare (Parker, 2018). Dikes serve as the primary defense from the North Sea; if failure were to occur, disaster would be catastrophic with the endless supply of water that has the potential to bury cities (K. Spaan, personal communication, June 6, 2017). The primary flood defense for Almere is the Oostvaardersdijk and the Knardijk; the Oostvaardersdijk acts as the perimeter for Almere and serves to compartmentalize and allow for new reclaimed land, while the Knardijk bisects the polder in half to serve as a secondary regional defense for the area (Pleijster & Van der Veeken, 2014). IJburg differs in that the whole town is above sea level, thus becoming

Fig. 3 New towns toolkit. *Retrieved from Parker, Alaina. (2018). Quantifying impacts of urbanization on flood risk under contrasting management strategies.*

(Continued)

Fig. 3—Cont'd In B. Kothuis, Y. Lee, S. Brody (Eds.), Authentic Learning and Transformative Education: Volume I (1st ed., pp. 151–161). NSF-PIRE Coastal Flood Risk Reduction Program.

an elevated polder at 2.5 m (K. Spaan, personal communication, June 6, 2017). Although already elevated, the perimeter of the islands is surrounded by an adaptable dike that provides additional flood protection as sea levels rise (K. Spaan, personal communication, June 6, 2017).

The threat of inundation from inland waterways is the secondary consideration that comes into play when planning new communities in the Netherlands (Parker, 2018). The secondary critical flood-risk infrastructure for Almere includes: pumping stations, sluices, major canals, canals, and dry canals (Fig. 3). Pumping stations, such as the Blocq van Kuffeler, will convey water out of the polder and have the ability to prepump in anticipation of storm events (M. Visser, personal communication, June 13, 2017). A greater network of sluices, major canals, canals, and dry canals makes up the framework of Almere and is strategically placed and designed to disrupt storm surge, hold additional water during surge events, and ultimately convey additional rainwater out of the polder system (Fig. 3). Maintenance and monitoring water levels is critical in Almere due to the nature of the polder system (M. Visser, personal communication, June 13, 2017), whereas in IJburg because of its elevated status the secondary critical flood-risk infrastructure is simplified into floating houses, and blue green grids (Fig. 3). Floating houses have the ability to fluctuate with surrounding water levels in the town, ensuring that the specific structures are not impacted during storm events (Andersson, 2014). Blue green grids refers to the strategic design of IJburg that allows for sheet flow conveyance of rainfall around developed land to be directed into surrounding water bodies (K. Spaan, personal communication, June 6, 2017). The blue green grid implementation ensures that stress is relieved from traditional stormwater infrastructure, ultimately reducing risk from inland inundation (Groenblauw, 2020). Monitoring IJburg's secondary critical flood-risk infrastructure may not be as critical to maintain constantly as for Almere; however, authority review of proposed land development should occur to ensure the continued functionality of floating houses and blue green grids.

The final threat of inundation occurs from extreme rainfall events. This tertiary critical flood-risk infrastructure category is characterized by the implementation of low-impact development best management practices (LID-BMPs). Almere utilizes retention lakes, which serve a dual purpose of providing visual aesthetics to residents while providing additional storage capacity during localized rain events (Fig. 3). In addition to the retention lakes, Almere has also dedicated a significant amount of land to be preserved as greenspace (J. Balkema, personal communication, June 6, 2017). The preservation of greenfields ensures that rainwater can permeate and infiltrate the landscape, reducing flood risk as opposed to traditional development in the United States that has high concentrated areas of impervious coverage. IJburg currently employs a variety of LID-BMP strategies to serve as tertiary critical flood-risk infrastructure, including: extensive green roofs, intensive green roofs, dike roof, and infiltration basin (Fig. 3). These mechanisms serve the larger goal for IJburg to maintain its own water cycle (K. Spaan, personal communication, June 6, 2017); future phases of IJburg are anticipated to also include the following strategies: embedded rain pipes, open gutters, rain gardens, swales, permeable pavement, etc. (Groenblauw, 2020). Implementing LID-BMP strategies allows for developments to ameliorate the effects of rainfall inundation (Groenblauw, 2020).

Ultimately, the documentation of the methods by which new towns Almere and IJburg mitigate flood risk serves as a reference for development within the greater Houston-Galveston Metropolitan Area. All sites will be unique in their environmental and cultural strengths, weaknesses, opportunities, and constraints. Before implementing solutions, ample site analysis should occur to ensure design solutions are suitable and sustainable by being socially equitable, environmentally sound, and economically feasible (Calkins, 2012). Although my initial research question was to determine what critical flood-risk reduction infrastructure *should be incorporated* into new towns, my research served to better document what critical flood-risk reduction infrastructure *can be incorporated* into new towns. More importantly, there is no stand-alone solution for reducing flood risk and developments should employ a variety of structural and nonstructural critical flood-risk infrastructure in order to ensure the protection of public health, safety, and welfare.

NSF PIRE and beyond

Three years have passed since my involvement as an active NSF PIRE participant; within this time I have graduated with an undergraduate degree in landscape architecture with a minor in urban and regional planning from Texas A&M University. I have gained professional experience as a landscape designer within a private landscape architecture firm in San Antonio, Texas, which specializes in civic and cultural landscapes, urban design, complete streets, low-impact development, etc.

The experiences I gained from being an NSF PIRE program participant has opened my perspective on design strategies that would otherwise seem infeasible, but are in fact possible with multidisciplinary collaboration and frame of mind. Although my day-to-day schedule does not revolve around protecting the San Antonio Region from large critical flood risks, I have found that with my background in the PIRE program I have become incentivized to continue my education on LID strategies and understanding the nuances of proposing and implementing these strategies in projects. The San Antonio River Authority offers multiple classes for professionals of varying backgrounds to receive additional training in technical design, construction, and maintenance of LID strategies. Urbanization and impacts from climate change are very real threats that San Antonio faces, as the city is located in the "Flash Flood Alley" region, spanning from Del Rio to the Dallas-Fort Worth Metropolitan area (San Antonio River Authority, 2020). Designers are encouraged to implement LID strategies to assist in ameliorating these impacts.

One of the challenges facing landscape architects and designers is the ability to implement sustainable design (including LID strategies) in projects with limited scope and funding, especially with projects in the private sector. Typically, features such as LID become add alternates or completely removed from project scope due to the economic feasibility of implementation. Instead landscape architects will be brought into the project too late to convince clients that implementing sustainable design is economically feasible and should be incorporated into project goals/vision for greater public health, wellness, and safety. It is difficult to accurately evaluate the economic benefits of certain implementations using sustainable strategies, and the cost to do so is often outside of the scope for design projects (Calkins, 2012). As designers we must get creative about our tactics to ensure we can propose sustainable design while providing financial benefits to our clients.

In our office, we have found that becoming involved in programs that offer financial incentives for implementing LID has been one of the most successful methods by which we are able to design sustainably. The San Antonio River Authority Watershed Wise Rebate Program serves to provide funding to LID-BMP projects that meet specified program requirements (Watershed wise rebate program, 2020). In my emerging professional experience, it seems that until cities and towns put in place stricter stormwater and development regulation requirements (such as local zoning ordinances, overlay districts, etc.), there will continue to be an up-hill battle for landscape architects and designers to strive for sustainable designs. As perceptions toward flood risk change, we must continue to empower ourselves with knowledge of how others are adapting to flood risk successfully, such as in the case of new towns Almere and IJburg, how to remain creative and utilize strategies such as financial incentives, and be involved in the front end of project visioning and strategizing.

Being a participant of the NSF PIRE Coastal Flood Risk Reduction Program has allowed me to become a well-rounded educated professional, develop a unique skill set and background, and ultimately transfer knowledge learned internationally to benefit projects that reduce flood risk within the United States.

References

Andersson, C. (2014). *Rising tides: resilient Amsterdam*. Honors Theses. Paper 25.
Calkins, M. (2012). *The sustainable sites handbook: a complete guide to the principles, strategies, and best practices for sustainable landscapes*. Hoboken, NJ: Wiley.
City of League City. (2013). *League City Comprehensive Plan 2035*. 19 November 2020. Retrieved from: https://www.leaguecity.com/1038/Comprehensive-Plan.
Claus, F., Dongen, F. V., Schaap, T., Burg, L. V., & Huizinga, N. (2001). *IJburg: Haveneiland en rieteilanden*. Rotterdam: Uitgeverij 010.
Climate Change Impacts. (2021). *Retrieved From United States Environmental Protection Agency*.
Groenblauw, A. (2020). *Bouw groen en blauw: inspiratie voor rainproof en natuurinclusief bouwen op Centrumeiland*.
IP Course ACCE. (2013). *Climate change in Almere*. (p. 0-87, Rep.). Larensteinselaan 26a Postbus 9001 6882 CT Velp: Van Hall Larenstein.
O'Driscoll, M., Clinton, S., Jefferson, A., Manda, A., & McMillan, S. (2010). Urbanization effects on watershed hydrology and in-stream processes in the southern United States. *Water, 2*(3), 605–648.
Parker, A. (2018). Quantifying impacts of urbanization on flood risk under contrasting management strategies. In B. Kothuis, Y. Lee, & S. Brody (Eds.), *Authentic Learning and Transformative Education: Volume I* (1st ed., pp. 151–161). NSF-PIRE Coastal Flood Risk Reduction Program.
Pleijster, E., & Van der Veeken, C. (2014). *Dutch dikes*. Rotterdam: NAi Uitgevers.
San Antonio River Authority. (2020). *Flood Risk: San Antonio River Authority*. Retrieved November 19, 2020, from: https://www.sariverauthority.org/be-river-proud/flood-risk.
Watershed Wise Rebate Program. (2020). Retrieved November 19, 2020, from: https://www.sariverauthority.org/be-river-proud/sustainability/rebates.
Williams, P. A. (1994). Flood control vs. flood management. *Civil Engineering—ASCE, 64*(5), 51–54.

CHAPTER 31

Conclusions

Samuel Brody[a,b], Baukje Bee Kothuis[c,d], and Yoonjeong Lee[a,b]
[a]Institute for a Disaster Resilient Texas, Texas A&M University, College Station, TX, United States
[b]Department of Marine and Coastal Environmental Science, Texas A&M University, Galveston Campus, Galveston, TX, United States
[c]Department of Hydraulic Engineering and Flood Risk, Faculty of Civil Engineering and Geosciences, Delft University of Technology, Delft, The Netherlands
[d]Netherlands Business Support Office, Houston, TX, United States

In the field of flood risk reduction, comparative international work is rare. Even more so are book-length treatments that look not only across international boundaries, but disciplines as well. *Coastal Flood Risk Reduction* does just that and more. This book is the most systematic, integrated, and detailed study of flooding ever produced between the Netherlands and the Upper Texas Coast in the United States. Based on an ongoing 6-year collaborative investigation, **43** authors representing numerous academic entities in both countries contributed **35** chapters addressing seemingly every angle of the problem and their potential solutions. Even more unique, this body of work is driven by a place-based research and learning approach to address floods, where students provide the inspiration for inquiry and innovation.

The book organized many different perspectives through five major thematic areas associated with flood risk reduction: (1) predicting floods based on environmental and physical characteristics, (2) investigating the socioeconomic and political drivers of flood risk, (3) examining the role of planning, design, and the built environment, (4) finding solutions that enhance flood resiliency, and (5) assessing the impact of place-based and authentic learning. These themes provide an analytical framework in which expositions move from tulips to tacos, and back again.

In telling this story, we span multiple disciplines, scales, methods, and solutions all aimed at increasing resiliency to floods not just within the two countries, but also around the world. For example, ocean engineers wonder how the movement of sand in coastal environments may form the first line of defense against flooding. Hazard economists assess the risks, benefits, and costs of decision-making. Environmental planners focus on the impacts of development and local plans in reducing future flood risk. Landscape architects probe the effectiveness of nature-based approaches and the incorporation of green infrastructure in urban design. Educators describe how to provide students with integrative research and authentic learning experiences. And, students tell their own research stories stemming from an immersive place-based approach to learning about local issues in both countries. Above all, this book is about leveraging research to find solutions that reduce the adverse impacts of floods—whether that entails constructing a major surge barrier, protecting open spaces or critical habitats, communicating risk to local residents, or some combination thereof.

Integrated comparisons of flood issues across international boundaries, combined with a collective focus on problem solving is what sets this book apart from others. Only by looking beyond borders and across disciplines can we fully understand and potentially solve such a complex, global problem. Moreover, the transfer of knowledge across sites, regions, and countries provides invaluable insights into flood risk reduction that cannot be gained by less holistic work. Readers can draw from lessons learned in the upper Texas coast and the Netherlands, and apply this knowledge to their own localities, contexts, or situations around the world.

However, this book focuses on just two countries in two areas of the world. More work is needed comparing and integrating findings from many more flood-prone regions coping with rapid development, environmental change, and inundation threats coming from both the sky and the sea. Additional studies are especially needed in Asian, African, and island nations. *Coastal Flood Risk Reduction* should be just the start of future work on flooding as a transboundary, multidisciplinary problem addressed through in-depth, place-based, comparative investigations.

Index

Note: Page numbers followed by *f* indicate figures, *t* indicate tables, and *b* indicate boxes.

A

Aboveground storage tanks (ASTs), 152–153, 397
Adaptation opportunities, 186
ADCIRC Surge Guidance System (ASGS), 24
ADvance CIRCulation (ADCIRC), 96–98
Amsterdam water forest, 222–223
Anderson, 168–170
Artificial Intelligence (AI), 343–344
ASGS. *See* ADCIRC Surge Guidance System (ASGS)
ASTs. *See* Aboveground storage tanks (ASTs)
Avoidance, flood events, 303–304

B

Beneficial use of dredged material (BUDM), 38–39
Benefit–cost analysis (BCA), 306–307
Benefit–cost ratio (BCR), 309–311
Big U, 82
Black gold, 193
Botlek, 399
BUDM. *See* Beneficial use of dredged material (BUDM)
Building with nature, 380
Built environment impacts, flood loss, 167
   built environment, obstacles, 171–172
   dutch, solutions, 172–174
   impervious surfaces, spread, 168–171, 169*f*
   inadequate, aging infrastructure, 172
   putting more people, harm's way, 167–168
BuyersBewhere.com, 344

C

Carrot-and-stick strategy, 362
CGE. *See* Computable general equilibrium (CGE)
Clear Creek Watershed, 309–311
Coastal
   areas, 6
   defense, 271
   dunes, design, implementation, flood risk reduction, 287
   dunes evolution, 291
   dunes, storm impacts, 289–292
      impact scale, 289–291, 290*t*
   engineered dunes, 292–298
   environment, 317
   flood risk reduction, 423
   galveston/seabale dunes, 297–298, 297*f*
   hondsbossche dunes, 295, 296*f*
   houston-Galveston coastal spine, 298
   katwijk dunes, 296–297, 296*f*
   natural dunes, 287–288, 288*f*
   recovery after a storm, 292
   regions, 131
   spine, 96, 283–284
Coastal Flood Risk Reduction Program (CFRRP), 351–352
Coastline, 77, 287–288
Collaboration, Dutch, Texas researchers, 14–15
Collaborative learning, 368
Collision, 289–291
Command-and-control approach, 305
Communication, flood-resilient behaviors, 343
Community resilience, 239
Compound flooding, 77
   galveston Island, 84–86, 85*f*
   managing coastal flood hazards, 82–86
   modeling coastal flood hazards, numerical, statistical approaches, 78–82
   noorderzijlvest, 83–84
   numerical modeling, 79–80
   rotterdam, 82–83, 83*f*
   statistical modeling, 80–82
   west side of Galveston Bay, 84
Comprehensive, integrative problem-solving approach, 371
Computable general equilibrium (CGE), 95
Cost-benefit analysis (CBA), 92
   direct, indirect impacts, 96–100, 97*f*, 98–99*t*
   dutch approach, coastal infrastructure BCA, United States, 100–104, 101*t*
   modeling approach, 93–100, 94*f*
   study area, 96
Costs, 137
Coupled models, 53–54, 53*f*
Cross sectional time series (CSTS) approach, 320

D

Data
   analytics, 342–343
   driven framework, 312–313
   visualization, 343

Index

Delta
  Committee, 72
  landscapes, 241
  works, 271–272
DEM. *See* Digital elevation models (DEM)
Demolition, properties, 303–304
Designers, storm surge barrier, 284
Development decisions, flood risk, 167
Digital elevation models (DEM), 96–98
Dike, 119
District-hazard zones (DHZs), 177–179, 185–186
Dune design, 265
Dutch
  American approaches, flood risk reduction, 5–7
  dike ring system, 338
  principles, Galveston Bay region, 11–13
  specific design-concept/building-with-nature, 361
  water management agency, 307

E

Educational effects, 366
  authentic problem/learning, 367–368
  methods, 368–369
  scoring rubric, NSF PIRE CFRRP student survey, 373–375
  transformative learning (TL), 366–368
Effective flood mitigation practices, Netherlands, 351–352
Engineering with nature, 380
Environmental gate, 13

F

Federal Emergency Management Agency (FEMA), 48, 61, 93, 241, 303–304
Field trips, 414–415
Flood, 19, 33
  damages, 5
  defense(s), 92, 242–243
  hazard, 77
    machine learning, alternative model, 62–63
    models, visualizations, importance, 61–63
    potential application, Netherlands, 72–73
Floodplain wetlands, reduce/delay floods, 317
Flood proof houses
  application areas, types, 330
  consequence reduction measures, application, 334–336

  flood proof plinth, 336
  shelters, 336
  zoning, 336
Netherlands, 329
technical practicalities, 336–337
types, 329–337
  amphibious houses, 334, 335*f*
  floating houses, 333–334
  mount houses, 330–331, 330*f*, 332*f*
  pole houses, 331–333, 333*f*
Floodproofing the Metropolitan Region Amsterdam (MRA), nature based approach, 209–223, 210*f*, 212*f*
climate change/sea level rise, increased rainfall scenarios, 211–213
demonstration projects, 215–223
  amsterdam forest/new water retaining forest, 220–223, 221–222*f*
  flevopolder/floodproof houses, amphibious street furniture, 217–219, 218–219*f*
  flevopolder/transferring subsiding grounds, 215–217, 216*f*
  heuvelrug/restoring the natural water infiltration system, 219–220
design approach, 214
extending the system, 213–214
integrated strategies, 214–215
landscape zones, 214
reflection, houston context, 224
regional water system, 211
Flood protection, 242
Flood resiliency, United States, 303
  buyouts, contextual differences, Netherlands, 305–307
  case study analysis, 308*f*, 309–311, 310*f*, 311–312*t*
  prioritizing ecological gains, model, 307–311, 308*f*
  property acquisition, 303–305
Flood-resilient communities, plan evaluation, 177, 178*f*
  transatlantic application, 179–180
    de Staart neighborhood, Dordrecht, 185–186
context, 185
findings, 186, 187*f*
process, 185–186
    Feijenoord, Rotterdam, 180–182
context, 180–182, 181*f*
findings, 182

process, 182
  Nijmegen, 183–185
context, 183
findings, 184–185, 184f
process, 183
Flood risk, 2, 5, 61, 109, 119, 137, 272, 370–371
Flood risk reduction, Galveston bay, coastal barrier system preliminary design, 257–258, 317
  Bolivar roads storm surge barrier, 261–263, 262f
    environmental section gates, 262–263
  coastal spine system, 259–266, 266t
    system, approach, 259–261, 260f, 261t
  land barrier, 263–265, 264f
  Netherlands, 115–117
  setting the scene/risk-based evaluation of strategies, 258, 259f
  strategy, 304
Flood risk reduction strategy, 304
Foredune, 289–291
Forerunner, 22
Fortified dune, 265
Frames of reference, 366

G

General Land Office (GLO), 14
Geodesign framework, 303
Global Flood Monitor, 65
Global petroleumscapes, 193
Great Storm, 9–10
Green infrastructure-based design, Texas coastal communities, 227
  flood mitigation/flood risk reduction, 228–229
  flood mitigation tools LID, 229–230
  GI application project, 232–238
    design strategy, layout, 233–236, 233–234f
    landscape performance, 235f, 236–238, 237f, 238t
    study area, 232–233, 232f
  moving forward, 238–239
  planning, design promoting GI, 231–232
Gross State Product (GSP), 100
Gulf of Mexico (GOM), 323–324

H

HAND. *See* Height Above Nearest Drainage (HAND)
Harris County Flood Control District (HCFCD), 173

Hazard data, 398–399
Hazards US Multi-Hazard (HAZUS-MH), 93
Heemraadschap Hollandsnoorderkwartier (HHNK), water authority, 381
Height Above Nearest Drainage (HAND), 54
Hemmati, Ellingwood, Mahmoud, 2020, 167
HHNK. *See* Heemraadschap Hollandsnoorderkwartier (HHNK), water authority
Historical loss, National Flood Insurance Program (NFIP), 113–115, 113–114f
Household(s), 1, 109
Housing and Urban Development (HUD), 112
Houston, Rotterdam beyond oil, ship channels, 193–194, 199t
  cracks, fractures, 198–202
  creating new perspectives, 202–207, 204–206f
  modern industrial urban landscape, dream building, 194–198, 195f, 197f
Houston-Ship Channel (HSC), 153–155
HUD. *See* Housing and Urban Development (HUD)
Human impact, 1
Hurricane, 19, 119
  events, 93
  Harvey, 61
Hydrologic system, 62

I

IDRT. *See* Institute for a Disaster Resilient Texas (IDRT)
IJburg, 416
Ike Dike, 11–12, 25, 96, 249
Improved knowledge, students' knowledge level, 370
Improving individual flood preparedness results, 128–129, 129f
Incentive, 111
Industrial operations, 92
Infrastructure impacts, vulnerability, coastal flood events, 151–152, 292–293
  international case studies, 152–158
    envisioning, design, management, 158–161
coastal engineering, advance performance, 158–160
smart resilience, 160–161
    storage tanks, coastal port, industrial complexes, 152–156, 154–155f
    transportation infrastructure, 156–158, 157f

Innovation, 423
Institute for a Disaster Resilient Texas (IDRT), 5
Insufficient demand, flood insurance, 121–126
   charity hazard, 121–122
   inertia bias, 123–125, 124f
   simplification, 122–123
   systematic biases, 122–125
Insurance, 109
Integrated water management approach, 305–306
International integrative research education program, 352
   background theories, program design concept, 352–354, 355–358t
      integrative case study research design, 354–356, 358–359f
   lessons learned, 362
   open, inclusive learning environment, 360–362
      group research meetings, presentations, 360–361
      multidisciplinary design workshop, 361–362
   student recruitment, securing diversity, 356–360
International Program on Climate Change (IPCC), 77–78
Inundation, 289–291
Investigation, 423
IPCC. *See* International Program on Climate Change (IPCC)
I-STORM, 271–272

J
JSU (2018) research, flood risk, 25–27, 26f

L
Lake IJssel, 330
Land acquisition, 304–305
Land use and land cover (LULC), 168–170, 324
Learning, flood risk, 343–344
Letter of intent (LOI), 358
Levels, flood insurance demand, 129–131, 130f
LiDAR-derived Digital Elevation Models (DEMs), 52
LISFLOOD-FP, 53
Local individual risk (LIR), 140
Low impact development (LID), 412
Low impact development best management practices (LID-BMPs), 419
Low-probability/high-impact (LPHI), 119–120
LULC. *See* Land use and land cover (LULC)

M
Machine learning workflow, flood hazard estimation, 63–71
   case study, Southeast Texas, 63–64
   flood hazard concept measurement, 65–66
   interpretation, flood hazard model outputs, 64
   interpreting, modeling results, 69–71
   model calibration, 68–69
   overview, machine learning models, flood hazard estimation, 66–68, 67t
Management, Maintenance and Operations (MMO), 271–272
Media outlets, 1
Mega-nourishment (MN), 38–39
Mezirow's transformative learning theory, 366
Mississippi River Gulf Outlet (MRGO), 82
Mitigation, 91–92
   examples, 371
   ideas, 371
   techniques, 2–3
Modeling, 19–20
   flood hazards, urban areas, 49–55
      alternative approaches, 54–55
      1D models (one dimensional models), 49–51, 50t, 51f
      2D models (two dimensional models), 51–53, 52f
      3D models, key research priorities, 55–56
      risks, urban flooding, 55
Morphodynamics, 37, 288
Movable storm surge barriers, 271–272
   characteristic, implications, 276–280
   maintenance windows, 278
   movable storm surge barriers, 272–275, 273f, 274–275t
   occurrence, intended functioning, 277
   prototype, 276
   reasoned design, 280–284
      bolivar roads barrier/preliminary design, 283–284, 283f
      maeslant barrier/replacing compression blocks, 280–281, 281f
      ramspol barrier/corroded nut, 282–283, 282f
   static object, dynamic environment, 278–280, 279t
Multilayer safety approach, 329

Multiperspective problem approach, integrated nature-based design, transdisciplinary workshops, 377–378, 378f
  method, 381, 382b
  results, 382–389
    codesign process, 382–385, 383b
    learning outcomes, 388f, 389–392, 390–392t
    substantive outcomes, 385–389, 386–388f
  theoretical background, 379–380
Murphy oil, 397

N

National Flood Insurance Program (NFIP), 65, 115, 324
National Oceanic and Atmospheric Administration (NOAA), 24, 91, 167–168
National research council (2014), 353
National Science Foundation (NSF), 351–352
National Science Foundation Partnerships International Research Education (NSF-PIRE), 179–180
Natural disaster, 119
Nature-based flood defenses (NBFD), 317

O

Out-of-bag (OOB), 69

P

Paired t-test/repeated measures t-test, 369
Palustrine, upland wetlands, 323–324
Partnership for International Research and Education (PIRE), 2, 351–352
Partnerships for International Research and Education Coastal Flood Risk Reduction (PIRE-CFRR) program, 377
PBL. See Problem-based learning (PBL)
Peak annual flow (PAF), 320
Performance-based coastal engineering (PBCE), 159–160
PIRE. See Partnerships for International Research and Education (PIRE)
PIRE-CFRR. See Partnerships for International Research and Education Coastal Flood Risk Reduction (PIRE-CFRR) program
Plan integration Resilience Scorecard (PIRS), 177
Pluvial flooding, critical research area, 48–49, 72
Policy analytic activities, 380
Policy recommendations, 131–133
Political drivers, flood risk, 423
Precipitation, 78
Present flood defenses, Dutch central coast, upper Texas coast, 10–11, 11f
Prevention, flood, 6
Proactive framework, 308–309
Problem-based learning (PBL) techniques, 24, 367
Prospect theory (PT), 125–126

R

Rainfall, 1
Razende Bol, 385, 387f
Reactive approach, 305–306
Repetitive loss (RL), 111–112
Research by Design, 258
Research, convergence, 352
Research outcomes, place based learning, 411–412, 412f
  case study research, results, 413–420, 416–418f
  NSF PIRE program, 420–421
  program methodology, 413
Resiliency, 423
RftR. See Room for the River program (RftR)
Rijkswaterstaat (RWS), 8
Risicokaart/risk card/risk map, 345, 345f
Risk analysis, safety standards, 137
  comparison, 147–148
    costs, benefits, 147–148
    excedance probability, design water level, 147
    flooding probability, 147
  economic optimization, 140
  examples, 145–147
    costs-benefits framework, 146–147
    exceedance probability framework, 145–146, 145–146f
    flooding probability framework, 146
  risk approaches, 137–140, 139f
  safety standards, 140–144, 141f, 144f
Risk communication tools, bridging knowledge gap
  challenges, opportunities, 346–347
  data-driven web, example, 344–346, 344f
  risk perception, role, 341–342
Room for the River program (RftR), 173, 306
RWS. See Rijkswaterstaat (RWS)

S

Sea level rise (SLR), 7
Sediment, 34, 287–288

Self-reported knowledge level, 371
Service desks, 381
Small Business Administration (SBA), 112
Social Accounting Matrix (SAM), 95
South Padre Island (SPI), 41
Spatial quality, 306–307
Special Flood Hazard Area (SFHA), 48, 110
State of engineering practice, modeling hurricane storm surge, 23–24
Stilt houses, 331–333
Storage tanks, flood risk management, Rotterdam port, 397–399, 398f
  AST inventory analysis, 399–400, 400f
  dutch flood risk management philosophy, 404–407
  PIRE program impact, 407
  storm surge hazard data, 400–401
  vulnerability analysis, 401–404, 402f, 403t, 404–405f
Storm, 19
  event, 303
  surge, extreme water levels, 19–20, 20t, 20f
  surge generation, northwestern Gulf, 21–23, 21f
  surge models, different applications, 24–25
  water, 78
STWAVE model, 24
Support vector machines (SVMs), 68
Surge, 19
Swash, 289–291

T

Texas, 22
  plan, 13–14
  state data center, 167–168
Third Industrial Revolution, 193–194
Threshold level of concern, probability neglect, willingness to pay for flood insurance, probability weighting results, 127–128
Total factor productivity (TFP), 100
Transition zone, 78–79
Tropical cyclone, 78
TX-LA floods, 47

U

Underweighting LPHI risk, 125–126
United Nations Educational, Scientific, and Cultural Organization (UNESCO), 366

United States Army Corps of Engineers (USACE), 7, 319–320
United States Geological Survey (USGS), 320
Urban design, 227
Urban flood design integration, United States, Netherlands, 241–242
  ecological system, reintegration, 245, 246f
  Galveston case study, 249–250, 250f
  spatial design approach, 242–245, 245t
  urban fabric integration, 245
  Vlissingen (Flushing) case study, 246–248, 247–248f
US Army Corps of Engineers (USACE), 96, 241
US Department of Agriculture (USDA), 112
US Geological Survey (USGS), 65
US National Flood Insurance Program (NFIP), 109–112

V

Vacant land, 304–305
Vertical evacuation, 336
Vlissings model, 246

W

Wadden Sea Education Research Center (TWERC), 385
WAM model, 24
Water discharge system, 211–213
Water marks, 65
Water-plot, 331–333
Water, sediment movement, coastal environments, 33
  bedload collector, 41–42
  combining coastal hydrodynamics, morphodynamics, 34–38
  mud motor, 40–41
  numerical modeling, 37–38
  physical modeling, 35–37, 35f
  sand engine, 39–40, 39f
  submerged nearshore feeder bar, 41
  utilizing sediment (transport), mitigate flooding, 38–42
Water surface elevations (WSE), 49–51
WCRP. *See* World Climate Research Program (WCRP)
Wetlands, ecological function, 317
  dutch experience, comparison, 325
  early comparative research, 317

observational research, 319–323
section 404 wetland permits, alteration, 319–323, 321*t*
simulation-based research, 318–319
type, shape, identification, flood loss reduction, 323–324

Withdrawal, 203
Working with nature, 380
World Climate Research Program (WCRP), 77–78
Write Your Own (WYO), 110
WSE. *See* Water surface elevations (WSE)

Printed in the United States
by Baker & Taylor Publisher Services